STATISTICAL MECHANICS

THE MATHEMATICAL PHYSICS MONOGRAPH SERIES

A. S. Wightman, EDITOR
Princeton University

Ralph Abraham, *Princeton University*
FOUNDATIONS OF MECHANICS

Freeman J. Dyson, *The Institute for Advanced Study*
SYMMETRY GROUPS IN NUCLEAR AND PARTICLE PHYSICS

Robert Hermann, *Argonne National Laboratory*
LIE GROUPS FOR PHYSICISTS

Rudolph C. Hwa, *State University of New York at Stony Brook*
Vigdor L. Teplitz, *Massachusetts Institute of Technology*
HOMOLOGY AND FEYNMAN INTEGRALS

John R. Klauder and E. C. G. Sudarshan, *Syracuse University*
FUNDAMENTALS OF QUANTUM OPTICS

André Lichnerowicz, *Collège de France*
RELATIVISTIC HYDRODYNAMICS AND MAGNETOHYDRODYNAMICS

George W. Mackey, *Harvard University*
THE MATHEMATICAL FOUNDATIONS OF QUANTUM MECHANICS

Roger G. Newton, *Indiana University*
THE COMPLEX j-PLANE

David Ruelle, *Institut des Hautes Etudes Scientifiques*
STATISTICAL MECHANICS: RIGOROUS RESULTS

R. F. Streater, *Imperial College of Science and Technology*
A. S. Wightman, *Princeton University*
PCT, SPIN AND STATISTICS, AND ALL THAT

STATISTICAL MECHANICS

Rigorous Results

DAVID RUELLE

Institut des Hautes Etudes Scientifiques

W. A. BENJAMIN, INC.

New York *Amsterdam* *1969*

STATISTICAL MECHANICS Rigorous Results

SBN 8053–8360–3 (Cloth) 8053–8361–1 (Paper)
Library of Congress Catalog Card Number 69–12566
Manufactured in the United States of America
12345R32109

The manuscript was put into production on June 10, 1968;
this volume was published on June 1, 1969

W. A. BENJAMIN, INC.
New York, New York 10016

*Richtiges Auffassen einer Sache und
Missverstehen der gleichen Sache
schliessen einander nicht vollständig
aus.*

F. Kafka

Preface

Physics is a human attempt at understanding a certain class of natural phenomena, using scientific methods. We shall not try to define scientific methods in general, but note that they have been very successful and very diverse. Mathematical physics is one of these methods or approaches in which a mathematical structure is somehow associated with the structure of physical phenomena. The mathematical structure is then studied by mathematical techniques to yield new statements with physical relevance. Provided it yields physically relevant statements, no mathematical technique is to be judged too sophisticated or too trivial.

Applying the tools of mathematics to the investigation of the physical world may be a very rewarding intellectual experience. On the one hand, the knowledge which we have of physical phenomena hints at new mathematical theorems and at ways of proving them. On the other hand, mathematical analysis gives to the physical world a new structure and meaning. The knowledge of this structure and meaning constitutes an understanding of the "nature of things" as deep as we can hope to be.

Not every field of physics yields interesting mathematical physics. Luckily, we live in a period with many unsolved problems that are interesting and appear amenable to treatment. An exception to this statement may be relativistic quantum mechanics, largely because of "overgrazing," but there are also vast areas of *terra incognita*.

The present monograph is devoted to the study of a certain class of rigorous results in equilibrium statistical mechanics. These results concern the limit of an infinite system and the nature of phases and phase transitions. The emphasis is on general methods, and special models are essentially disregarded. (Exactly soluble models would deserve a book in their own right.)

The understanding of phase transitions remains one of the major unsolved problems of mathematical physics. The following chapters give an introduction to this problem, where some fundamental progress has taken place in

recent years. After some generalities on ensembles (Chapter 1), we analyze the infinite system limit of thermodynamic functions for lattice systems (Chapter 2) and continuous systems (Chapter 3). We proceed with a detailed study of dilute systems (Chapter 4) and with a description of what is known on phase transitions (Chapter 5). We then discuss the nature of the states of physical systems with an invariance group (Chapter 6), and we analyze in particular the equilibrium states of infinite systems in statistical mechanics (Chapter 7). An appendix describes some mathematical tools used in Chapters 6 and 7, as these chapters are mathematically more sophisticated than the preceding ones. (The progress of mathematical physics could be significantly promoted, in the author's opinion, by the availability of the results of important mathematical theories in concise form and without proofs, in the spirit of Bourbaki's "Fascicules de Résultats." The appendix of this monograph is an attempt in that direction.) The logical relations between the chapters are roughly as follow:

The people who have developed the subject matter of this monograph may be working in mathematics, physics, or chemistry departments. This diversity is one of the charms of the topic, but it creates a problem of language; the reader should remember this if he finds the text too mathematical, or too unmathematical.

The field of rigorous results in statistical mechanics extends beyond what is included in this volume (general results about equilibrium systems). A few pieces of work that are not treated include the proof of stability of quantum Coulomb systems by Dyson and Lenard [1], the study of the two-phase region for the Ising model by Minlos and Sinai [1], and the analysis of the time evolution of infinite classical systems by Lanford [1].

I am much indebted to G. Gallavotti, C. Gruber, O. Lanford, S. Miracle-Sole, and A. S. Wightman for suggesting many corrections and improvements to the manuscript of this book.

David Ruelle

Bures-sur-Yvette
August 1968

Contents

CHAPTER 1

THERMODYNAMIC BEHAVIOR.
ENSEMBLES

In this chapter, we review the systems and ensembles which will be used in the rest of the monograph. We discuss the concepts of thermodynamic behavior and thermodynamic limit (limit of an infinite system).

1.1 THERMODYNAMIC SYSTEMS

Statistical mechanics has originated from the desire to obtain a mathematical understanding of a class of physical systems of the following nature.

(a) The system is an assembly of identical subsystems.[1]

(b) The number of subsystems is large.

(c) The interactions between the subsystems are such as to produce a thermodynamic behavior of the system.

At this stage the notion of thermodynamic behavior is phenomenological and refers to a macroscopic[2] description of the system, obtained by idealizing actual observations. Typically, thermodynamic behavior may be described as follows.

(a′) Equilibrium states are operationally definable. The state of an isolated system tends to an equilibrium state as time tends to $+\infty$ ("approach to equilibrium").

(b′) An equilibrium state of a system consists of one or more macroscopically homogeneous regions (called phases).

[1] In this preliminary discussion we disregard the more general situation where the subsystems may belong to several distinct species.

[2] If a system exhibits thermodynamic behavior, the notion of "macroscopic observation" is operationally definable. It refers to a situation where the subsystems are considered as small and not individually observed; statistical correlations between the subsystems may, however, be observed (e.g., by x-ray diffraction).

(c') Equilibrium states can be described by thermodynamics, in particular they can be parametrized by a finite number of thermodynamic parameters which determine all thermodynamic functions.[3] It is believed that the thermodynamic functions depend piecewise analytically (or smoothly) on the parameters, the singularities corresponding to changes in the phase structure of the system (phase transitions).

(d') "Transport coefficients" may be defined from the first-order response of the system to small perturbations away from equilibrium.

The art of the mathematical physicist consists in finding a mathematical justification for statements like (a')–(d'). We can already guess (from the fact that we have a "large" number of "small" subsystems) that this will involve considering the limit of an infinite system, the *thermodynamic limit*. The understanding which we have at present of (a')–(d') is extremely incomplete, most investigations being based on uncontrolled approximations. We shall in this book concentrate our attention on (b') and (c'), i.e., on equilibrium theory. Unfortunately, our understanding of (a') is entirely inadequate to provide a sound basis for the notion of equilibrium, we shall instead for a moment turn to a special class of systems and appeal to ergodic theory.

We consider a classical system with time-independent Hamiltonian H and phase space Ω. Time evolution preserves the Liouville measure $d\omega$ and also[4]

$$\delta(H(\omega) - E)\, d\omega \tag{1.1}$$

for each E, at least if this expression is well defined. For the systems which we shall consider, the measures (1.1) are bounded (i.e., have finite total mass). Let $t \to \omega_t$ describe the time evolution of a point of Ω. An observation of the system at time t will yield a value $f(\omega_t)$ where f is some suitable function on Ω. Now, if the system exhibits thermodynamic behavior, $f(\omega_t)$ will be practically independent of t, at least[5] for most t. Under these circumstances it is reasonable to identify the result of an observation "at equilibrium" corresponding to the function f on Ω with the average over t of $f(\omega_t)$. If the measures (1.1) are ergodic, this time average is given explicitly by

$$\lim_{T \to \infty} \frac{1}{T} \int_0^T dt\, f(\omega_t)$$

$$= \left[\int d\omega\, \delta(H(\omega) - E) \right]^{-1} \int d\omega\, \delta(H(\omega) - E) f(\omega) \tag{1.2}$$

[3] Pressure, energy, volume, temperature, entropy,... are thermodynamic functions, some of which are chosen as thermodynamic parameters.

[4] δ denotes here Dirac's δ measure.

[5] Due to Poincaré recurrence we expect that a large (but finite) system which is not in equilibrium at time 0 will reach an apparent equilibrium from which it will deviate for relatively short periods at long time intervals. This behavior is of course not exhibited by small systems [for which (b) does not hold].

with

$$E = H(\omega_0) \tag{1.3}$$

for almost every initial condition $\omega_0 \in \Omega$ (with respect to Liouville measure). The result $\langle f \rangle$ of an observation corresponding to the function f is thus given by

$$\langle f \rangle = \left[\int d\omega \, \delta(H(\omega) - E) \right]^{-1} \int d\omega \, \delta(H(\omega) - E) f(\omega) \tag{1.4}$$

if (1.1) is ergodic.

Some version of (1.4) is needed as a starting point for equilibrium statistical mechanics and much effort has therefore been spent in investigating the ergodicity of (1.1) for specific systems. This is a difficult problem and only recently could Sinaï prove that certain systems with realistic interactions are ergodic.[6] In spite of the importance attached historically to the justification of (1.4) by ergodicity, it should be stressed that a more satisfactory argument should involve the fact that we deal with a large system. In particular a large system may have from the physical viewpoint a completely normal thermodynamic behavior without being, strictly speaking, ergodic.[7] In any case we shall now accept, with some adjustments, the description of equilibrium given by (1.4) and corresponding to the measure (1.1); this description appears indeed to be adequate in the cases of physical interest.

Other measures may be used instead of (1.1); they correspond to the well-known *ensembles* of Gibbs. It is believed that the various ensembles yield equivalent descriptions in the thermodynamic limit, i.e., for large systems; to prove this is the problem of *equivalence of ensembles*, which is only incompletely solved.

Equilibrium statistical mechanics is not restricted to classical systems of the type discussed above. By suitably modifying (1.1) and the other ensembles it can be extended to various types of systems. This is discussed in the next two sections.

1.2 CLASSICAL SYSTEMS AND ENSEMBLES

In this section we review the various types of systems and ensembles which will be studied in the rest of this book. The treatment is formal and the convergence of the integrals and series which occur will be investigated only later.

[6] These are some systems of particles interacting through pair potentials with a hard core and enclosed in a box. The proof (see Sinaï [1]) is not yet entirely published, but an outline of the argument is given by Arnold and Avez [1], p. 65.

[7] We shall see in Section 3.1.4 an example of a system where the main connected component of phase space describes a gas, but there are also small connected components with abnormal behavior.

1.2.1 CLASSICAL CONTINUOUS SYSTEMS. The subsystems which form a system of statistical mechanics are often called particles. For continuous systems these particles move in a v-dimensional space (where usually $v = 3$). Apart from translations, the particles may have degrees of freedom corresponding to rotations, and so forth, and several different species of particles may be present. For simplicity we shall here restrict our discussion to systems composed of only one species of particles having only translational degrees of freedom.

Consider a classical continuous system containing n particles with positions $x_i \in \mathbf{R}^v$ and momenta p_i; we assume that it is described by a Hamiltonian H of the form

$$H(p_1, \ldots, p_n, x_1, \ldots, x_n) = \sum_{i=1}^{n} \frac{p_i^2}{2m} + U(x_1, \ldots, x_n) \qquad (2.1)$$

where U is the (total) potential energy. It is typical of statistical mechanics that the systems have to be enclosed in a container; we assume that the bounded region $\Lambda \subset \mathbf{R}^v$ is a suitable container and we supplement (2.1) with $x_i \in \Lambda$ for $i = 1, \ldots, n$ and the condition that the particles be reflected elastically at the boundary of Λ.

(a) *Classical Continuous Systems, Microcanonical Ensemble.* The average (1.4) is called the microcanonical average. Its existence requires that the measure (1.1) be well defined but this problem can be avoided if (1.1) is replaced by one of the following measures[8] on $(\mathbf{R}^v)^n \times \Lambda^n$:

$$\frac{1}{n!} \delta^\Delta(H(\omega) - E)\, d\omega = \frac{1}{n!} \delta^\Delta(H(p_1, \ldots, p_n, x_1, \ldots, x_n) - E)\, dp_1 \cdots dp_n$$

$$\times\, dx_1 \cdots dx_n \qquad (2.2)$$

$$\frac{1}{n!} \delta^-(H(\omega) - E)\, d\omega = \frac{1}{n!} \delta^-(H(p_1, \ldots, p_n, x_1, \ldots, x_n) - E)\, dp_1 \cdots dp_n$$

$$\times\, dx_1 \cdots dx_n \qquad (2.3)$$

Here δ^Δ and δ^- are, respectively, the characteristic functions of the intervals $(-\Delta, 0)$ and $(-\infty, 0)$. Since only macroscopic observations of our system are to be made, it should not be possible to give an absolutely precise value[9] for its total energy E; hence, the replacement of δ by δ^Δ is heuristically justifiable. The replacement of δ by δ^- is justified a posteriori[10] by the fact that for a

[8] The measures in (2.2), (2.3) occur with a new factor $1/n!$ which has no significance at present since only the normalized measures are used in defining the ensemble averages. The normalization of the measures will however be given a physical meaning later (partition function, see Section 1.4), and we give for it the conventional expression.

[9] The same argument could be applied to the particle number n, but it is technically convenient here to keep n precisely known.

[10] See Section 3.3.14.

large system most of the mass of (2.3) corresponds to values of $H(\omega)$ close to E.

Because of the special form (2.1) of the Hamiltonian, the momenta p_i play a rather trivial role in classical equilibrium statistical mechanics. (This will appear more clearly in the case of the canonical ensemble.) A consequence of this fact is that it is interesting to introduce a configurational micro-canonical ensemble defined by a measure on the configuration space Λ^n of the system. We may take either

$$\frac{1}{n!}\,\delta^\Delta(U(x_1,\ldots,x_n) - E)\,dx_1\cdots dx_n \tag{2.4}$$

or

$$\frac{1}{n!}\,\delta^-(U(x_1,\ldots,x_n) - E)\,dx_1\cdots dx_n \tag{2.5}$$

by analogy with Eqs. (2.2) and (2.3).

(b) *Classical Continuous Systems, Canonical Ensemble.* In passing from the microcanonical to the canonical ensemble, the energy E is replaced as thermodynamic parameter by the *inverse temperature* $\beta > 0$. The measure defining the canonical ensemble is

$$\frac{1}{n!}\exp(-\beta H(\omega))\,d\omega$$

$$= \prod_{i=1}^{n}\left[\exp\left(-\beta\frac{p_i^2}{2m}\right)dp_i\right] \times \left[\frac{1}{n!}\exp(-\beta U(x_1,\ldots,x_n))\,dx_1\cdots dx_n\right] \tag{2.6}$$

This expression factorizes into a kinetic part

$$\exp\left(-\beta\frac{p_i^2}{2m}\right)dp_i \tag{2.7}$$

for each particle and a configurational part

$$\frac{1}{n!}\exp(-\beta U(x_1,\ldots,x_n))\,dx_1\cdots dx_n \tag{2.8}$$

which defines a configurational canonical ensemble.

(c) *Classical Continuous Systems, Grand Canonical Ensemble.* In the micro-canonical and canonical ensembles the particle numbers are fixed. The grand canonical ensemble on the other hand is defined by a measure on the sum over n of the phase spaces $(\mathbf{R}^\nu)^n \times \Lambda^n$. The number of particles is replaced as thermodynamic parameter by the *chemical potential* μ. The restriction to $(\mathbf{R}^\nu)^n \times \Lambda^n$ of the measure defining the grand canonical

ensemble is[11]

$$\frac{1}{n!}e^{n\beta\mu}\exp(-\beta H(\omega))\,d\omega = e^{n\beta\mu}\prod_{i=1}^{n}\left[\exp\left(-\beta\frac{p_i^2}{2m}\right)dp_i\right]$$

$$\times\left[\frac{1}{n!}\exp(-\beta U(x_1,\ldots,x_n))\,dx_1\cdots dx_n\right] \qquad (2.9)$$

A configurational grand canonical ensemble is defined by a measure on $\sum_{n=0}^{\infty}\Lambda^n$ with restriction to Λ^n given by

$$\frac{z^n}{n!}\exp(-\beta U(x_1,\ldots,x_n))\,dx_1\cdots dx_n \qquad (2.10)$$

Here it is convenient to consider that (2.10) is obtained from (2.9) by integration over the momenta, so that

$$z = e^{\beta\mu}\left(\frac{2\pi m}{\beta}\right)^{\nu/2} \qquad (2.11)$$

The parameter $z > 0$ is the "activity."

1.2.2 CLASSICAL CONTINUOUS SYSTEMS WITH HARD CORES. In a system of particles with hard cores certain configurations are excluded from the configuration space Λ^n. It is usually required that the Euclidean distance $|x_j - x_i|$ of each pair of particles remains larger than some constant $R > 0$. If we define

$$S_R^n = \{(x_1,\ldots,x_n)\in(\mathbf{R}^\nu)^n:|x_j - x_i| < R \quad \text{for some} \quad i \neq j\} \qquad (2.12)$$

the allowed configuration space is thus the complement $\Lambda^n\backslash S_R^n$. Together with this restriction of phase space, the Hamiltonian (2.1) has to be supplemented with the condition that two particles undergo an elastic collision when their mutual distance becomes equal to R. The introduction of hard cores brings no essential modification in the treatment of the various ensembles as described above.

1.2.3 CLASSICAL LATTICE SYSTEMS. Lattice systems are obtained most naturally by the idealization of crystals constituted of subsystems which may exist in a finite number of (classical or quantum) states. We assume that the subsystems (=lattice sites) are parametrized by ν-tuples of integers, so that the lattice is identified with \mathbf{Z}^ν. We restrict ourselves now to the classical case and let the states of the subsystem at $x\in\mathbf{Z}^\nu$ be numbered by $n_x = 0, 1,\ldots, N$. We may interpret n_x as describing the species of particle

[11] For $n = 0$, $(\mathbf{R}^\nu)^n \times \Lambda^n$ (resp. Λ^n) is reduced to a point, to which (2.9) [resp. (2.10)] attributes the weight 1.

present at x ("alloys") or as an occupation number ("lattice gases"[12]) or we may interpret $s_x = n_x - \frac{1}{2}N$ as a spin component ("spin systems").

A bounded region Λ will be here a finite subset of $\mathbf{Z}^v : \Lambda = \{x_1, \dots, x_V\}$. An ensemble for a classical lattice system will be a measure on the set of all possible configurations of the system within the region Λ. There are $(N + 1)^V$ such configurations parametrized by the V-tuples $(n_{x_1}, \dots, n_{x_V})$, and an ensemble will attribute a certain weight to each of these. We describe only a grand canonical ensemble for which the weight is

$$\exp\left[\beta \sum_{i=1}^{V} \mu(n_{x_i}) \right] \exp[-\beta U(n_{x_1}, \dots, n_{x_V})] \tag{2.13}$$

In this expression $U(n_{x_1}, \dots, n_{x_V})$ is the potential energy corresponding to the configuration $(n_{x_1}, \dots, n_{x_V})$. In the "alloy" interpretation, $\mu(n_x)$ is the chemical potential for the species n_x. In the "lattice gas" interpretation we should take $\mu(n_x) = n_x \mu$ and (2.13) becomes

$$\exp\left[\left(\sum_{i=1}^{V} n_{x_i} \right) \beta \mu \right] \exp[-\beta U(n_{x_1}, \dots, n_{x_V})]$$

$$= z^n \exp[-\beta U(n_{x_1}, \dots, n_{x_V})] \tag{2.14}$$

where we have put

$$z = e^{\beta \mu} \qquad n = \sum_{i=1}^{V} n_{x_i} \tag{2.15}$$

The right-hand side of (2.14) is clearly analogous to the expression (2.10) for the configurational grand canonical ensemble in the continuous case. In the "spin system" interpretation the spins may interact with a "magnetic field" h and we choose $\mu(n_x)$ to be minus the interaction energy: $\mu(n_x) = s_x h = (n_x - \frac{1}{2}N)h$. Using (2.15) also, (2.13) becomes[13]

$$\exp[(n - \tfrac{1}{2}NV)\beta h - \beta U(n_{x_1}, \dots, n_{x_V})] \tag{2.16}$$

1.3 QUANTUM SYSTEMS AND ENSEMBLES

While the observable quantities in classical systems can be described by real functions on phase or configuration space, they are given for quantum systems by self-adjoint operators on the Hilbert space of the system. We have seen that in classical statistical mechanics the ensemble averages are defined by a probability measure on phase or configuration space. In

[12] "Lattice gases" are the discrete version of the systems studied in Section 1.2.1, but only the configurational version of the ensembles makes sense. Usually we take $N = 1$ so that every lattice site may be occupied only by zero or one particle; this is effectively a "hard core" condition. A "lattice gas" need not be a "gas" in any sense of the word.

[13] There is no kinetic energy associated with spin.

quantum statistical mechanics they are given by a density matrix on the Hilbert space of the system. We call unnormalized density matrix a positive operator of trace class. To each quantum ensemble there will correspond such an operator T and the ensemble average of a quantity A will be given by

$$\langle A \rangle = (\mathrm{Tr}\ T)^{-1}(\mathrm{Tr}\ TA) \tag{3.1}$$

1.3.1 QUANTUM CONTINUOUS SYSTEMS. As in the classical case we restrict our attention to systems in \mathbf{R}^v composed of only one species of particles having only translational degrees of freedom. We describe a system of n such particles by a Hamiltonian of the form

$$H_n = \sum_{i=1}^{n} \left(-\frac{1}{2m}\Delta_i \right) + U(x_1,\ldots,x_n) \tag{3.2}$$

where Δ_i is the Laplace operator for the coordinate x_i of the ith particle and U acts multiplicatively on the wave function. If the system is enclosed in a suitable bounded region Λ of \mathbf{R}^v, H_n acts on $L^2(\Lambda^n)$, the Hilbert space of (Lebesgue-) square integrable functions of n arguments in Λ. Furthermore we supplement (3.2) with the condition that the boundary of Λ is "infinitely repulsive"[14]. In this way H_n becomes a self-adjoint operator $H_n(\Lambda)$, bounded below, with spectrum consisting of isolated eigenvalues of finite multiplicity (see Section 2.10).

In the quantum description of a system containing identical particles, it is required that the wave function be either symmetric (if the particles are "bosons") or antisymmetric (if the particles are "fermions") under permutations of the arguments corresponding to these particles.[15] Let $L_\varepsilon^2(\Lambda^n)$, with $\varepsilon = \pm$, be the subspace of $L^2(\Lambda^n)$ consisting of the functions which are either symmetric (for $\varepsilon = +$) or antisymmetric (for $\varepsilon = -$) under permutations of their n arguments. The Hamiltonian $H_n(\Lambda)$ is symmetric under permutations of the particles and may thus be restricted to $L_+^2(\Lambda^n)$ in the case of bosons (Bose–Einstein or B.–E. statistics) or to $L_-^2(\Lambda^n)$ in the case of fermions (Fermi–Dirac or F.–D. statistics). It is sometimes interesting to discuss also the case where no symmetry restriction is imposed on the wave functions and $L^2(\Lambda^n)$ is kept as Hilbert space of the system (Maxwell–Boltzmann or M.–B. statistics).

[14] This condition is usually expressed by requiring that the eigenfunctions of H vanish when some argument reaches the boundary of Λ; we shall give in Section 2.10 a prescription of wider applicability which uses the Friedrichs extension.

[15] It happens that a particle is a boson or a fermion depending on whether its spin is an integer or half-odd number. This rule (the "connection between spin and statistics") can be justified in relativistic field theory, but has no logical necessity in our nonrelativistic description. There is therefore no logical need for us to introduce spin components even for fermions. Actually the reader can easily supply indices of spin and particle species which we have omitted for notational convenience.

(a) *Quantum Continuous Systems, Microcanonical Ensemble*. The original formulation of the classical microcanonical ensemble given by (1.1) and (1.4) has no satisfactory quantum counterpart due to the fact that the spectrum of $H_n(\Lambda)$ is discrete. On the other hand, to the measures (2.2) and (2.3) correspond the following unnormalized density matrices on $L^2(\Lambda^n)$

$$\frac{1}{n!}\delta^\Delta(H_n(\Lambda) - E) = \frac{1}{n!}\sum_{E_{n\alpha}\in(E-\Delta,\,E)} P_{n\alpha} \tag{3.3}$$

$$\frac{1}{n!}\delta^-(H_n(\Lambda) - E) = \frac{1}{n!}\sum_{E_{n\alpha}<E} P_{n\alpha} \tag{3.4}$$

where $E_{n\alpha}$ are the eigenvalues of $H_n(\Lambda)$ and $P_{n\alpha}$ the projections on the corresponding eigenspaces.[16] In the case of B.–E. or F.–D. statistics, $H_n(\Lambda)$ and $P_{n\alpha}$ are restricted to $L_\varepsilon^2(\Lambda^n)$ and (3.3), (3.4) are replaced by

$$\delta^\Delta(H_n(\Lambda) - E) = \sum_{E_{n\alpha}\in(E-\Delta,\,E)} P_{n\alpha} \tag{3.5}$$

$$\delta^-(H_n(\Lambda) - E) = \sum_{E_{n\alpha}<E} P_{n\alpha} \tag{3.6}$$

(b) *Quantum Continuous Systems, Canonical Ensemble*. The unnormalized density matrix defining the canonical ensemble is

$$\frac{1}{n!}\exp[-\beta H_n(\Lambda)] = \frac{1}{n!}\sum_\alpha \exp[-\beta E_{n\alpha}]P_{n\alpha} \tag{3.7}$$

acting on $L^2(\Lambda)$ or

$$\exp[-\beta H_n(\Lambda)] = \sum_\alpha \exp[-\beta E_{n\alpha}]P_{n\alpha} \tag{3.8}$$

acting on $L_\varepsilon^2(\Lambda)$, depending on the statistics.

(c) *Quantum Continuous Systems, Grand Canonical Ensemble*. Since the number of particles is not fixed in the grand canonical ensemble, the corresponding density matrix will act on the (Hilbert space) direct sum of the spaces $L^2(\Lambda^n)$:

$$\mathcal{H}(\Lambda) = \bigoplus_{n=0}^\infty L^2(\Lambda^n) \tag{3.9}$$

[16] We cannot justify the introduction of the quantum ensembles from ergodic theory and we are just using analogy with the classical situation. Notice however that the time invariance of the ensemble average requires the corresponding density matrix ρ to commute with $H_n(\Lambda)$. If $H_n(\Lambda)$ has nondegenerate eigenvalues (general situation), ρ must be a function of $H_n(\Lambda)$: $\rho = f(H_n(\Lambda))$. It turns out that for a large system the nature of the ensembles obtained is very insensitive to the form of f (equivalence of ensembles).

For B.–E. or F.–D. statistics, this is restricted to

$$\mathcal{H}_\varepsilon(\Lambda) = \bigoplus_{n=0}^{\infty} L_\varepsilon^2(\Lambda^n) \tag{3.10}$$

with $\varepsilon = +$ or $\varepsilon = -$, respectively; \mathcal{H}_ε is the Fock space[17] associated with the region Λ. The self-adjoint unbounded operator on $\mathcal{H}(\Lambda)$ or $\mathcal{H}_\varepsilon(\Lambda)$ which has eigenspaces $L^2(\Lambda^n)$ or $L_\varepsilon^2(\Lambda^n)$ with corresponding eigenvalue n will be denoted by $n(\Lambda)$ and called particle number operator; notice that $n(\Lambda) \geq 0$. The self-adjoint operator $H(\Lambda)$ on $\mathcal{H}(\Lambda)$ or $\mathcal{H}_\varepsilon(\Lambda)$, which has the restriction $H_n(\Lambda)$ to $L^2(\Lambda^n)$ or $L_\varepsilon^2(\Lambda^n)$, is called the total Hamiltonian. Clearly $n(\Lambda)$ and $H(\Lambda)$ commute. With this notation, the grand canonical ensemble is defined by the unnormalized density matrix

$$(n(\Lambda)!)^{-1} \exp[\beta(\mu n(\Lambda) - H(\Lambda))] = \sum_n \frac{e^{n\beta\mu}}{n!} \sum_\alpha [\exp(-\beta E_{n\alpha})]P_{n\alpha} \tag{3.11}$$

acting on $\mathcal{H}(\Delta)$ or

$$\exp[\beta(\mu n(\Lambda) - H(\Lambda))] = \sum_n e^{n\beta\mu} \sum_\alpha [\exp(-\beta E_{n\alpha})]P_{n\alpha} \tag{3.12}$$

acting on $\mathcal{H}_\varepsilon(\Lambda)$, depending on the statistics.

1.3.2 QUANTUM CONTINUOUS SYSTEMS WITH HARD CORES. If we consider, as in the classical case, spherical hard cores of diameter $R > 0$, we have to restrict the n-particle Hilbert space to $L^2(\Lambda^n \backslash S_R^n)$ or $L_\varepsilon^2(\Lambda^n \backslash S_R^n)$. The elasticity of the collisions between the particles is imposed in the same manner as for the collisions with the boundary of Λ (see Section 2.10).

1.3.3 QUANTUM LATTICE SYSTEMS. In the quantum version of the lattice systems discussed in Section 1.2.3, we associate with each lattice site $x \in \mathbf{Z}^\nu$ a Hilbert space \mathcal{H}_x of finite dimension $N + 1$ describing the states of the corresponding subsystem. If $\Lambda = \{x_1, \ldots, x_V\}$ is a finite subset of \mathbf{Z}^ν, we associate with it the Hilbert space

$$\mathcal{H}(\Lambda) = \bigotimes_{x \in \Lambda} \mathcal{H}_x \tag{3.13}$$

of dimension $(N + 1)^V$. The Hamiltonian $H(\Lambda)$ for the region Λ of the lattice system will be a self-adjoint operator on $\mathcal{H}(\Lambda)$. An ensemble of the grand canonical type is now described by the unnormalized density matrix

$$\exp[-\beta H(\Lambda)] \tag{3.14}$$

on $\mathcal{H}(\Lambda)$ (where we have omitted a chemical potential or magnetic field

[17] That is, the Hilbert space of the Fock representation of the canonical commutation ($\varepsilon = +$) or anticommutation ($\varepsilon = -$) relations which we shall describe in Section 7.1.4.

term). Let $(\psi_{xn})_{0 \le n \le N}$ be an orthonormal basis of \mathscr{H}_x for each $x \in \mathbf{Z}^\nu$. Then the number operators n_x, $n(\Lambda)$, are defined on $\mathscr{H}(\Lambda)$ by

$$n_{x_k} \bigotimes_{i=1}^{\nu} \psi_{x_i n_i} = n_k \bigotimes_{i=1}^{\nu} \psi_{x_i n_i} \qquad (3.15)$$

$$n(\Lambda) = \sum_{x \in \Lambda} n_x \qquad (3.16)$$

and we can write instead of (3.14) the expression

$$\exp\left[\beta \sum_{i=1}^{\nu} \mu(n_{x_i}) \right] \exp[-\beta H(\Lambda)] \qquad (3.17)$$

which is analogous to (2.13) and can be similarly specialized.[18] Notice that if we choose

$$H(\Lambda) \bigotimes_{x \in \Lambda} \psi_{xn_x} = U(n_{x_1}, \ldots, n_{x_\nu}) \bigotimes_{x \in \Lambda} \psi_{xn_x} \qquad (3.18)$$

then (3.17) reduces to (2.13), showing that classical lattice systems are a special case of quantum lattice systems.

1.4 EQUILIBRIUM STATES AND THERMODYNAMIC FUNCTIONS

Let us call *state* of a physical system an expectation value functional on the observable quantities for this system. The averages corresponding to the ensembles introduced in Section 1.2 and 1.3 are such states. We shall say that the systems for which the ensemble averages are defined are *finite systems*, but we may also consider the corresponding *infinite systems* which contain an infinity of subsystems and extend throughout \mathbf{R}^ν or \mathbf{Z}^ν. The general discussion in Section 1.1 makes it likely that if a system exhibits thermodynamic behavior the states defined by the ensemble averages for large[19] finite systems approach in some sense states of the corresponding infinite systems. Besides the existence of the limit states (existence of the thermodynamic limit), we would also like to prove that they are not dependent on the choice of ensemble (equivalence of ensembles). We shall call the limit states *equilibrium states* (for the infinite systems).

We shall develop the point of view that the main problem of equilibrium statistical mechanics is to study the infinite system equilibrium states and their relation to the interactions which give rise to them. Because we are interested in systems with thermodynamic behavior, we expect that the equilibrium states describe a situation where

[18] In the "lattice gas" situation a "kinetic energy" may be simulated by a difference operator.
[19] We shall make precise, in specific cases, in which sense the system has to become large.

(a) in a bounded region of space the number of subsystems remains finite, and
(b) each subsystem has negligible interaction with the subsystems which are far away.

This implies of course that the interactions between subsystems cannot be arbitrary: they must be weak between distant subsystems, and they must also prevent the collapse of an infinite number of subsystems into a bounded region of space. Another property which is assumed for the interactions is their invariance under translations in \mathbf{R}^ν or \mathbf{Z}^ν. This invariance is not very interesting for finite systems; it is broken by the fact that the system is enclosed in a bounded region Λ. For infinite systems, however, the translation invariance[20] of the equilibrium states will be essential in their study.

We expect that if we restrict an infinite system equilibrium state to a finite region Λ, it will yield a state analogous to those defined by the ensemble averages, described therefore by a probability measure or density matrix on the appropriate space. This provides a possible mathematical description of equilibrium states: a collection of measures or density matrices corresponding to all bounded regions of \mathbf{R}^ν or \mathbf{Z}^ν. We shall also use more conventional descriptions: correlation functions for classical systems and reduced density matrices for quantum systems.

It is now time to consider the problem of the determination of thermodynamic functions from statistical mechanics. In the traditional approach we introduce for each ensemble a *partition function*, which is the total mass of the measure defining the ensemble (classical systems) or the trace of the unnormalized density matrix defining the ensemble (quantum systems). It is then accepted that the logarithm of the partition function divided by the volume of the region Λ containing the system has a limit when this system becomes large, and this limit is identified with a thermodynamic function.[21] This is the approach which we shall adopt in the next chapter. We remark however that if the equilibrium states introduced above give a complete description of systems in the thermodynamic limit, they should also determine the thermodynamic functions. It will turn out later that this is true, although it is not apparent when thermodynamic functions are defined from the logarithm of normalization factors. In fact, equilibrium states give a statistical description of systems and the amount of randomness present in this description will be reflected in the entropy per unit volume (see Section 7.2).

[20] Or the invariance under a larger group.

[21] This thermodynamic function depends on the ensemble, and the thermodynamic functions obtained from the various ensembles should be related by the usual thermodynamic relations (equivalence of ensembles).

THE THERMODYNAMIC LIMIT FOR THERMODYNAMIC FUNCTIONS: LATTICE SYSTEMS

In Chapters 2 and 3 we study the thermodynamic limit for thermodynamic functions. In this chapter we treat the simple case of lattice systems (classical and quantum) in the grand canonical ensemble. We introduce also some notation and results needed later concerning the limit of an infinite volume (Section 2.1) and inequalities for quantum systems (Section 2.5).

2.1 LIMIT OF AN INFINITE VOLUME

All the ensembles which we have considered describe systems enclosed in a bounded region Λ of \mathbf{R}^v or \mathbf{Z}^v. Taking the thermodynamic limit involves letting Λ "tend to infinity"; we have to describe how.

Let $a = (a^1, \ldots, a^v) \in \mathbf{R}^v$ (resp. $\in \mathbf{Z}^v$), if $a^1 > 0, \ldots, a^v > 0$ we define

$$\Lambda(a) = \{x \in \mathbf{R}^v : 0 \le x^i < a^i \quad \text{for} \quad i = 1, \ldots, v\} \tag{1.1}$$

respectively

$$\Lambda(a) = \{x \in \mathbf{Z}^v : 0 \le x^i < a^i \quad \text{for} \quad i = 1, \ldots, v\} \tag{1.2}$$

The "volume" of $\Lambda(a)$ is

$$V(a) = \prod_{i=1}^{v} a^i \tag{1.3}$$

We write $a \to \infty$ if $a^1 \to +\infty, \ldots, a^v \to +\infty$. When $a \to \infty$, $\Lambda(a)$ tends to infinity in a manner suitable for the study of the thermodynamic limit.

There are, however, more general and physically more satisfactory notions.

If $\Lambda(a)$ is translated by a vector $na = (n^1 a^1, \ldots, n^\nu a^\nu)$ with $n \in \mathbf{Z}^\nu$ the result is a set

$$\Lambda_n := \Lambda(a) + na \qquad (1.4)$$

and the family $(\Lambda_n)_{n \in \mathbf{Z}^\nu}$ forms a partition of \mathbf{R}^ν or \mathbf{Z}^ν. For any subset Λ of \mathbf{R}^ν or \mathbf{Z}^ν we let $N_a^+(\Lambda)$, resp. $N_a^-(\Lambda)$, be the number of sets Λ_n such that $\Lambda_n \cap \Lambda \neq \varnothing$, resp. $\Lambda_n \subset \Lambda$.

2.1.1 DEFINITION. *The sets Λ tend to infinity in the sense of Van Hove[1] if*

$$\lim N_a^-(\Lambda) = +\infty \qquad \lim N_a^-(\Lambda)/N_a^+(\Lambda) = 1 \qquad (1.5)$$

for all a.

This notion is invariant under inhomogeneous linear transformations of \mathbf{R}^ν or \mathbf{Z}^ν.

Suppose that the sets $\Lambda \subset \mathbf{R}^\nu$ are bounded and Lebesgue measurable. Let $V(\Lambda)$ be the measure of Λ and $V_h(\Lambda)$ the measure of the set of points with distance less than or equal to h to the boundary of Λ. The notion of Van Hove convergence to infinity is equivalent to

$$\lim V(\Lambda) = +\infty \qquad \lim V(\Lambda)^{-1} V_h(\Lambda) = 0 \qquad (1.6)$$

It has been necessary in some cases to use a stronger notion in order to prove the existence of the thermodynamic limit.

2.1.2 DEFINITION. *We denote by $d(\Lambda)$ the diameter[2] of Λ. The sets Λ tend to infinity in the sense of Fisher[3] if*

$$\lim V(\Lambda) = +\infty \qquad (1.7)$$

and there exists a "shape function" π such that

$$\lim_{\alpha \to 0} \pi(\alpha) = 0 \qquad (1.8)$$

and for sufficiently small α and all Λ

$$V_{\alpha d(\Lambda)}(\Lambda)/V(\Lambda) \leq \pi(\alpha) \qquad (1.9)$$

This notion is again invariant under linear transformations. Since $V_{\alpha d(\Lambda)}$ is larger than the volume of a sphere with radius $\alpha d(\Lambda)$, (1.9) implies that there exists $C > 0$ such that

$$V(\Lambda) \geq C\, d(\Lambda)^\nu \qquad (1.10)$$

The sets $\Lambda((t^2, t))$ in \mathbf{R}^2 (see Fig. 1) tend to infinity in the sense of Van Hove but not in the sense of Fisher when $t \to +\infty$.

[1] See Van Hove [1].
[2] The diameter of a set is the supremum of the distance between pairs of points.
[3] See Fisher [1].

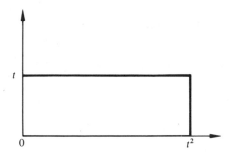

FIGURE 1 The sets $\Lambda((t^2, t))$ in \mathbf{R}^2.

2.2 INTERACTIONS OF QUANTUM LATTICE SYSTEMS

We shall start our study of the thermodynamic limit with quantum lattice systems. For such a system a Hamiltonian $H(\Lambda)$ is associated with every finite $\Lambda \subset \mathbf{Z}^\nu$. The Hamiltonian $H(\Lambda)$ is a self-adjoint operator on the Hilbert space $\mathscr{H}(\Lambda)$ defined by Eq. (3.13) of Chapter 1; if $\Lambda = \varnothing$, we put $H(\Lambda) = 0$.

Let $\Lambda_1 \cap \Lambda_2 = \varnothing$, $\mathscr{H}(\Lambda_1 \cup \Lambda_2)$ can be identified naturally with $\mathscr{H}(\Lambda_1) \otimes \mathscr{H}(\Lambda_2)$. We shall also identify any operator A_1 on $\mathscr{H}(\Lambda_1)$ with the operator[4] $A_1 \otimes \mathbf{1}$ on $\mathscr{H}(\Lambda_1 \cup \Lambda_2)$. In particular, for any $\Lambda \supset \Lambda_1$, $H(\Lambda_1)$ is identified with an operator on $\mathscr{H}(\Lambda)$.

In order to express the translation invariance of the theory we first introduce for each $a \in \mathbf{Z}^\nu$ and each $x \in \mathbf{Z}^\nu$ a unitary mapping $V_x(a): \mathscr{H}_x \to \mathscr{H}_{x+a}$ such that $V_x(0)$ is the identity in \mathscr{H}_x and $V_x(a_1 + a_2) = V_{x+a_2}(a_1)V_x(a_2)$; we also define

$$V_\Lambda(a) = \bigotimes_{x \in \Lambda} V_x(a) \tag{2.1}$$

We postulate now that

$$H(\Lambda + a) = V_\Lambda(a)H(\Lambda)V_{a+\Lambda}(-a) \tag{2.2}$$

or, omitting the index[5] of V,

$$H(\Lambda + a) = V(a)H(\Lambda)V(a)^{-1} \tag{2.3}$$

We may write

$$H(\Lambda) = \sum_{X \subset \Lambda} \Phi(X) \tag{2.4}$$

This relation determines a $\Phi(X)$ recursively, as a self adjoint operator on

[4] We use the symbol $\mathbf{1}$ for the identity operator; the space on which it acts is obvious from the context.
[5] This omission is compatible with the identification $H(\Lambda_1) \to H(\Lambda_1) \otimes \mathbf{1}$ mentioned above.

$\mathcal{H}(X)$ (identified with an operator on $\mathcal{H}(\Lambda)$ for any $\Lambda \supset X$) and (2.3) yields

$$\Phi(X + a) = V(a)\Phi(X)V(a)^{-1} \tag{2.5}$$

The operator-valued set function Φ is the *potential*[6] describing the interactions of the subsystems in the theory. Equation (2.5) expresses the translation invariance of the potential. We now describe classes of potentials which will be used in the sequel.

We say that the potential Φ has *finite range* if the sets X, such that $X \ni 0$ and $\Phi(X) \neq 0$, are finite in number. Let Δ be their union. Obviously Δ is finite, $\Delta \ni 0$ and $\Delta = -\Delta$[7]; Δ may be called the "range" of Φ. If $x, y \in X$ and $x - y \notin \Delta$, then $\Phi(X) = 0$ because $X - x \ni 0$, $X - x \notin \Delta$ imply $\Phi(X - x) = 0$, and hence $\Phi(X) = 0$. Let

$$a \notin \Lambda_1 - \Lambda_2 + \Delta \tag{2.6}$$

or equivalently $[\Lambda_1 - (\Lambda_2 + a)] \cap \Delta = \emptyset$. If $x \in \Lambda_1$ and $y \in \Lambda_2 + a$, we have $x - y \notin \Delta$ and hence, if $x, y \in X$, $\Phi(X) = 0$. In particular, (2.6) implies

$$H(\Lambda_1 \cup (\Lambda_2 + a)) = \sum_{X \subset \Lambda_1 \cup (\Lambda_2 + a)} \Phi(X)$$

$$= \sum_{X \subset \Lambda_1} \Phi(X) + \sum_{X \subset \Lambda_2 + a} \Phi(X) = H(\Lambda_1) + H(\Lambda_2 + a) \tag{2.7}$$

This shows that the regions Λ_1, $\Lambda_2 + a$ "do not interact if they are sufficiently far apart."

In what follows we shall restrict our attention to the potentials Φ such that

$$\|\Phi\| = \sum_{X \ni 0} \frac{\|\Phi(X)\|}{N(X)} < +\infty \tag{2.8}$$

where $N(X)$ is the number of points in X. The space \mathcal{B} of these potentials is a real Banach space with respect to the norm (2.8). The space \mathcal{B}_0 of potentials with finite range is dense in \mathcal{B}. From (2.4) and (2.8) we obtain

$$\|H(\Lambda)\| = \left\| \sum_{x \in \Lambda} \sum_{\substack{X \ni x \\ X \subset \Lambda}} \frac{\Phi(X)}{N(X)} \right\| \leq \sum_{x \in \Lambda} \sum_{X \ni x} \frac{\|\Phi(X)\|}{N(X)}$$

$$= N(\Lambda)\|\Phi\| \tag{2.9}$$

We have introduced in Section 1.3.3 an ensemble described by the unnormalized density matrix $\exp[-\beta H(\Lambda)]$; the corresponding partition

[6] We adopt here the point of view that the Hamiltonian contains no kinetic energy. This would apply for instance to a quantum spin system; in this interpretation $\Phi(\{x\})$ is the interaction of the spin at $x \in \mathbf{Z}^\nu$ with an external magnetic field.

[7] We have $\Delta = \{x : \text{there exists } X \text{ such that } x \in X, 0 \in X, \text{ and } \Phi(X) \neq 0\}$ hence $x \in X$ implies $-x \in X - x$; $0 \in X - x$ and $\Phi(X - x) \neq 0$ imply $x \in -\Delta$.

function is

$$\Xi(\Lambda, \beta) = \mathrm{Tr}_{\mathcal{H}(\Lambda)} \exp[-\beta H(\Lambda)] \tag{2.10}$$

and the corresponding thermodynamic function[8] is

$$\lim N(\Lambda)^{-1} \log \Xi(\Lambda, \beta) \tag{2.11}$$

when Λ tends to infinity in a suitable manner. It is apparent that the factor β plays a trivial role in these formulas; for notational clarity we shall absorb it in the interaction potential Φ (which amounts to setting $\beta = 1$). We write

$$Z_\Lambda(\Phi) = \mathrm{Tr}_{\mathcal{H}(\Lambda)} \exp[-H(\Lambda)] \tag{2.12}$$

so that $\Xi(\Lambda, \beta) = Z_\Lambda(\beta\Phi)$, and we define

$$P_\Lambda(\Phi) = N(\Lambda)^{-1} \log Z_\Lambda(\Phi) \tag{2.13}$$

Notice that by translation invariance

$$Z_{\Lambda+a}(\Phi) = Z_\Lambda(\Phi) \qquad P_{\Lambda+a}(\Phi) = P_\Lambda(\Phi) \tag{2.14}$$

2.2.1 LEMMA. *If A and B are self-adjoint operators on \mathbf{C}^n, then*

$$|\log \mathrm{Tr}(e^A) - \log \mathrm{Tr}(e^B)| \le \|A - B\| \tag{2.15}$$

Let $h \to A(h)$ be linear from \mathbf{R} to the self-adjoint operators on \mathbf{C}^n, then

$$\left| \frac{d}{dh} \log \mathrm{Tr}(e^{A(h)}) \right| = \left| \frac{\mathrm{Tr}[H(dA(h)/dh)(\exp A(h))]}{\mathrm{Tr}(\exp A(h))} \right|$$

$$\le \left\| \frac{d}{dh} A(h) \right\|$$

and (2.15) follows by taking $A(h) = B + h(A - B)$ and integrating from zero to one.

2.2.2 PROPOSITION. (a) *If $\Phi, \Psi \in \mathcal{B}$, then*

$$|P_\Lambda(\Phi) - P_\Lambda(\Psi)| \le \|\Phi - \Psi\| \tag{2.16}$$

(b) *The function $P_\Lambda(\cdot)$ is convex on \mathcal{B}.*
Using (2.13), (2.15), and (2.9) we get

$$|P_\Lambda(\Phi) - P_\Lambda(\Psi)| = N(\Lambda)^{-1} |\log \mathrm{Tr} \exp H_\Phi(\Lambda) - \log \mathrm{Tr} \exp H_\Psi(\Lambda)|$$

$$\le N(\Lambda)^{-1} \|H_\Phi(\Lambda) - H_\Psi(\Lambda)\| = N(\Lambda)^{-1} \|H_{\Phi-\Psi}(\Lambda)\|$$

$$\le \|\Phi - \Psi\|$$

which proves (a). Property (b) results immediately from Proposition 2.5.5 which we shall establish later.

[8] In the spin system interpretation this is up to sign a spin "free energy" multiplied by β.

2.2.3 PROPOSITION. *If* $\Phi \in \mathscr{B}$, *then*

$$|P_\Lambda(\Phi) - \log(N + 1)| < \|\Phi\| \qquad (2.17)$$

This follows from Proposition 2.2.2(a), with $\Psi = 0$:

$$P_\Lambda(0) = N(\Lambda)^{-1} \log \mathrm{Tr}_{\mathscr{H}(\Lambda)} 1 = N(\Lambda)^{-1} \log(N + 1)^{N(\Lambda)} = \log(N + 1)$$

2.3 THERMODYNAMIC LIMIT FOR QUANTUM LATTICE SYSTEMS

2.3.1 PROPOSITION. *Let the potential* Φ *have finite range* ($\Phi \in \mathscr{B}_0$) *and* Δ *be the associated finite subset of* \mathbf{Z}^ν. *Let* $\Lambda_1 \cap \Lambda_2 = \varnothing$ *and* $N(\Lambda_1, \Lambda_2)$ *be the number of points* $a \in \mathbf{Z}^\nu$ *such that* $\Lambda_1 \cap (\Delta + a) \neq \varnothing$ *and* $\Lambda_2 \cap (\Delta + a) \neq \varnothing$; *then*

$$|N(\Lambda_1 \cup \Lambda_2)P_{\Lambda_1 \cup \Lambda_2}(\Phi) - N(\Lambda_1)P_{\Lambda_1}(\Phi) - N(\Lambda_2(\Phi)|)P_{\Lambda_2} \leq N(\Lambda_1, \Lambda_2)\|\Phi\| \quad (3.1)$$

We have by (2.4)

$$\|H(\Lambda_1 \cup \Lambda_2) - H(\Lambda_1) - H(\Lambda_2)\| = \|\sum_X{}^* \Phi(X)\|$$

$$= \left\| \sum_{x \in \Lambda_1 \cup \Lambda_2} \sum_{X \ni x}{}^* \frac{\Phi(X)}{N(X)} \right\| \leq N(\Lambda_1, \Lambda_2)\|\Phi\|$$

$$(3.2)$$

where \sum^* extends over those $X \subset \Lambda_1 \cup \Lambda_2$ such that $X \cap \Lambda_1 \neq \varnothing$ and $X \cap \Lambda_2 \neq 0$. Notice that $\Phi(X)$ vanishes unless $X - x \subset \Delta$; hence unless $\Lambda_1 \cap (\Delta + x) \neq \varnothing$, $\Lambda_2 \cap (\Delta + x) \neq \varnothing$ which justifies the last inequality.

Using Lemma 2.2.1 with $A = -H(\Lambda_1 \cup \Lambda_2)$ and $B = -[H(\Lambda_1) + H(\Lambda_2)]$ on $\mathscr{H}(\Lambda_1 \cup \Lambda_2)$ we obtain

$$|N(\Lambda_1 \cup \Lambda_2)P_{\Lambda_1 \cup \Lambda_2}(\Phi) - N(\Lambda_1)P_{\Lambda_1}(\Phi) - N(\Lambda_2)P_{\Lambda_2}(\Phi)|$$

$$= |\log \mathrm{Tr} \exp[-H(\Lambda_1 \cup \Lambda_2)] - \log \mathrm{Tr} \exp[-H(\Lambda_1) - H(\Lambda_2)]|$$

$$\leq \|(H(\Lambda_1 \cup \Lambda_2) - H(\Lambda_1) - H(\Lambda_2)\| \leq N(\Lambda_1, \Lambda_2)\|\Phi\|$$

which concludes the proof.

2.3.2 PROPOSITION. *If* $\Phi \in \mathscr{B}_0$ *the following limit exists and is finite*

$$P(\Phi) = \lim_{\Lambda \to \infty} P_\Lambda(\Phi) \qquad (3.3)$$

when Λ *tends to infinity in the sense of Van Hove.*

According to Section 2.1 we write $\Gamma_a^- \subset \Lambda$ where Γ_a^- is the union of $N_a^-(\Lambda)$ translates of $\Lambda(a)$ contained in Λ. Let Λ_j be the translates which

constitute $\Gamma_a^-, j = 1, \ldots, N_a^-(\Lambda)$, and let

$$\Gamma(j) = (\Lambda \setminus \Gamma_a^-) \cup \bigcup_{i=1}^{j} \Lambda_i$$

so that $\Gamma(N_a^-(\Lambda)) = \Lambda$. Proposition 2.3.1 gives

$$|N(\Gamma(j))P_{\Gamma(j)}(\Phi) - N(\Gamma(j-1))P_{\Gamma(j-1)}(\Phi) - V(a)P_{\Lambda(a)}(\Phi)|$$
$$\leq N(\Gamma(j-1), \Lambda_j)\|\Phi\|$$

and, summing over j

$$|N(\Lambda)P_\Lambda(\Phi) - N(\Lambda \setminus \Gamma_a^-)P_{\Lambda \setminus \Gamma_a^-}(\Phi) - N_a^-(\Lambda)V(a)P_{\Lambda(a)}(\Phi)|$$
$$\leq \|\Phi\| \sum_j N(\Gamma(j-1), \Lambda_j) \tag{3.4}$$

By definition $N(\Gamma(j-1), \Lambda_j)$ is less than the number of points b such that $\Delta + b$ intersects both Λ_j and its complement. Since these points b are "close to the boundary" of Λ_j, given $\varepsilon > 0$, we can choose a so large that

$$V(a)^{-1}N(\Gamma(j-1), \Lambda_j) < \varepsilon/(2\|\Phi\|)$$

Inserting this in (3.4) yields

$$|N(\Lambda)[P_\Lambda(\Phi) - P_{\Lambda(a)}(\Phi)] - N(\Lambda \setminus \Gamma_a^-)[P_{\Lambda \setminus \Gamma_a^-}(\Phi) - P_{\Lambda(a)}(\Phi)]|$$
$$\leq N_a^-(\Lambda)V(a)(\varepsilon/2) \tag{3.5}$$

From (2.17) we get

$$|P_{\Lambda \setminus \Gamma_a^-}(\Phi) - P_{\Lambda(a)}(\Phi)| \leq 2\|\Phi\| \tag{3.6}$$

On the other hand, in view of Van Hove convergence, we may assume that

$$N(\Lambda)^{-1}N(\Lambda \setminus \Gamma_a^-) \leq N_a^+(\Lambda)^{-1}[N_a^+(\Lambda) - N_a^-(\Lambda)] < \varepsilon/(4\|\Phi\|) \tag{3.7}$$

Inserting Eqs. (3.6) and (3.7) in (3.5) yields

$$|P_\Lambda(\Phi) - P_{\Lambda(a)}(\Phi)| < \varepsilon$$

if a and Λ are sufficiently large, which concludes the proof.

2.3.3 THEOREM.[9] *If* $\Phi \in \mathscr{B}$, *the following limit exists and is finite*

$$P(\Phi) = \lim_{\Lambda \to \infty} P_\Lambda(\Phi) \tag{3.8}$$

when Λ *tends to infinity in the sense of Van Hove. Furthermore*
(a) *if* $\Phi, \Psi \in \mathscr{B}$, *then*

$$|P(\Phi) - P(\Psi)| \leq \|\Phi - \Psi\| \tag{3.9}$$

(b) *the function* $P(\cdot)$ *is convex on* \mathscr{B}.

[9] See Robinson [1], I.

The functions $P_\Lambda : \mathscr{B} \to \mathbf{R}$ are equicontinuous[10] by (2.16); since they converge on the dense set \mathscr{B}_0 by Proposition 2.3.2, they converge everywhere, proving (3.8). The rest follows from Proposition 2.2.2.

2.3.4 PERIODIC BOXES. It is often convenient to consider systems in a periodic box obtained by "glueing together opposite faces" of $\Lambda(a)$. Let $T(a)$ be the quotient of \mathbf{Z}^ν by the subgroup $\mathbf{Z}^\nu a = \{na : n \in \mathbf{Z}^\nu\}$. If $\alpha \in T(a)$, let $[\alpha]$ be its representative in $\Lambda(a)$. We define for each $\alpha \in T(a)$ and $x \in \Lambda(a)$ a unitary mapping $\hat{V}_x(\alpha) : \mathscr{H}_x \to \mathscr{H}_{[x+\alpha]}$ by

$$\hat{V}_x(\alpha) = V_x([x + \alpha] - x)$$

and let $\hat{V}(\alpha)$ be the unitary transformation of $\mathscr{H}(\Lambda(a))$ given by

$$\hat{V}(\alpha) = \bigotimes_{x \in \Lambda(a)} \hat{V}_x(\alpha)$$

Now let Φ have finite range ($\Phi \in \mathscr{B}_0$); if a is sufficiently large and $X \subset \Lambda(a)$, the following definition is unambiguous

(a) $\hat{\Phi}(X) = \Phi(X)$ if $\Phi(X) \neq 0$

(b) $\hat{\Phi}(X) = \hat{V}(\alpha)^{-1}\hat{\Phi}([X + a])\hat{V}(\alpha)$
 if $\hat{\Phi}([X + \alpha])$ is already defined by (a)

(c) $\hat{\Phi}(X) = 0$ otherwise

and for all $\alpha \in T(a)$

$$\hat{\Phi}([X + \alpha]) = \hat{V}(\alpha)\hat{\Phi}(X)\hat{V}(\alpha)^{-1} \tag{3.10}$$

We can define \hat{H}, $\hat{Z}_{\Lambda(a)}(\Phi)$, $\hat{P}_{\Lambda(a)}(\Phi)$ by formulas analogous to (2.4), (2.12), and (2.13). We check by an argument analogous to that of Proposition 2.3.2 that

$$\lim_{a \to \infty} \hat{P}_{\Lambda(a)}(\Phi) = P(\Phi) \tag{3.11}$$

2.4 THERMODYNAMIC LIMIT FOR CLASSICAL LATTICE SYSTEMS

We remarked in Section 1.3.3 that classical lattice systems are a special case of quantum lattice systems. The connection is obtained by taking a basis $(\psi_{xn})_{0 \leq n \leq N}$ in \mathscr{H}_x for each $x \in \mathbf{Z}^\nu$ so that

$$\psi_{x+a,n} = V_x(a)\psi_{xn} \tag{4.1}$$

and inserting (3.18) into (3.17) of Chapter 1.[11] We shall consider here the special case of a lattice gas with $N = 1$, i.e., where each lattice site can be

[10] See Section A.1.2.

[11] In Sections 2.2 and 2.3 we have used (3.14) rather than (3.17) of Chapter 1, but the one goes into the other if $H(\Lambda)$ is redefined as $H(\Lambda) - \sum_{i=1}^\nu \mu(n_{x_i})$.

occupied only by either zero or one particle. We write

$$U(n_{x_1}, \ldots, n_{x_V}) = U_\Lambda(X) \qquad (4.2)$$

where $X = \{x \in \Lambda : n_x = 1\}$. Using this notation we assume that

$$U_\Lambda(X) = \sum_{Y \subset X} \Phi(Y) \qquad (4.3)$$

where the "potential" Φ is a function of finite subsets of \mathbf{Z}^v. Equation (4.3) amounts to assuming that empty sites do not contribute to the interaction[12]; in particular, $U_\Lambda(X)$ does not depend upon the index Λ, which we shall drop henceforth. Notice that the invariance relation (2.3) becomes

$$U(X + a) = U(X) \qquad (4.4)$$

Using the results of Sections 2.2 and 2.3 we now give a general description of the lattice gas obtained as just indicated.

Let Φ be a real function of the finite subsets of \mathbf{Z}^v which is invariant under translations, i.e.,

$$\Phi(X + a) = \Phi(X) \qquad (4.5)$$

for all $a \in \mathbf{Z}^v$, and such that $\Phi(\varnothing) = 0$. We call \mathscr{B} the space of such functions for which

$$\|\Phi\| = \sum_{X \ni 0} \frac{|\Phi(X)|}{N(X)} < +\infty \qquad (4.6)$$

where $N(X)$ is the number of points in X. With respect to the norm (4.6), \mathscr{B} is a real Banach space. We let \mathscr{B}_0 be the subspace of \mathscr{B} constituted by those Φ such that $\Phi(X) \neq 0$ only for a finite number of sets $X \ni 0$ (finite range potentials). Define

$$U_\Phi(X) = \sum_{Y \subset X} \Phi(Y) \qquad (4.7)$$

then if $\Phi \in \mathscr{B}$

$$|U_\Phi(X)| \leq N(X)\|\Phi\| \qquad (4.8)$$

Let Λ be a finite subset of \mathbf{Z}^v; a configuration of a classical lattice gas in Λ is described by the set $X \subset \Lambda$ of occupied sites. An ensemble is given by attributing to each configuration X a weight

$$\exp[-\beta U_\Phi(X)] \qquad (4.9)$$

Let \mathscr{B}^k be the subspace of \mathscr{B} consisting of those Φ such that $\Phi(X) = 0$ unless $N(X) = k$ (k-body potentials). Because of (4.5), \mathscr{B}^1 is one-dimensional,

[12] The operator $\Phi(X)$ defined by (2.4) is diagonal in the present situation; its only non-vanishing eigenvalue corresponds to $n_x = 1$ for all $x \in X$ and is here called $\Phi(X)$ again.

and we may write

$$\mathcal{B} = \mathcal{B}^1 \oplus \tilde{\mathcal{B}} \tag{4.10}$$

where \mathcal{B}^1 is identified with the real numbers and $\tilde{\mathcal{B}}$ consists of those Φ such that $\Phi(X) = 0$ if $N(X) = 1$. Corresponding to (4.10), if $\Phi \in \mathcal{B}$ we write $\Phi = (\Phi^1, \tilde{\Phi})$ with $\Phi^1 \in \mathcal{B}^1$ and $\tilde{\Phi} \in \tilde{\mathcal{B}}$; then

$$U_\Phi(X) = N(X)\Phi^1 + U_{\tilde{\Phi}}(X) \tag{4.11}$$

If we write $\Phi^1 = -\mu$, and $z = e^{\beta\mu}$, (4.9) becomes

$$\exp(N(X)\beta\mu)\exp[-\beta U_{\tilde{\Phi}}(X)] = z^{N(X)}\exp[-\beta\mu_{\tilde{\Phi}}(X)] \tag{4.12}$$

which is of the grand canonical form (2.14) of Chapter 1. The "grand partition function" corresponding to (4:12) is

$$\Xi(\Lambda, z, \beta) = \sum_{n \geq 0} z^n \sum_{X \subset \Lambda,\, N(X)=n} \exp[-\beta U_{\tilde{\Phi}}(X)] \tag{4.13}$$

We also introduce

$$Z_\Lambda(\Phi) = \sum_{X \subset \Lambda} \exp[-U_\Phi(X)] \tag{4.14}$$

so that

$$\Xi(\Lambda, e^{\beta\mu}, \beta) = Z_\Lambda(\beta(-\mu, \tilde{\Phi})) \tag{4.15}$$

and we define

$$P_\Lambda(\Phi) = N(\Lambda)^{-1}\log Z_\Lambda(\Phi) \tag{4.16}$$

2.4.1 THEOREM.[13] *If* $\Phi \in \mathcal{B}$, *the following limit exists and is finite*

$$P(\Phi) = \lim_{\Lambda \to \infty} P_\Lambda(\Phi) \tag{4.17}$$

when Λ *tends to infinity in the sense of Van Hove. Furthermore*
 (a) *if* $\Phi, \Psi \in \mathcal{B}$, *then*

$$|P(\Phi) - P(\Psi)| \leq \|\Phi - \Psi\| \tag{4.18}$$

 (b) *the function* $P(\cdot)$ *is convex and positive on* \mathcal{B}.
 We have

$$Z_\Lambda(\Phi) = \sum_{X \subset \Lambda} \exp[-U_\Phi(X)] \geq \exp[-U_\Phi(\varnothing)] = 1 \tag{4.19}$$

Theorem 2.4.1 follows from Theorem 2.3.3 of which it is a special case, except for the positivity of $P(\cdot)$ which follows from (4.19).

[13] See Gallavotti and Miracle [1].

If $\Phi \in \mathscr{B}$ we let

$$C_\Phi = \sum_{X \ni 0} \frac{\Phi(X)}{N(X)} \tag{4.20}$$

Clearly, $|C_\Phi| \le \|\Phi\|$.

2.4.2 PERIODIC BOXES. We may, as in Section 2.3.4 enclose a system with interaction $\Phi \in \mathscr{B}_0$ in a periodic box obtained by glueing together opposite faces of $\Lambda(a)$ for a sufficiently large. The potential energy of $X \subset \Lambda(a)$ becomes

$$\hat{U}_\Phi(X) = \sum_{Y \subset X} \hat{\Phi}(Y) \tag{4.21}$$

where $\hat{\Phi}$ is such that, for all $\alpha \in T(a)$,

$$\hat{\Phi}([X + \alpha]) = \hat{\Phi}(X) \tag{4.22}$$

and we have, if $x \in \Lambda(a)$,

$$C_\Phi = \sum_{X \ni x} \frac{\hat{\Phi}(X)}{N(X)} \tag{4.23}$$

We can define $\hat{Z}_{\Lambda(a)}$, $\hat{P}_{\Lambda(a)}(\Phi)$ by formulas analogous to (4.14), (4.16) and, as in the quantum case,

$$\lim_{a \to \infty} \hat{P}_{\Lambda(a)}(\Phi) = P(\Phi) \tag{4.24}$$

2.4.3 CONNECTION WITH SPIN SYSTEMS.[14] We shall say that $\Phi \in \mathscr{B}$ is a *pair interaction* if $N(X) > 2$ implies $\Phi(X) = 0$, i.e., if $\Phi = (\Phi^1, \Phi^2) \in \mathscr{B}^1 \oplus \mathscr{B}^2$. Let Φ also have finite range; then using (4.23) the potential energy for a periodic box may be written as

$$\hat{U}_\Phi(X) = N(X)\Phi^1 + \sum_{\{x,y\} \subset X} \hat{\Phi}^2(\{x, y\})$$

$$= N(X)\Phi^1 + \frac{1}{2} \sum_{x \in X} \left[\sum_{y \in \Lambda(a),\, y \neq x} \hat{\Phi}^2(\{x, y\}) - \sum_{y \in \Lambda(a) \setminus X} \hat{\Phi}^2(\{x, y\}) \right]$$

$$= N(X)(\Phi^1 + C_{\Phi^2}) - \frac{1}{2} \sum_{x \in X} \sum_{y \in \Lambda(a) \setminus X} \hat{\Phi}^2(\{x, y\}) \tag{4.25}$$

We may thus interpret \hat{U}_Φ as the potential energy of a spin system with interaction potential $-\frac{1}{2}\hat{\Phi}^2(\{x, y\})$ between the sites x, y if they have different spins, zero otherwise.[15]

[14] See Lee and Yang [1].
[15] The term $N(X)(\Phi^1 + C_{\Phi^2})$ is, apart from the constant $\frac{1}{2}V(a)(\Phi^1 + C_{\Phi^2})$, the interaction of the spins with a magnetic field.

The potential energy corresponding to the right-hand side of (4.25) for a region $\Lambda \subset \mathbf{Z}^\nu$ (instead of a periodic box) is

$$U'_\Lambda(X) = N(X)(\Phi^1 + C_{\Phi^2}) - \tfrac{1}{2} \sum_{x \in X} \sum_{y \in \Lambda \setminus X} \Phi^2(\{x, y\}) \qquad (4.26)$$

From the classical version of Section 2.3.4 and from (4.24), (4.25) it follows that

$$\lim_{\Lambda \to \infty} N(\Lambda)^{-1} \log \sum_{X \subset \Lambda} \exp[-U'_\Lambda(X)]$$

$$= \lim_{a \to \infty} V(a)^{-1} \log \sum_{X \subset \Lambda(a)} \exp[-\hat{U}_\Phi(X)]$$

$$= P(\Phi)$$

We may omit the condition $\Phi \in \mathscr{B}_0$ in this result because of the density of finite range interactions in $\mathscr{B}^1 \oplus \mathscr{B}^2$ and of the equicontinuity property (2.16):

For all $\Phi = (\Phi^1, \Phi^2) \in \mathscr{B}^1 \oplus \mathscr{B}^2$,

$$\lim_{\Lambda \to \infty} N(\Lambda)^{-1} \log \sum_{X \subset \Lambda} \exp\left[-N(X)(\Phi^1 + C_{\Phi^2}) + \tfrac{1}{2} \sum_{x \in X} \sum_{y \in \Lambda \setminus X} \Phi^2(\{x, y\}) \right]$$

$$= P(\Phi) \qquad (4.27)$$

when Λ *tends to infinity in the sense of Van Hove.*

2.4.4 A SYMMETRY PROPERTY.[16] Equation (4.25) shows that for a pair interaction the potential energy of a lattice gas may be put in a form where occupied and empty sites play symmetrical roles. We can go further and replace the interaction Φ originally present in (4.3) between occupied sites by an interaction $\mathscr{L}\Phi$ between empty sites. We assume that there exists k such that $\Phi(X) = 0$ if $N(X) > k$ (this is satisfied if Φ is a pair interaction or has finite range) and we define $\mathscr{L}\Phi$ by

$$\mathscr{L}\Phi(\varnothing) = 0 \qquad (4.28)$$

$$\mathscr{L}\Phi(X) = (-1)^{N(X)} \sum_{Y \supset X} \Phi(Y) \qquad \text{if} \quad N(X) \geq 1 \qquad (4.29)$$

Clearly \mathscr{L} maps $\mathscr{B}^1 \oplus \mathscr{B}^2 \oplus \cdots \oplus \mathscr{B}^k$ linearly into itself, and we readily check that \mathscr{L} is an involution, i.e.,

$$\mathscr{L}(\mathscr{L}\Phi) = \Phi \qquad (4.30)$$

[16] See Gallavotti, Miracle, and Robinson [1].

Let now $\Phi \in \mathscr{B}_0$ and consider again a system enclosed in a sufficiently large periodic box; if $X \subset \Lambda(a)$, we find

$$\hat{U}_{\mathscr{L}\Phi}(X) = \sum_{S \subset X} \mathscr{L}\widehat{\Phi}(S) = \sum_{S \subset X,\, S \neq \varnothing} (-1)^{N(S)} \sum_{T \supset S} \hat{\Phi}(T)$$

$$= \sum_{T:\, T \cap X \neq \varnothing} \hat{\Phi}(T) \sum_{S \subset T \cap X,\, S \neq \varnothing} (-1)^{N(S)} = - \sum_{T:\, T \cap X \neq \varnothing} \hat{\Phi}(T),$$

and on the other hand

$$\hat{U}_\Phi(\Lambda(a) \backslash X) - \hat{U}_\Phi(\Lambda(a)) = - \sum_{S \not\subset \Lambda(a) \backslash X} \hat{\Phi}(S) = - \sum_{S:\, S \cap X \neq \varnothing} \hat{\Phi}(S)$$

Therefore, using (4.23),

$$\hat{U}_\Phi(\Lambda(a) \backslash X) - \hat{U}_{\mathscr{L}\Phi}(X) = \hat{U}_\Phi(\Lambda(a))$$

$$= \sum_{x \in \Lambda(a)} \sum_{X \ni x} \frac{\hat{\Phi}(X)}{N(X)} = V(a)C_\Phi \qquad (4.31)$$

From (4.31) we get

$$\hat{Z}_{\Lambda(a)}(\Phi) = \sum_{X \subset \Lambda(a)} \exp[-\hat{U}_\Phi(\Lambda(a) \backslash X)]$$

$$= \sum_{X \subset \Lambda(a)} \exp[-\hat{U}_{\mathscr{L}\Phi}(X) - V(a)C_\Phi] = \exp[-V(a)C_\Phi]\hat{Z}_{\Lambda(a)}(\mathscr{L}\Phi)$$

and therefore

$$P(\mathscr{L}\Phi) = P(\Phi) + C_\Phi \qquad (4.32)$$

We have proved (4.32) for $\Phi \in \mathscr{B}_0$ but this identity remains true by continuity for all $\Phi \in \mathscr{B}^1 \oplus \mathscr{B}^2 \oplus \cdots \oplus \mathscr{B}^k$.

Notice that by changing Φ to $\mathscr{L}\Phi$ in (4.32) we get

$$C_{\mathscr{L}\Phi} = -C_\Phi \qquad (4.33)$$

2.5 SOME INEQUALITIES FOR QUANTUM SYSTEMS[17]

As we shall see in Chapter 3, the study of the thermodynamic limit for quantum systems can be done in close analogy with the treatment of classical systems. To make this analogy apparent we shall need Proposition 2.5.1 below. We prove also a few other inequalities relative to quantum systems which are needed for various parts of this book. The reader may study these inequalities at his convenience.

2.5.1 PROPOSITION (minimax principle). Let A be an unbounded self-adjoint operator on \mathscr{H}. For every finite-dimensional subspace M in the domain

[17] The author learned most of the contents of this section verbally from R. Jost, D. W. Robinson, and A. Huber.

of A let

$$\lambda(M) = \sup_{\varphi \in M, \, \|\varphi\| = 1} (\varphi, A\varphi) \qquad (5.1)$$

For every integer $m \geq 1$, let

$$\lambda_m = \inf_{\dim M = m} \lambda(M) \qquad (5.2)$$

(a) $\lambda_1 \neq -\infty$ *if and only if A is bounded below.*
(b) *If $\lambda_1 \neq -\infty$ and $\lambda_m \to \infty$ then the spectrum of A consists of discrete eigenvalues with finite multiplicity.*
(c) *If $\lambda_1 \neq -\infty$ and if the spectrum of A consists of discrete eigenvalues with finite multiplicity, then these, in increasing order and repeated according to multiplicity, are the λ_m.*

Part (a) is obvious. To prove (b) we notice that if the spectrum is not as described we can find a sequence of M such that $\dim M \to \infty$ and $\lambda(M)$ is bounded.

To prove (c) let $(\tilde{\lambda}_m)$ be the increasing sequence of the eigenvalues of A repeated according to multiplicity and $(\tilde{\varphi}_m)$ a corresponding orthonormal family of eigenvectors. It suffices to show that if $\dim M = m$, then $\lambda(M) \geq \tilde{\lambda}_m$ (that $\lambda_m \leq \tilde{\lambda}_m$ is clear). Given $\varepsilon > 0$, it is easy to approximate M by \tilde{M} such that $\lambda(\tilde{M}) < \lambda(M) + \varepsilon$ and \tilde{M} is contained in the subspace of \mathcal{H} spanned by $\tilde{\varphi}_1, \ldots, \tilde{\varphi}_N$ for sufficiently large N. It suffices now to show that $\lambda(\tilde{M}) \geq \tilde{\lambda}_m$. Counting dimensions we see that the space \tilde{M} and the space spanned by $\tilde{\varphi}_m, \tilde{\varphi}_{m+1}, \ldots, \tilde{\varphi}_N$ contain a common vector φ with $\|\varphi\| = 1$; therefore

$$\lambda(\tilde{M}) \geq (\varphi, A\varphi) = \sum_{i=m}^{N} |\varphi, \tilde{\varphi}_i|^2 \tilde{\lambda}_i \geq \tilde{\lambda}_m$$

which concludes the proof.

2.5.2 A CONVEXITY INEQUALITY (O. KLEIN). *Let A, B be self-adjoint (unbounded) operators with spectrum in the domain of definition of the convex function f; then*

$$\mathrm{Tr}(f(A) - f(B) - (A - B)f'(B)) \geq 0 \qquad (5.3)$$

We refrain from giving more detailed indications about the spectrum of A, B and the sense to be given to f': the proof of (5.3) is simple and can be referred to in specific cases. Let (φ_i) [resp. (ψ_i)] be a complete orthonormal family of eigenvectors of A (resp. B) and (a_i) [resp. (b_i)] the corresponding families of eigenvalues. Let $c_{ij} = (\varphi_i, \psi_j)$; then

$$(\varphi_i, [f(A) - f(B) - (A - B)f'(B)]\varphi_i)$$
$$= f(a_i) - \sum_j |c_{ij}|^2 f(b_j) - \sum_j |c_{ij}|^2 (a_i - b_j) f'(b_j)$$

$$= \sum_j |c_{ij}|^2 [f(a_i) - f(b_j) - (a_i - b_j) f'(b_j)]$$

$$\geq 0 \tag{5.4}$$

which proves (5.3).

2.5.3 PROPOSITION. *Let A, B be positive, self-adjoint, trace class operators;* then

$$\mathrm{Tr}\, A \log A - \mathrm{Tr}\, A \log B \geq \mathrm{Tr}(A - B) \tag{5.5}$$

Proof. Take $f(t) = t \log t$ in (5.3).

2.5.4 PROPOSITION (*Peierls' theorem*). *Let A be an (unbounded) self-adjoint operator and (φ_i) an orthonormal family of vectors in the domain of A:* then

$$\sum_i \exp[-(\varphi_i, A\varphi_i)] \leq \mathrm{Tr}\, e^{-A} \tag{5.6}$$

We may assume that (φ_i) is complete, i.e., a basis. Let B be the self-adjoint operator with eigenvectors φ_i and corresponding eigenvalues $(\varphi_i, A\varphi_i)$. Inserting in (5.3) yields

$$\sum_i f((\varphi_i, A\varphi_i)) \leq \mathrm{Tr}\, f(A) \tag{5.7}$$

and (5.6) is obtained by taking $f(t) = e^{-t}$.

2.5.5 PROPOSITION. *Let $\mathcal{B}(\mathcal{H})$ be the real vector space of self-adjoint operators on the finite-dimensional Hilbert space \mathcal{H}; then the function $A \to \log \mathrm{Tr}\, e^A$ is convex and increasing on $\mathcal{B}(\mathcal{H})$.*
First, using (5.6), the Hölder inequality, and again (5.6) we have

$$\mathrm{Tr}\, e^{\alpha A + (1 - \alpha)B} = \sup_{(\varphi_i)} \sum_i \exp(\varphi_i, [\alpha A + (1 - \alpha)B]\varphi_i)$$

$$= \sup_{(\varphi_i)} \sum_i [\exp(\varphi_i, A\varphi_i)]^\alpha [\exp(\varphi_i, B\varphi_i)]^{1-\alpha}$$

$$\leq \sup_{(\varphi_i)} \left[\sum_i \exp(\varphi_i, A\varphi_i) \right]^\alpha \left[\sum_i \exp(\varphi_i, B\varphi_i) \right]^{1-\alpha}$$

$$\leq (\mathrm{Tr}\, e^A)^\alpha (\mathrm{Tr}\, e^B)^{1-\alpha}$$

which proves convexity. Using again (5.6) we see that $A \leq B$ implies

$$\mathrm{Tr}\, e^A = \sup_{(\varphi_i)} \sum_i \exp(\varphi_i, A\varphi_i) \leq \sup_{(\varphi_i)} \sum_i \exp(\varphi_i, B\varphi_i) = \mathrm{Tr}\, e^B$$

which concludes the proof.

2.5.6 PROPOSITION. *Let σ be a density matrix (positive operator of trace 1) on $\mathcal{H} = \mathcal{H}_1 \otimes \mathcal{H}_2$, and let σ_1 and σ_2 be the partial traces of σ with respect to*

\mathscr{H}_2 and \mathscr{H}_1 so that σ_1, σ_2 are density matrices on $\mathscr{H}_1, \mathscr{H}_2$; then

$$\mathrm{Tr}_{\mathscr{H}} \, \sigma \log \sigma \geq \mathrm{Tr}_{\mathscr{H}_1} \, \sigma_1 \log \sigma_1 + \mathrm{Tr}_{\mathscr{H}_2} \, \sigma_2 \log \sigma_2 \tag{5.8}$$

Proof. Let $A = \sigma$, $B = \sigma_1 \otimes \sigma_2$ in (2.9.5) and use $\log(\sigma_1 \otimes \sigma_2) = \log \sigma_1 \otimes \mathbf{1} + \mathbf{1} \otimes \log \sigma_2$.

2.5.7 PROPOSITION. *The function* $\sigma \to \mathrm{Tr} \, \sigma \log \sigma$ *is convex on the convex set of density matrices.*

Let $\sigma = \alpha\sigma_1 + (1 - \alpha)\sigma_2$; then using (5.5) with $B = \sigma$ and $A = \sigma_1$ or σ_2 we have

$$\mathrm{Tr} \, \sigma \log \sigma = \alpha \, \mathrm{Tr} \, \sigma_1 \log \sigma + (1 - \alpha) \, \mathrm{Tr} \, \sigma_2 \log \sigma$$

$$\leq \alpha \, \mathrm{Tr} \, \sigma_1 \log \sigma_1 + (1 - \alpha) \, \mathrm{Tr} \, \sigma_2 \log \sigma_2$$

2.5.8 PROPOSITION.[18] *Let* A, B *be self-adjoint bounded operators with bounded inverses such that*

$$0 < A \leq B \tag{5.9}$$

then

$$B^{-1} \leq A^{-1} \tag{5.10}$$

$$\log A \leq \log B \tag{5.11}$$

Let $C = B^{-1/2} A B^{-1/2}$, then (5.9) gives $0 \leq C \leq \mathbf{1}$ and hence $\mathbf{1} \leq C^{-1} = B^{1/2} A^{-1} B^{1/2}$, which proves (5.10). We obtain (5.11) from (5.10) and the representation

$$\log B - \log A = \int_0^\infty dx[(x\mathbf{1} + A)^{-1} - (x\mathbf{1} + B)^{-1}] \tag{5.12}$$

[18] This proposition is a special result in the general theory of monotone operator functions; see Bendat and Sherman [1].

CHAPTER 3

THE THERMODYNAMIC LIMIT FOR THERMODYNAMIC FUNCTIONS: CONTINUOUS SYSTEMS

In this chapter we extend to continuous systems (classical and quantum) the study of the thermodynamic limit started in Chapter 2. First we introduce conditions on the interaction such that thermodynamic behavior can be expected (Sections 3.1 and 3.2). Then we prove the existence of the thermodynamic limit for various ensembles and establish the equivalence of these ensembles.

3.1 INTERACTIONS OF CLASSICAL CONTINUOUS SYSTEMS

In Sections 2.2–2.4 we have studied the thermodynamic limit for classical and quantum lattice systems, restricting ourselves to one type of ensemble (essentially grand canonical). The situation envisaged was particularly favorable and we were able to prove not only the existence of the thermodynamic limit for the thermodynamic function P, but also that P is a continuous convex function on the Banach space \mathscr{B} of interactions. A similar result may also be proved for a class of continuous systems with hard cores, using the grand canonical ensemble,[1] but the general investigation of classical continuous systems involves some completely new problems, among which are the following.

(a) As a rule the microcanonical ensemble is more difficult to study than the canonical, which is more difficult than the grand canonical. In particular

[1] See Exercise 3.H.

29

some convexity properties which occur for finite systems if the grand canonical ensemble is used [cf. Proposition 2.2.2,(b)], appear only after the thermodynamic limit has been taken if we use the canonical or micro-canonical ensemble.

(b) We shall see in Section 3.2 that changing the sign of a reasonable inter-action of a continuous system usually yields a catastrophic interaction (not leading to thermodynamic behavior). Catastrophic behavior is possible because for a continuous system without hard cores, a bounded region may be occupied by an arbitrarily large number of particles.

Given a classical continuous system, for each integer $n \geq 0$ and $x_1, \ldots, x_n \in \mathbf{R}^\nu$ the potential energy $U(x_1, \ldots, x_n)$ of n particles located at x_1, \ldots, x_n is defined. If $n = 0$, we set $U = 0$. We assume that U takes its values in $\mathbf{R} \cup \{\infty\}$, i.e., the value ∞ is permitted for the potential energy. Notice that the configurations of particle positions such that $U = \infty$ are given the weight zero by the various ensembles of 1.2.1, and are thus in effect "forbidden." In particular, systems with hard cores (see Section 1.2.2) are obtained by setting $U = \infty$ for the excluded configurations. We assume that $U(x_1, \ldots, x_n)$ is invariant under permutation of its arguments, in accordance with its physical meaning (potential energy). Furthermore we assume that U is invariant under translations: if $a \in \mathbf{R}^\nu$, then

$$U(x_1 + a, \ldots, x_n + a) = U(x_1, \ldots, x_n) \tag{1.1}$$

If U does not take the value ∞, we may write

$$U(x_1, \ldots, x_n) = \sum_k \sum_{1 \leq i_1 < \cdots < i_k \leq n} \Phi^k(x_{i_1}, \ldots, x_{i_k}) \tag{1.2}$$

and (1.2) determines uniquely (by induction on k) the k-body potential Φ^k. The Φ^k are real-valued functions, invariant under permutations of their arguments and under translations. Even if U is allowed to take the value ∞, we shall assume the form (1.2) where Φ^k is now also allowed to take the value ∞. We shall always assume that Φ^k is Lebesgue measurable. Because of translation invariance Φ^1 is a constant; conventionally, we take $\Phi^1 = 0$. This is really no loss of generality as Φ^1 reappears, for instance, in the guise of chemical potential in the grand canonical ensemble. *Altogether, we assume that U is defined by*

$$U(x_1, \ldots, x_n) = \sum_{k \geq 2} \sum_{1 \leq i_1 < \cdots < i_k \leq n} \Phi^k(x_{i_1}, \ldots, x_{i_k}) \tag{1.3}$$

where the k-body potential Φ^k is Lebesgue measurable with values in $\mathbf{R} \cup \{\infty\}$, is invariant under permutation of its k arguments and under translations:

$$\Phi^k(x_1 + a, \ldots, x_k + a) = \Phi^k(x_1, \ldots, x_k) \tag{1.4}$$

A sequence $(\Phi^k)_{k \geq 2}$ of k-body potentials will be called an *interaction*. The 2-body potential Φ^2 is also known as a *pair potential* and an interaction $(\Phi^k)_{k \geq 2}$ such that $\Phi^k = 0$ for $k > 2$ is a *pair interaction*. Notice that because of translation invariance $\Phi^2(x_1, x_2)$ depends only upon $x_2 - x_1$; we shall write $\Phi^2(x_1, x_2) = \Phi(x_2 - x_1)$. Because of invariance under permutation of x_1 and x_2, we have thus

$$\Phi(-x) = \Phi(x) \tag{1.5}$$

We say that a pair potential has *finite range* if Φ has compact support, i.e., if there exists R_0 such that

$$\Phi(x) = 0 \qquad \text{for} \quad |x| \geq R_0 \tag{1.6}$$

The hard core condition corresponding to the exclusion of $S_R{}^n$ [see (2.12) of Chapter 1] from configuration space is expressed by setting $\Phi(x) = \infty$ if $|x| < R$.

An interaction is invariant under Euclidean transformations if, for every Euclidean rotation O of \mathbf{R}^ν, we have

$$\Phi^k(Ox_1, \ldots, Ox_k) = \Phi^k(x_1, \ldots, x_k) \tag{1.7}$$

For $k = 2$, (1.7) expresses that the pair potential Φ is spherically symmetric; it follows that $\Phi(x)$ depends only upon the Euclidean norm $|x|$ of its argument. The interactions between atoms of the rare gases are reasonably well represented by pair interactions with a spherically symmetric pair potential like the one shown in Fig. 2.

As discussed already in Section 1.4, certain conditions have to be imposed on the interaction if the system is to exhibit thermodynamic behavior:

(a) the interaction between distant particles must be negligible;

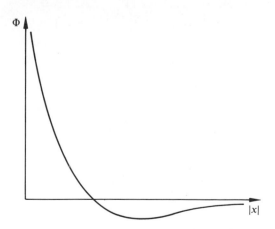

FIGURE 2 A typical pair potential.

(b) the interaction must not cause the collapse of infinitely many particles into a bounded region of \mathbf{R}^ν.

For the purposes of this chapter, (a) will be ensured by the *temperedness* condition 3.1.1, and (b) will be ensured by the *stability* condition discussed at length in Section 3.2.

Let a mutual potential energy $W_{n_1 n_2}$ be defined by

$$W_{n_1 n_2}(x'_1, \ldots, x'_{n_1}; x''_1, \ldots, x''_{n_2})$$

$$= \sum_{k \geq 1} \sum_{l \geq 1} \sum_{1 \leq i_1 < \cdots < i_k \leq n_1} \sum_{1 \leq j_1 < \cdots < j_l \leq n_2} \Phi(x'_{i_1}, \ldots, x'_{i_k}, x''_{j_1}, \ldots, x''_{j_l}) \quad (1.8)$$

so that

$$U(x'_1, \ldots, x'_{n_1}, x''_1, \ldots, x''_{n_2})$$

$$= U(x'_1, \ldots, x'_{n_1}) + U(x''_1, \ldots, x''_{n_2}) + W_{n_1 n_2}(x'_1, \ldots, x'_{n_1}, x''_1, \ldots, x''_{n_2}) \quad (1.9)$$

3.1.1 DEFINITION.[2] *We say that the interaction $(\Phi^k)_{k \geq 2}$ is tempered if there exist $\lambda > \nu$ and $R_0 > 0$, $A \geq 0$ such that*

$$W_{n_1 n_2}(x'_1, \ldots, x'_{n_1}; x''_1, \ldots, x''_{n_2}) \leq A n_1 n_2 r^{-\lambda} \quad (1.10)$$

whenever $|x''_j - x'_i| \geq r \geq R_0$ for all $i = 1, \ldots, n_1$, $j = 1, \ldots, n_2$.

This notion is invariant under linear transformations of \mathbf{R}^ν (in spite of the use of the Euclidean norm). We could, in (1.10), take $A = 1$ by appropriate choice of λ and R_0. Temperedness implies that the positive part of $W_{n_1 n_2}$ tends to zero when $r \to \infty$ but leaves the negative part arbitrary; however, the combination of temperedness and stability (Definition 3.2.1) will be sufficient to prove the existence of the thermodynamic limit.

Temperedness is ensured if

$$W_{n_1 n_2}(x'_1, \ldots, x'_{n_1}; x''_1, \ldots, x''_{n_2}) \leq 0 \quad (1.11)$$

whenever $|x''_j - x'_i| \geq R_0$ for all $i = 1, \ldots, n_1$, $j = 1, \ldots, n_2$; this is simply the case $A = 0$ of Definition 3.1.1. Another case of interest is that of pair interactions: temperedness holds if and only if the pair potential satisfies

$$\Phi(x) \leq A|x|^{-\lambda} \quad \text{for} \quad |x| \geq R_0 \quad (1.12)$$

We obtain an idea of the meaning of temperedness if we estimate with (1.12) the energy of interaction of a particle with other particles distributed randomly with density d at a distance greater than $D \geq R_0$:

$$d \int_{|x| > D} \Phi(x)\, dx \leq Ad \int_{|x| > D} |x|^{-\lambda}\, dx$$

$$= C \int_D^\infty t^{\nu - \lambda - 1}\, dt \quad (1.13)$$

[2] See Fisher [1].

where C is a constant. The right-hand side of (1.13) converges since $\lambda > \nu$ and tends to zero when $D \to \infty$. Temperedness implies thus that the positive part of the interaction energy between particles at large distances is negligible [the negative part of the interaction energy is controlled by the stability condition (2.1)]. Temperedness as we have defined it is certainly not the best possible condition to impose for the proof of the existence of the thermodynamic limit, but it has the advantage of simplicity, and it is satisfied for most interactions usually considered.

A particular case of (1.12) is

$$\Phi(x) \leq 0 \qquad \text{for} \quad |x| \geq R_0 \tag{1.14}$$

Condition (1.14) will be satisfied by realistic intermolecular potentials (attractive character of the "long range" van der Waals forces) and by finite range pair potentials.

3.2 STABLE INTERACTIONS

This section may be omitted on first reading, except for the following definition.

3.2.1 DEFINITION. *We say that the interaction* $(\Phi^k)_{k \geq 2}$ *is stable if there exists* $B \geq 0$ *such that*

$$U(x_1, \ldots, x_n) \geq -nB \tag{2.1}$$

for all $n \geq 0$ *and* $x_1, \ldots, x_n \in \mathbf{R}^\nu$.

The condition (2.1) is clearly satisfied (with $B = 0$) if $\Phi^k \geq 0$ for all k (positive interactions).[3] The sum of two stable interactions is stable. The equivalent of stability for the classical lattice systems considered in Section 2.4 is automatically satisfied in view of (4.8) of Chapter 2.

We shall give below useful criteria for the stability of pair interactions. Before doing this we want to convince the reader of the fact that an interaction which violates stability is likely to lead to nonthermodynamic behavior (catastrophic interaction). Let Λ be a bounded Lebesgue measurable subset of \mathbf{R}^ν with volume ($=$ Lebesgue measure) $V(\Lambda)$; the *grand partition function* Ξ_Λ is the partition function corresponding to the grand canonical ensemble [see eqs. (2.9) and (2.10) of Chapter 1] and is given by

$$\Xi_\Lambda = 1 + \sum_{n=1}^{\infty} \frac{z^n}{n!} \int_{\Lambda^n} dx_1 \cdots dx_n \exp[-\beta U(x_1, \ldots, x_n)] \tag{2.2}$$

with $\beta > 0$, $z > 0$. *Stability implies the convergence of the grand partition*

[3] Notice however that $U \geq 0$ does not imply $\Phi^k \geq 0$.

function as the following term by term majorization shows

$$1 + \sum_{n=1}^{\infty} \frac{z^n}{n!} \int_{\Lambda^n} dx_1 \cdots dx_n \exp[-\beta U(x_1, \ldots, x_n)]$$

$$\leq 1 + \sum_{n=1}^{\infty} \frac{z^n}{n!} V(\Lambda)^n \exp(n\beta B)$$

$$= \exp[zV(\Lambda)\exp(\beta B)] \tag{2.3}$$

On the other hand the divergence of the series (2.2) indicates a nonthermo-dynamic behavior. We now show that for a large class of pair interactions the series (2.2) diverges if and only if the stability condition (2.1) is violated

3.2.2 PROPOSITION.[4] *Let* Φ *be a real upper semicontinuous*[5] (u.s.c.) *function on* \mathbf{R}^v *such that* $\Phi(-x) = \Phi(x)$, *and define*

$$U(x_1, \ldots, x_n) = \sum_{1 \leq i < j \leq n} \Phi(x_j - x_i) \tag{2.4}$$

The following properties are equivalent:

(a) *the inequality*

$$\sum_{i=1}^{n} \sum_{j=1}^{n} \Phi(x_j - x_i) \geq 0 \tag{2.5}$$

holds for all $n \geq 0$ *and all* $x_1, \ldots, x_n \in \mathbf{R}^v$,

(b) *there exists* $B \geq 0$ *such that*

$$U(x_1, \ldots, x_n) \geq -nB \tag{2.6}$$

for all n, x_1, \ldots, x_n,

(c) *the following series converges for all bounded Lebesgue measurable* $\Lambda \subset \mathbf{R}^v$

$$\Xi_\Lambda = 1 + \sum_{n=1}^{\infty} \frac{z^n}{n!} \int_{\Lambda^n} dx_1 \cdots dx_n \exp[-\beta U(x_1, \ldots, x_n)] \tag{2.7}$$

That (a) \Rightarrow (b) with $B = \frac{1}{2}\Phi(0)$ follows from the identity

$$\sum_{i=1}^{n} \sum_{j=1}^{n} \Phi(x_j - x_i) = n\Phi(0) + 2U(x_1, \ldots, x_n) \tag{2.8}$$

That (b) \Rightarrow (c) follows from (2.3). We show now that if (a) does not hold, then (c) does not hold. Indeed let n and $\tilde{x}_1, \ldots, \tilde{x}_n$ be such that

$$\sum_{i=1}^{n} \sum_{j=1}^{n} \Phi(\tilde{x}_j - \tilde{x}_i) = -2\varepsilon < 0 \tag{2.9}$$

[4] See Ruelle [5].
[5] See Section A.1.3.

The function $(y_1, \ldots, y_n, z_1, \ldots, z_n) \to \sum_{i=1}^{n} \sum_{j=1}^{n} \Phi(z_j - y_i)$ is u.s.c. on \mathbf{R}^{2nv} and takes the value -2ε at $(\tilde{x}_1, \ldots, \tilde{x}_n, \tilde{x}_1, \ldots, \tilde{x}_n)$; there exists therefore $\delta > 0$ such that

$$|y_1 - \tilde{x}_1| < \delta, \ldots, |y_n - \tilde{x}_n| < \delta, |z_1 - \tilde{x}_1| < \delta, \ldots, |z_n - \tilde{x}_n| < \delta$$

implies

$$\sum_{i=1}^{n} \sum_{j=1}^{n} \Phi(z_j - y_i) < -\varepsilon \qquad (2.10)$$

Let $\Lambda = \bigcup_{i=1}^{n} \{x \in \mathbf{R}^v : |x - \tilde{x}_i| < \delta\}$ and

$$M_s = \{(x_1, \ldots, x_{sn}) \in \mathbf{R}^{snv} : |x_{pn+i} - \tilde{x}_i| < \delta$$
$$\text{for} \quad p = 0, \ldots, s - 1 \quad \text{and} \quad i = 1, \ldots, n\}$$

By (2.10), if $(x_1, \ldots, x_{sn}) \in M_s$ we have

$$U(x_1, \ldots, x_{sn}) = \frac{1}{2} \left[\sum_{p=1}^{s} \sum_{q=1}^{s} \sum_{i=1}^{n} \sum_{j=1}^{n} \Phi(x_{qn+j} - x_{pn+i}) - sn\Phi(0) \right]$$
$$\leq \frac{1}{2}[-s^2\varepsilon - sn\Phi(0)] \qquad (2.11)$$

Therefore, if V_δ is the volume of the sphere of radius δ in v dimensions

$$\Xi_\Lambda \geq \sum_{s=1}^{\infty} \frac{z^{sn}}{(sn)!} \int_{M_s} dx_1 \cdots dx_{sn} \exp[-\beta U(x_1, \ldots, x_{sn})]$$
$$\geq \sum_{s=1}^{\infty} \frac{z^{sn}}{(sn)!} V_\delta^{sn} \exp\left[\frac{\beta}{2}(s^2\varepsilon + sn\Phi(0))\right] = +\infty \qquad (2.12)$$

This concludes the proof.[6]

3.2.3 CATASTROPHIC POTENTIALS. The catastrophic behavior which occurs in the situation of Proposition 3.2.2 if stability is violated arises from the fact that a large weight is attributed to configurations where many particles are crowded into a bounded region of \mathbf{R}^v. Using 3.2.2(a) with $n = 1$ we check that a negative pair potential like that of Fig. 3 is catastrophic. The potential given by Fig. 4 with $v = 3$ appears more honest; we show that it is nevertheless also catastrophic. We remind the reader of the fact that in the face-centered cubic lattice in three dimensions every site has 12 nearest neighbors. Let R be the distance between nearest neighbors, and let x_1, \ldots, x_n be the sites of the lattice contained inside of a sphere so large that the number of nearest neighbor bonds between x_1, \ldots, x_n is $N > 11n/2$;

[6] The general term of the series in (2.12) diverges due to the presence of $\exp[(\beta\varepsilon/2)s^2]$ (use Stirling's formula).

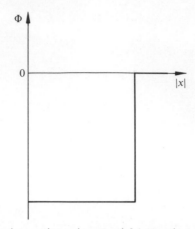

FIGURE 3 A negative pair potential (attractive square well).

then

$$\sum_{i=1}^{n} \sum_{j=1}^{n} \Phi(x_j - x_i) = n\Phi(0) + 2N\Phi(R)$$

$$= 11n - 2N < 11n - 11n = 0$$

and therefore, by Proposition 3.2.2, Φ is catastrophic.

Other examples of catastrophic pair potentials are provided by the following result.

FIGURE 4 A catastrophic pair potential ($v = 3$).

3.2.4 PROPOSITION.[7] *Let the pair potential Φ be absolutely integrable. If*

$$\int dx \Phi(x) < 0 \tag{2.13}$$

then we may choose Λ bounded such that Ξ_Λ diverges.

From the convexity of the exponential function, we have

$$V(\Lambda)^{-n} \int_{\Lambda^n} dx_1 \cdots dx_n \exp[-\beta U(x_1, \ldots, x_n)]$$

$$\geq \exp\left[-\beta V(\Lambda)^{-n} \int_{\Lambda^n} dx_1 \cdots dx_n U(x_1, \ldots, x_n)\right]$$

$$= \exp\left[-\beta \frac{n(n-1)}{2} V(\Lambda)^{-2} \int_{\Lambda^2} dx_1 \, dx_2 \Phi(x_2 - x_1)\right]$$

It follows from (2.13) that if we choose for Λ a sufficiently large cube we have

$$-\varepsilon = V(\Lambda)^{-2} \int_{\Lambda^2} dx_1 \, dx_2 \Phi(x_2 - x_1) < 0$$

hence

$$\Xi_\Lambda \geq \sum_{n=1}^{\infty} \frac{z^n}{n!} V(\Lambda)^n \exp\left[\beta \frac{n(n-1)}{2}\varepsilon\right] = +\infty$$

We indicate now classes of pair interactions for which the stability condition is satisfied.

3.2.5 PAIR INTERACTIONS WITH HARD CORE. Let Φ be a pair potential with hard core:

$$\Phi(x) = \infty \qquad \text{if} \quad |x| < R$$

Suppose furthermore that

$$\sum_{i=1}^{n} \Phi(x_i) \geq -2B \tag{2.14}$$

for all n and all sequences (x_1, \ldots, x_n) such that $|x_j - x_i| \geq R$ for $i \neq j$. In that case stability is satisfied because

$$U(x_1, \ldots, x_n) = \frac{1}{2} \sum_{i=1}^{n} \left[\sum_{j \neq i} \Phi(x_j - x_i)\right] \geq -nB \tag{2.15}$$

The condition (2.14) holds in particular if there exists a positive decreasing

[7] See Dobrushin [1].

(bounded) function φ on $[R, +\infty)$ such that

$$\int_R^\infty \varphi(t)t^{\nu-1}\, dt < +\infty \qquad (2.16)$$

and

$$\Phi(x) \geq -\varphi(|x|) \qquad \text{if}\quad |x| \geq R \qquad (2.17)$$

3.2.6 POSITIVE-TYPE PAIR POTENTIALS. A complex continuous function f on \mathbf{R}^ν is said to be of positive type if, for all $x_1, \ldots, x_n \in \mathbf{R}^\nu$ and $z_1, \ldots, z_n \in \mathbf{C}$, the following inequality is satisfied

$$\sum_{i=1}^n \sum_{j=1}^n z_i z_j^* f(x_j - x_i) \geq 0 \qquad (2.18)$$

It is known (Bochner's theorem) that a function is of positive type if and only if it is the Fourier transform of a positive measure of finite total mass on \mathbf{R}^ν. In particular the Fourier transform of a positive integrable function is of positive type. If a function f of positive type is real then $f(-x) = f(x)$ and, taking $z_1 = \cdots = z_n = 1$ in (2.18) we obtain

$$\sum_{i=1}^n \sum_{j=1}^n f(x_j - x_i) \geq 0$$

i.e., precisely condition (a) of Proposition 3.2.2. It follows that f is a stable pair potential.

3.2.7 PROPOSITION.[8] *If the pair potential Φ can be written in the form*

$$\Phi = \Phi_1 + \Phi_2 \qquad (2.19)$$

where

(a) *Φ_1 is positive*
(b) *Φ_2 is a real continuous function of positive type,*

then Φ is stable.

As we shall see this criterion is very effective.[9]

3.2.8 PROPOSITION.[10] *Let $0 < a_1 < a_2$ and let φ_1, φ_2 be positive decreasing functions on $(0, a_1), (a_2, +\infty)$, respectively, and such that*

$$\int_0^{a_1} \varphi_1(t)t^{\nu-1}\, dt = +\infty \qquad \int_{a_2}^\infty \varphi_2(t)t^{\nu-1}\, dt < +\infty \qquad (2.20)$$

[8] See Ruelle [1].
[9] It seems to be an open problem to construct a stable potential which is not of the form (2.19).
[10] See Dobrushin [1], the proof outlined here is due to Fisher and Ruelle [1].

If the pair potential Φ is bounded below and satisfies

$$\Phi(x) \geq \varphi_1(|x|) \qquad \text{for} \quad |x| \leq a_1 \qquad (2.21)$$

$$\Phi(x) \geq -\varphi_2(|x|) \qquad \text{for} \quad |x| \geq a_2 \qquad (2.22)$$

then Φ is stable.

We indicate only the idea of a proof, which consists in constructing a positive-type minorant to Φ and then applying Proposition 3.2.7.

(a) Construct a bounded function $\psi_1 \geq 0$ on \mathbf{R}^ν such that

$$\int dx \psi_1(x) < +\infty$$

$$-\psi_1(x) \leq \Phi(x') \qquad \text{if} \quad |x - x'| < 1$$

(b) By regularization[11] of ψ_1 obtain a bounded function $\psi_2 \geq 0$ such that

$$\int dx \psi_2(x) < +\infty$$

$$-\psi_2(x) \leq \Phi(x) \qquad \text{for all} \quad x$$

and the Fourier transform $\hat{\psi}_2$ of ψ_2 tends to zero rapidly at infinity, so that for some $C_1 > 0$

$$\hat{\psi}_2(p) \leq C_1(1 + p^2)^{-\nu} \qquad (2.23)$$

(c) Let χ_1 be such that its Fourier transform is $\hat{\chi}_1(p) = (1 + p^2)^{-\nu}$, and let $\chi_2 = \chi_1 \cdot \alpha$ where $\alpha \geq 0$ is C^∞, of positive type with compact support. The Fourier transform of χ_2 is the convolution product of $(1 + p^2)^{-\nu}$ with $\hat{\alpha} \geq 0$; hence, there exists $C_2 > 0$ such that

$$\hat{\chi}_2(p) \geq C_2(1 + p^2)^{-\nu}$$

(d) $\Phi + \psi_2$ is positive and has a nonintegrable divergence at the origin. On the other hand χ_2 is bounded with compact support, and therefore we can find $\lambda_1, \ldots, \lambda_N > 1$ and $\alpha_1, \ldots, \alpha_N > 0$ so that

$$\sum_{i=1}^{N} \alpha_i \chi_2(\lambda_i x) \leq \Phi(x) + \psi_2(x) \qquad (2.24)$$

$$\sum_{i=1}^{N} \alpha_i \lambda_i^{-\nu} \geq C_1/C_2 \qquad (2.25)$$

[11] Convolution with a C^∞ function $\alpha \geq 0$ of suitably small support such that $\int \alpha(x)\,dx = 1$.

Using (2.25) and (2.23) we obtain for the Fourier transform of $\Sigma_{i=1}^{N} \alpha_i \chi_2(\lambda_i x)$

$$\sum \alpha_i \lambda_i^{-\nu} \hat{\chi}_2(\lambda_i^{-1} p) \geq C_2 \sum \alpha_i \lambda_i^{-\nu}(1 + \lambda_i^{-2} p^2)^{-\nu} \geq C_2 \sum_{i=1}^{N} \alpha_i \lambda_i^{-\nu}(1 + p^2)^{-\nu}$$

$$\geq C_1(1 + p^2)^{-\nu} \geq \hat{\psi}_2(p) \tag{2.26}$$

Therefore $\Sigma \, \alpha_i \chi_2(\lambda_i x) - \psi_2$ is a positive-type minorant to Φ.

3.2.9 SUPERSTABLE POTENTIALS.[12] We mention without proof some refinements of the stability condition (2.1). Let $\Phi = \Phi' + \Phi''$ where Φ' is a stable pair potential while Φ'' is continuous greater than or equal to zero and $\Phi''(0) > 0$. There exist then constants $B, C > 0$ such that if Λ is a cube of sufficiently large volume V and $x_1, \ldots, x_n \in \Lambda$, then

$$U(x_1, \ldots, x_n) \geq n\left[-B + C\frac{n}{V}\right] \tag{2.27}$$

If we take now $\Phi''(x) = \alpha|x|^{-\lambda}$ with $\alpha > 0$, $\lambda > \nu$, (2.27) can be replaced by

$$U(x_1, \ldots, x_n) \geq n\left[-B + C\left(\frac{n}{V}\right)^{\lambda/\nu}\right] \tag{2.28}$$

The potential energy per particle therefore goes up linearly (2.27) or faster (2.28) when the density becomes sufficiently high. It may be noted that potentials satisfying Proposition 3.2.8 are superstable.

3.2.10 POTENTIALS OF THE LENNARD–JONES TYPE. We say that a pair potential Φ is of Lennard–Jones type if it is bounded below and satisfies Eqs. (2.21) and (2.22) with $\varphi_1(|x|) = \varphi_2(|x|) = |x|^{-\lambda}$, $\lambda > \nu$. Stability and temperedness are then automatically satisfied, as well as (2.28).

3.2.11 CASE OF SEVERAL SPECIES OF PARTICLES.[13] The results obtained for one species of particles extend to the situation where several species are present. Of particular interest is the case of electric charge distributions ρ_1, \ldots, ρ_n in \mathbf{R}^3; the Coulomb (= electrostatic) interaction energy between these is bounded below by $-\Sigma_{i=1}^{n} E_i$ where E_i is the self-energy of ρ_i:

$$E_i = \tfrac{1}{2} \int dx \int dx' \frac{\rho_i(x)\rho_i(x')}{|x - x'|}$$

Notice that systems of classical point charges do not behave thermo-dynamically; the configuration integral for two particles of opposite signs

[12] See Ruelle [1].
[13] See Fisher and Ruelle [1].

diverges

$$\int_\Lambda dx_1 \int_\Lambda dx_2 \exp\left[\beta\frac{1}{|x_2 - x_1|}\right] = +\infty$$

3.3 THERMODYNAMIC LIMIT FOR THE CONFIGURATIONAL MICROCANONICAL ENSEMBLE

In this section a fixed interaction $(\Phi^k)_{k\geq 2}$ is chosen and U is computed with the use of (1.3).

Again let Λ be a bounded Lebesgue measurable subset of \mathbf{R}^v with volume $V(\Lambda)$. We introduce the *configurational microcanonical partition function*.[14]

$$\Omega(\Lambda, n, E) = \frac{1}{n!}\int_{\Lambda^n} \delta^-(U(x_1,\ldots,x_n) - E)\,dx_1\cdots dx_n$$

$$= \frac{1}{n!}\operatorname{meas}\{(x_1,\ldots,x_n) \subset \Lambda^n: U(x_1,\ldots,x_n) \leq E\} \qquad (3.1)$$

corresponding to the ensemble defined by Eq. (2.5) of Chapter 1. We further define the *configurational entropy*

$$S(\Lambda, n, E) = \log \Omega(\Lambda, n, E) \qquad (3.2)$$

where S may take the value $-\infty$ (if $\Omega = 0$). The quantity $S(\Lambda, n, E)$ is an increasing function of Λ and E, and as a function of E it may be inverted[15] to yield $E(\Lambda, n, S)$ which is an increasing function of S.

3.3.1 PROPOSITION. *For fixed n, $E(\Lambda, n, S)$ is*

(a) *an increasing function of S,*

(b) *a decreasing function of Λ.*

Part (b) follows from (a) and the fact that $S(\Lambda, n, E)$ is an increasing function of Λ.

Notice that if the stability condition (2.1) is satisfied then

$$E(\Lambda, n, S) \geq -nB \qquad (3.3)$$

We shall now work toward proving that $V(\Lambda)^{-1}S(\Lambda, n, E)$ has a limit when $\Lambda \to \infty$ in the sense of Fisher, and when $V(\Lambda)^{-1}n$, $V(\Lambda)^{-1}E$ tend to finite values. We shall proceed via the study of the thermodynamic limit for $V(\Lambda)^{-1}E(\Lambda, n, S)$, and the following inequality will play a basic role.

[14] In Eq. (3.1), meas denotes the Lebesgue measure.
[15] The inverted function is allowed to take the values $-\infty$ and $+\infty$, and ambiguities which may arise in its definition are lifted by making it continuous to the left.

3.3.2 PROPOSITION. *Let the temperedness condition 3.1.1 be satisfied by the interaction.*

(a) *If Λ_1 and Λ_2 are disjoint with distance $r \geq R_0$, then*

$$E(\Lambda_1 \cup \Lambda_2, n_1 + n_2, S_1 + S_2) \leq E(\Lambda_1, n_1, S_1) + E(\Lambda_2, n_2, S_2) + An_1n_2r^{-\lambda}$$

(3.4)

(b) *In the case of several regions $\Lambda_1, \ldots, \Lambda_m$ with mutual distances $\geq r \geq R_0$ we have*

$$E\left(\bigcup_{i=1}^{m} \Lambda_i, \sum_{i=1}^{m} n_i, \sum_{i=1}^{m} S_i\right) \leq \sum_{i=1}^{m} E(\Lambda_i, n_i, S_i) + \tfrac{1}{2}A\left(\sum_{i=1}^{m} n_i\right)^2 r^{-\lambda}$$ (3.5)

Let indeed $x'_1, \ldots, x'_{n_1} \in \Lambda_1$, $x''_1, \ldots, x''_{n_2} \in \Lambda_2$ be such that

$$U(x'_1, \ldots, x'_{n_1}) \leq E_1 \qquad U(x''_1, \ldots, x''_{n_2}) \leq E_2$$

then

$$U(x'_1, \ldots, x'_{n_1}, x''_1, \ldots, x''_{n_2})$$
$$= U(x'_1, \ldots, x'_{n_1}) + U(x''_1, \ldots, x''_{n_2}) + W(x'_1, \ldots, x'_{n_1}; x''_1, \ldots, x''_{n_2})$$
$$\leq E_1 + E_2 + An_1n_2r^{-\lambda}$$

Therefore

$$\frac{1}{(n_1 + n_2)!} \text{meas}\{(x_1, \ldots, x_{n_1+n_2}) \in \Lambda_1 \cup \Lambda_2$$

$$: U(x_1, \ldots, x_{n_1+n_2}) \leq E_1 + E_2 + An_1n_2r^{-\lambda}\}$$

$$\geq \frac{1}{n_1!} \text{meas}\{(x'_1, \ldots, x'_{n_1}) \in \Lambda_1 : U(x'_1, \ldots, x'_{n_1}) \leq E_1\}$$

$$\times \frac{1}{n_2!} \text{meas}\{(x''_1, \ldots, x''_{n_2}) \in \Lambda_2 : U(x''_1, \ldots, x''_{n_2}) \leq E_2\}$$

and taking the logarithm yields

$$S(\Lambda_1 \cup \Lambda_2, n_1 + n_2, E_1 + E_2 + An_1n_2r^{-\lambda})$$

$$\geq S(\Lambda_1, n_1, E_1) + S(\Lambda_2, n_2, E_2)$$ (3.6)

(3.4) follows immediately from (3.6), and (3.5) follows from repeated application of (3.4), using

$$\sum_{i<j} n_in_j \leq \tfrac{1}{2}\left(\sum_{i} n_i\right)^2$$

3.3.3 A SPECIAL SEQUENCE OF CUBES. From now on we assume tempered-ness of the interaction. We first prove the existence of a thermodynamic limit for a special sequence (Λ_N) of regions of \mathbf{R}^ν. Let θ satisfy

$$1 < 2^{\nu/\lambda} < \theta < 2 \tag{3.7}$$

and let R_0 be as in the definition 3.1.1 of temperedness. Also let

$$L > R = \frac{R_0}{2 - \theta} \tag{3.8}$$

and write

$$L_N = 2^N L - \theta^N R \tag{3.9}$$

for integer $N \geq 0$. Finally, define Λ_N as a cube with side L_N in \mathbf{R}^ν. It is possible (see Fig. 5) to place 2^ν translates $\Lambda_N^{(i)}$ of Λ_N in Λ_{N+1} at mutual distances at least

$$R_N = L_{N+1} - 2L_N = \theta^N(2 - \theta)R \geq R_0 \tag{3.10}$$

Applying successively Proposition 3.3.1,(b) and Proposition 3.3.2,(b) we get

$$E\left(\Lambda_{N+1}, \sum_{i=1}^{2^\nu} n_i, \sum_{i=1}^{2^\nu} S_i\right) \leq E\left(\bigcup_{i=1}^{2^\nu} \Lambda_N^{(i)}, \sum_{i=1}^{2^\nu} n_i, \sum_{i=1}^{2^\nu} S_i\right)$$

$$\leq \sum_{i=1}^{2^\nu} E(\Lambda_N, n_i, S_i) + \frac{A}{2}\left(\sum_{i=1}^{2^\nu} n_i\right)^2 R_N^{-\lambda} \tag{3.11}$$

In particular

$$E(\Lambda_{N+1}, 2^\nu n, 2^\nu S) \leq 2^\nu E(\Lambda_N, n, S) + \frac{A}{2}(2^\nu n)^2 R_N^{-\lambda} \tag{3.12}$$

For real δ and σ such that $2^{N\nu}\delta$ is an integer greater than or equal to zero

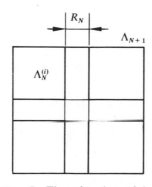

FIGURE 5 The cubes Λ_N and Λ_{N+1}.

we let

$$c_N(\delta, \sigma) = 2^{-N\nu} E(\Lambda_N, 2^{N\nu}\delta, 2^{N\nu}\sigma) - \frac{A}{2}\delta^2 \sum_{m=0}^{N-1} 2^{(m+1)\nu} R_m^{-\lambda} \qquad (3.13)$$

then (3.12) yields

$$c_{N+1}(\delta, \sigma) \le c_N(\delta, \sigma)$$

The sequence (c_N), being decreasing, has a limit:

$$\lim c_N(\delta, \sigma) = c(\delta, \sigma)$$

On the other hand the following limit exists

$$\lim_{N \to \infty} \sum_{m=0}^{N-1} 2^{(m+1)\nu} R_m^{-\lambda} = (2-\theta)^{-\lambda} R_0^{-\lambda} 2^\nu \sum_{m=0}^{\infty} (2^\nu \theta^{-\lambda})^m$$

because $2^\nu \theta^{-\lambda} < 1$ by (3.7). We have thus shown that *if δ is rational of the form $2^{-q}p \ge 0$ and σ is real then the following limit exists*

$$\lim_{N \to \infty} 2^{-N\nu} E(\Lambda_N, 2^{N\nu}\delta, 2^{N\nu}\sigma) = \eta(\delta, \sigma) \qquad (3.14)$$

$$\eta(\delta, \sigma) = c(\delta, \sigma) + \frac{A}{2}\delta^2 \sum_{m=0}^{\infty} 2^{(m+1)\nu} R_m^{-\lambda} \qquad (3.15)$$

and $\eta(\delta, \sigma) = +\infty$ if and only if $E(\Lambda_N, 2^{N\nu}\delta, 2^{N\nu}\sigma) = +\infty$ for all N.

If we divide (3.11) by $2^{(N+1)\nu}$ and let $N \to \infty$ we obtain as a special case

$$\eta(\tfrac{1}{2}\delta_1 + \tfrac{1}{2}\delta_2, \tfrac{1}{2}\sigma_1 + \tfrac{1}{2}\sigma_2) \le \tfrac{1}{2}\eta(\delta_1, \sigma_1) + \tfrac{1}{2}\eta(\delta_2, \sigma_2) \qquad (3.16)$$

If $\alpha = 2^{-q}p, 0 \le \alpha \le 1$, we get by repeated application of (3.16) also

$$\eta(\alpha\delta_1 + (1-\alpha)\delta_2, \alpha\sigma_1 + (1-\alpha)\sigma_2) \le \alpha\eta(\delta_1, \sigma_1) + (1-\alpha)\eta(\delta_2, \sigma_2) \qquad (3.17)$$

Since (c_N) is decreasing we have for $\delta^* = 0, 1,$

$$\eta(\delta^*, \sigma) \le \frac{A}{2}(\delta^*)^2 \sum_{m=0}^{\infty} 2^{(m+1)\nu} R_m^{-\lambda} + c_0(\delta^*, \sigma)$$

$$= \frac{A}{2}(\delta^*)^2 \sum_{m=0}^{\infty} 2^{(m+1)\nu} R_m^{-\lambda} + E(\Lambda_0, \delta^*, \sigma) \qquad (3.18)$$

We check from (3.1) and (3.2) that $E(\Lambda_0, 0, \sigma) = 0$ if $\sigma \le 0$ and that $E(\Lambda_0, 1, \sigma) = 0$ if $\sigma \le \log V(\Lambda_0)$. Inserting $\delta_1 = 0$, $\delta_2 = 1$, $\alpha = 1 - \delta$, $\sigma_1 = \sigma - \delta \log V(\Lambda_0)$, $\sigma_2 = \sigma + (1-\delta)\log V(\Lambda_0)$ in (3.17) we have therefore

$$\eta(\delta, \sigma) \le \delta \frac{A}{2} \sum_{m=0}^{\infty} 2^{(m+1)\nu} R_m^{-\lambda} \qquad (3.19)$$

provided (see Fig. 6)

$$0 \leq \delta \leq 1 \qquad \sigma \leq \delta \log V(\Lambda_0) \tag{3.20}$$

3.3.4 PROPOSITION. *Let the stability condition be satisfied. There exists a nonempty open convex set* $\Delta \subset \mathbf{R}^2$ *(Fig. 6) such that* η *extends by continuity to a convex function in* Δ, *and* $\eta(\delta, \sigma) = +\infty$ *if* (δ, σ) *does not belong to the closure* $\bar{\Delta}$ *of* Δ. *Furthermore* $\eta(\delta, \sigma)$ *is an increasing function of* σ.

Notice first that $\eta(\delta, \sigma)$ is an increasing function of σ by Proposition 3.3.1,(a). The proof that η is a convex function of its arguments is somewhat delicate because $\eta(\delta, \sigma)$ is initially defined only when δ is of the form $2^{-q}p$. We write

$$\Gamma = \{(\delta, \sigma): \eta(\delta, \sigma) \neq +\infty\} \tag{3.21}$$

and let $\bar{\Gamma}$ be the closure of this set in \mathbf{R}^2; by (3.16) $\bar{\Gamma}$ is convex. We denote by Δ the interior of $\bar{\Gamma}$, which is a convex open set.

It follows from (3.19) that η is bounded from above (where defined) in the region (3.20). This shows in particular that Δ is not empty. If (d, σ) is any point of Δ, we can find $(\delta_1, \sigma_1) \in \Gamma$ and $\alpha = 2^{-q}p$, $0 \leq \alpha \leq 1$, such that

$$\alpha(\delta_1, \sigma_1) + (1 - \alpha)(\delta_2, \sigma_2)$$

covers a neighborhood of (d, σ) if (δ_2, σ_2) varies over (3.20). By (3.17) this implies that η is also bounded from above in a neighborhood of (d, σ). On the other hand stability [see (3.3)] yields a lower bound

$$\eta(\delta, \sigma) \geq -\delta B \tag{3.22}$$

so that $|\eta|$ is bounded in a neighborhood of every point of Δ.

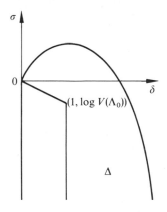

FIGURE 6 The convex region Δ.

From (3.17) and the boundedness of η we prove[16] now that η has a continuous extension to Δ. Let $D \subset \Delta$ be a disk such that for some $K > 0$

$$|\eta(\delta, \sigma)| < K \qquad \text{if} \quad (\delta, \sigma) \in D \tag{3.23}$$

and let the disk D' be concentric with D with half the radius. Since disks of the form D' cover Δ it is sufficient to show that η has a continuous extension to D'. This will now be done by proving that η is uniformly continuous where originally defined in D'.

For simplicity we let $(\delta, \sigma) = \xi$. We obtain from (3.17) in particular

$$\eta(\xi') \leq 2^{-q}\eta(\xi'' + 2^q(\xi' - \xi'')) + (1 - 2^{-q})\eta(\xi'') \tag{3.24}$$

Therefore

$$\eta(\xi') - \eta(\xi'') \leq 2^{-q}[\eta(\xi'' + 2^q(\xi' - \xi'')) - \eta(\xi'')] \tag{3.25}$$

and similarly

$$\eta(\xi'') - \eta(\xi') \leq 2^{-q}[\eta(\xi' + 2^q(\xi'' - \xi')) - \eta(\xi')] \tag{3.26}$$

If $\xi', \xi'' \in D'$, and $\xi' - \xi'' \to 0$, we may assume that $\xi'' + 2^q(\xi' - \xi'') \in D$, $\xi' + 2^q(\xi'' - \xi') \in D$ with $q \to \infty$. From (3.23), (3.25), and (3.26) we obtain

$$|\eta(\xi') - \eta(\xi'')| \leq 2^{1-q}K \tag{3.27}$$

where $q \to \infty$. This shows the uniform continuity of η and completes the proof.

The extension by continuity in Δ of the function η will again be denoted by η.

3.3.5 REMARK. *Let* $(\delta, \sigma_1), (\delta, \sigma_2) \in \Delta$ *with* $\delta = 2^{-q}p$; *the convergence of* $2^{-N\nu}E(\Lambda_N, 2^{N\nu}\delta, 2^{N\nu}\sigma)$ *to* $\eta(\delta, \sigma)$ *is uniform for* $\sigma \in [\sigma_1, \sigma_2]$ *because* pointwise convergence of monotone functions to a continuous function on a compact interval implies uniform convergence.

3.3.6 GENERAL DOMAINS. We shall now prove the existence of the thermodynamic limit for general domains Λ tending to ∞ in the sense of Fisher, instead of the special sequence of cubes used above.

3.3.7 PROPOSITION. *If* $\Lambda \to \infty$ *(Fisher) and if*

$$V(\Lambda)^{-1}n \to L^{-\nu}\delta \qquad V(\Lambda)^{-1}S \to L^{-\nu}\sigma \tag{3.28}$$

where $(\delta, \sigma) \in \Delta$, *then*

$$\lim_{\Lambda \to \infty} V(\Lambda)^{-1}E(\Lambda, n, S) = L^{-\nu}\eta(\delta, \sigma) \tag{3.29}$$

[16] This is a standard argument due to Jensen; see Polya and Szegö [1] Vol. I, Part 2, **124**, pp. 63 and 224.

First we show that

$$\limsup_{\Lambda \to \infty} V(\Lambda)^{-1} E(\Lambda, n, S) \le L^{-\nu} \eta(\delta, \sigma) \tag{3.30}$$

Let δ_0, close to but strictly larger than δ, be of the form $2^{-N\nu} n_0$. We write

$$n = m n_0 + r_0 \qquad 0 \le r_0 < n_0$$

Keeping δ_0 fixed, we choose N as a function of n such that

$$n^{-1} L_N^\nu N \to 0 \qquad n^{-1} L_N^\lambda \to \infty \tag{3.31}$$

It is easy to verify from Definition 2.1.2 that, for "large enough" Λ (i.e., as $\Lambda \to \infty$), Λ contains $m + r_0$ disjoint cubes with side ξL_N where

$$1 < \xi < \left(\frac{\delta_0}{\delta}\right)^{1/\nu}$$

Therefore, Λ contains translates $\Lambda_N^{(i)}$, $i = 1, \ldots, m + r_0$, of Λ_N with mutual distances at least $(\xi - 1) L_N$. Using Propositions 3.3.1,(b), 3.3.2,(b), and the fact that $E(\Lambda_N, 1, \log V(\Lambda_N)) = 0$ we have

$$E(\Lambda, n, S) \le E\left(\bigcup_{i=1}^{m + r_0} \Lambda_N^{(i)}, m n_0 + r_0, S\right)$$

$$\le m E\left(\Lambda_N, n_0, \frac{S - r_0 \log V(\Lambda_N)}{m}\right) + \tfrac{1}{2} A n^2 (\xi - 1)^{-\lambda} L_N^{-\lambda} \tag{3.32}$$

We may rewrite (3.32) as

$$V(\Lambda)^{-1} E(\Lambda, n, S)$$

$$\le \frac{n}{V(\Lambda)} \cdot \frac{m n_0}{n} \cdot \frac{1}{\delta_0} \cdot 2^{-N\nu} E\left(\Lambda_N, n_0, \frac{S - r_0 \log V(\Lambda_N)}{m}\right)$$

$$+ \tfrac{1}{2} A(\xi - 1)^{-\lambda} \cdot \frac{n}{V(\Lambda)} \cdot \frac{n}{L_N^\lambda}$$

Letting $\Lambda \to \infty$ yields thus

$$\limsup_{\Lambda \to \infty} V(\Lambda)^{-1} E(\Lambda, n, S) \le L^{-\nu} \cdot \frac{\delta}{\delta_0} \cdot \eta\left(\delta_0, \frac{\delta_0}{\delta} \sigma\right) \tag{3.33}$$

where we have used (3.31) and Remark 3.3.5. Finally, letting δ_0 approach δ from above, we get (3.30) from (3.33).

We conclude the proof of (3.30) by showing that

$$\liminf_{\Lambda \to \infty} V(\Lambda)^{-1} E(\Lambda, n, S) \ge L^{-\nu} \eta(\delta, \sigma) \tag{3.34}$$

In view of Fisher convergence [see (1.10) of Chapter 2] there exists $C' > 0$

and an integer N' such that $\Lambda_{N'}$ contains a translate Λ' of Λ and

$$C' \leq V(\Lambda_{N'})^{-1} V(\Lambda) \leq \tfrac{1}{2} \tag{3.35}$$

Let δ_1, close to δ, be of the form $2^{-q}p$; if $N'v \geq q$ we may write $\delta_1 = 2^{-N'v} n_1$. Also, let Λ'' be the set of points of $\Lambda_{N'}$ which has a distance to Λ' larger than $\theta^{N'} R$. Since $\Lambda \to \infty$ (Fisher), we have $\Lambda'' \to \infty$ (Fisher) and

$$\lim_{\Lambda \to \infty} V(\Lambda_{N'})^{-1}(V(\Lambda) + V(\Lambda'')) = 1 \tag{3.36}$$

Propositions 3.3.1,(b) and 3.3.2,(a) give

$$E(\Lambda_{N'}, 2^{N'v} \delta_1, 2^{N'v} \sigma)$$

$$\leq E(\Lambda, n, S) + E(\Lambda'', n_1 - n, 2^{N'v}\sigma - S) + \tfrac{1}{2} A n_1^{2} (\theta^{N'} R)^{-\lambda}$$

Dividing by $V(\Lambda)$ and letting $\Lambda \to \infty$ (which implies $N' \to \infty$) we get by (3.36)

$$\liminf_{\Lambda \to \infty} V(\Lambda)^{-1} E(\Lambda, n, S)$$

$$\geq \liminf_{\Lambda \to \infty} \left[\frac{V(\Lambda_{N'})}{V(\Lambda)} (L^{-v} 2^{-N'v} E(\Lambda_{N'}, 2^{-N'v} \delta_1, 2^{N'v} \sigma)) \right.$$

$$\left. - \frac{V(\Lambda_{N'}) - V(\Lambda)}{V(\Lambda)} (V(\Lambda'')^{-1} E(\Lambda'', n_1 - n, 2^{N'v}\sigma - S)) \right]$$

Using then in the right-hand side of (3.14), (3.30), and (3.35), we obtain the desired inequality (3.34) when $\delta_1 \to \delta$.

It is possible to supplement Proposition 3.3.7 as follows.

3.3.8 PROPOSITION. *Let* $\Lambda \to \infty$ *(Fisher) and*

$$V(\Lambda)^{-1} n \to L^{-v} \delta \qquad V(\Lambda)^{-1} S \to L^{-v} \sigma \tag{3.37}$$

(a) *If* (δ, σ) *is not in the closure* $\bar{\Delta}$ *of* Δ, *then*

$$\lim_{\Lambda \to \infty} V(\Lambda)^{-1} E(\Lambda, n, S) = +\infty \tag{3.38}$$

(b) *If* (δ, σ) *belongs to the boundary of* Δ, *then*

$$\liminf_{\Lambda \to \infty} V(\Lambda)^{-1} E(\Lambda, n, S) \geq \eta^*(\delta, \sigma) \tag{3.39}$$

where $\eta^*(\delta, \sigma)$ *is* $\liminf \eta(\delta^*, \sigma^*)$ *when* $(\delta^*, \sigma^*) \in \Delta, (\delta^*, \sigma^*) \to (\delta, \sigma)$.

Let $V(\Lambda)^{-1} E(\Lambda, n, S)$ tend to a finite value violating (a) or (b). We define $\Lambda_{N'}, \Lambda', \Lambda''$ as in the second part of 3.3.7 and (going to a subsequence) assume that $V(\Lambda_{N'})^{-1} V(\Lambda) \to \alpha$. Let also $V(\Lambda'')^{-1} n'' \to \delta''$, $V(\Lambda'')^{-1} S'' \to \sigma''$, with $(\delta'', \sigma'') \in \Delta$ and $n + n'' = 2^{N'v} \delta_1$, $S + S'' = 2^{N'v} \sigma_1$. Proposition 3.3.2,(a) yields

then

$$\alpha \liminf_{\Lambda \to \infty} V(\Lambda)^{-1} E(\Lambda, n, S) + (1 - \alpha) L^{-\nu} \eta(\delta'', \sigma'')$$

$$\geq L^{-\nu} \lim_{N' \to \infty} 2^{-N'\nu} E(\Lambda_{N'}, 2^{N'\nu}\delta_1, 2^{N'\nu}\sigma_1)$$

In case (a) we may choose $(\delta_1, \sigma_1) \notin \bar{\Delta}$ and obtain a contradiction to $E(\Lambda_N, 2^{N\nu}\sigma_1, 2^{N\nu}\sigma_1) = +\infty$ (see 3.3.3). In case (b) a contradiction is obtained if (δ'', σ'') tends to (δ, σ) along a straight line.

3.3.9 PROPOSITION. *The function η is continuous on the union of Δ and the half-line $\{(\delta, \sigma): \delta = 0, \sigma < 0\}$. If $\Lambda \to \infty$ (Fisher) and*

$$V(\Lambda)^{-1} n \to 0 \qquad V(\Lambda)^{-1} S \to L^{-\nu}\sigma < 0 \tag{3.40}$$

then

$$\lim_{\Lambda \to \infty} V(\Lambda)^{-1} E(\Lambda, n, S) = L^{-\nu}\eta(0, L^{-\nu}\sigma) = 0 \tag{3.41}$$

It follows from (3.19) and (3.22) that if $\delta^* \to 0$ and $\sigma^* \to \sigma < 0$, then $\eta(\delta^*, \sigma^*) \to 0 = \eta(0, \sigma)$. Finally, (3.41) results from (3.39) and (3.30) because the proof of the latter equation extends to the present situation.

3.3.10 ENTROPY AS FUNCTION OF ENERGY AND DENSITY. In our study of the thermodynamic limit we have up to now, for technical reasons, considered E as a function of n, S. We shall now regard S as a function of n, E. It is easy to obtain a theorem on the thermodynamic limit for these new variables. This follows from the fact that we can formulate the results obtained above in a manner which is symmetrical with respect to S, n, E. Let Σ^* be the subset of \mathbf{R}^3 which consists of

(a) the graph of η over Δ, i.e., $\{(\delta, \sigma, \eta): (\delta, \sigma) \in \Delta, \eta = \eta(\delta, \sigma)\}$,
(b) the union of the half-lines $\{(\delta, \sigma, \eta): \delta = \delta^*, \sigma = \sigma^*, \eta \geq \eta^*\}$ where (δ^*, σ^*) is a point of the boundary of Δ, not of the form $(\delta = 0, \sigma < 0)$ and η^* is $\liminf \eta(\delta, \sigma)$ when $(\delta, \sigma) \in \Delta, (\delta, \sigma) \to (\delta^*, \sigma^*)$,
(c) the half-line $\{(\delta, \sigma, \eta): \delta = 0, \sigma < 0, \eta = 0\}$.

Before stating our main result we change, for obvious reasons, the normalization of the variables, writing

$$\rho = L^{-\nu}\delta \qquad \varepsilon = L^{-\nu}\eta \qquad s = L^{-\nu}\sigma \tag{3.42}$$

and we call Σ the surface corresponding to Σ^* in these new variables. Inspection of 3.3.7–3.3.9 yields the following result.

3.3.11 PROPOSITION. *If $\Lambda \to \infty$ in the sense of Fisher and if $V(\Lambda)^{-1} n \to \rho$, $V(\Lambda)^{-1} E \to \varepsilon$, $V(\Lambda)^{-1} S(n, E) \to s$ where ρ, ε, s are finite, then $(\rho, \varepsilon, s) \in \Sigma$.*

We note that if $(\rho, \varepsilon, s) \in \Sigma$, then

$$\varepsilon \geq -\rho B \qquad s \leq \rho - \rho \log \rho \qquad (3.43)$$

because of (3.22) and

$$S(\Lambda, n, E) \leq \log \frac{V(\Lambda)^n}{n!} \qquad (3.44)$$

We define $\rho_{cp} = \sup_{(\rho,\varepsilon,s)\in\Sigma} \rho$, i.e., ρ_{cp} is the highest density which appears in Σ. In the absence of hard cores (i.e., if we may assume that U does not take the value $+\infty$) we have[17] $\rho_{cp} = +\infty$. If hard cores are present, ρ_{cp} is entirely due to them and is the "*close-packing*" density. If $0 \leq \rho < \rho_{cp}$, let $\varepsilon_0(\rho) = \inf_{(\rho,\varepsilon,s)\in\Sigma} \varepsilon$; also let $\varepsilon_0(0) = 0$. We now have all the ingredients of the following theorem (see Fig. 7).

3.3.12 THEOREM. *Let $S(\Lambda, n, E)$ be defined by (3.1), (3.2) for a stable, tempered interaction. There exist*

(a) $\rho_{cp} > 0$ or $\rho_{cp} = +\infty$,

(b) *a convex continuous function ε_0 on the interval $[0, \rho_{cp})$ such that $\varepsilon_0(0) = 0$ and $\varepsilon_0(\rho) \geq -\rho B$,*

(c) *a concave continuous function s on the region $\Theta = \{(\rho, \varepsilon): 0 \leq \rho < \rho_{cp}, \varepsilon > \varepsilon_0(\rho)\}$, increasing in ε for fixed ρ and such that $s(0, \varepsilon) = 0$ for $\varepsilon > 0$.*

Let $\Lambda \to \infty$ in the sense of Fisher and

$$\lim_{\Lambda \to \infty} V(\Lambda)^{-1} n = \rho \qquad \lim_{\Lambda \to \infty} V(\Lambda)^{-1} E = \varepsilon$$

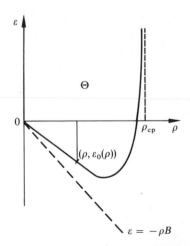

FIGURE 7 The convex region Θ.

[17] Notice that $\Omega(\Lambda, n, E)$ can be made unequal to zero by taking E sufficiently large, and use the fact 3.3.3 that $(\eta(\delta, \sigma) = +\infty$ if and only if $E(\Lambda_N, 2^{Nv}\delta, 2^{Nv}\sigma) = +\infty$ for all N.

FIGURE 8 Entropy versus energy.

(a) *If* $(\rho, \varepsilon) \in \Theta$ *then* $\lim_{\Lambda \to \infty} V(\Lambda)^{-1} S(\Lambda, n, E) = s(\rho, \varepsilon)$.

(b) *If* (ρ, ε) *belongs to the boundary of* Θ *then*

$$\limsup_{\Lambda \to \infty} V(\Lambda)^{-1} S(\Lambda, n, E) \leq s^*(\rho, \varepsilon)$$

where $s^*(\rho, \varepsilon)$ *is* $\limsup s(\rho^*, \varepsilon^*)$ *when* $(\rho^*, \varepsilon^*) \in \Theta, (\rho^*, \varepsilon^*) \to (\varepsilon, \rho)$.

(c) *If* (ρ, ε) *belongs to the complement of the closure of* Θ *then*

$$\lim_{\Lambda \to \infty} V(\Lambda)^{-1} S(\Lambda, n, E) = -\infty$$

3.3.13 UNIFORMITY OF THE CONVERGENCE. Let K be compact, $K \subset \Theta$ and $\Lambda \to \infty$ (Fisher). The convergence of $|V(\Lambda)^{-1} S(\Lambda, n, E) - s(V(\Lambda)^{-1} n, V(\Lambda)^{-1} E)|$ to zero is uniform for $(V(\Lambda)^{-1} n, V(\Lambda)^{-1} E) \in K$. (Otherwise we could find $C > 0$ and Λ_i, n_i, E_i such that $\Lambda_i \to \infty$, $V(\Lambda_i)^{-1} n_i \to \rho$, $V(\Lambda_i)^{-1} E_i \to \varepsilon, (\rho, \varepsilon) \in K$ and $|V(\Lambda)^{-1} S(\Lambda, n, E) - s(\rho, \varepsilon)| \geq C$, in contradiction with Theorem 3.3.12.)

3.3.14 THE ENSEMBLE (2.4) OF CHAPTER 1. We have introduced in Chapter 1 [Eqs. (2.4) and (2.5)] two measures describing configurational microcanonical ensembles:

$$\frac{1}{n!} \delta^\Delta [U(x_1, \ldots, x_n) - E] \, dx_1 \cdots dx_n \tag{3.45}$$

$$\frac{1}{n!} \delta^- [U(x_1, \ldots, x_n) - E] \, dx_1 \cdots dx_n \tag{3.46}$$

We shall now compare these ensembles.

Figure 8 shows a typical curve $\varepsilon \to s(\rho, \varepsilon)$ for fixed ρ, $0 < \rho < \rho_{cp}$. Because of (3.43), $\bar{s}(\rho) = \sup s(\rho, \varepsilon) < +\infty$. Let $\varepsilon_1(\rho)$ be the infimum of those ε for

which $s(\rho, \varepsilon) = \bar{s}(\rho); \varepsilon_1(\rho)$ is a real number or $+\infty$. Notice that (by convexity of s) $\varepsilon \to s(\rho, \varepsilon)$ is strictly increasing in the interval $(\varepsilon_0(\rho), \varepsilon_1(\rho))$. This fact permits a comparison between the ensemble defined by (3.46) which we have used up to now and the ensemble defined by (3.45). If Ω^Δ is defined by analogy with (3.1) for the ensemble (3.45) we have

$$\Omega(\Lambda, n, E) = \Omega(\Lambda, n, E - \Delta) + \Omega^\Delta(\Lambda, n, E) \qquad (3.47)$$

Let $\Lambda \to \infty$ (Fisher), $V(\Lambda)^{-1}n \to \rho$ where $0 < \rho < \rho_{cp}$, and $V(\Lambda)^{-1}E \to \varepsilon$ where $\varepsilon \in (\varepsilon_0(\rho), \varepsilon_1(\rho))$; the strict increase of s and Theorem 3.3.12 show that

$$\lim_{\Lambda \to \infty} V(\Lambda)^{-1} \log \Omega^\Delta(\Lambda, n, E) = s(\rho, \varepsilon) \qquad (3.48)$$

Furthermore it is easy to see that the normalized measures corresponding to (3.45) and (3.46) become close in norm as $\Lambda \to \infty$. For $\varepsilon + \delta > \varepsilon_1(\rho)$, the two ensembles may deviate (concave dotted line on Fig. 8), but this is unimportant for the transition to the classical ensembles of the next section. The ensemble (3.45) has the interesting feature that it may yield a nontrivial thermodynamic limit for nonstable interactions.

3.4 THERMODYNAMIC LIMIT FOR THE CLASSICAL ENSEMBLES

All interactions considered in this section are stable and tempered. Now that we have detailed results (Proposition 3.3.11 or Theorem 3.3.12) on the thermodynamic limit for the configurational microcanonical ensemble, it will be easy to treat the other ensembles. The essential idea in going from one ensemble to another is to isolate in the partition function of the latter one contribution which is shown to be dominant in the thermodynamic limit and which can be expressed in terms of the former ensemble. For each ensemble we shall obtain a thermodynamic function which is convex (or concave) in its arguments; this convexity is known as the *thermodynamic stability* property. The transition from microcanonical to canonical and grand canonical ensembles amounts, in the thermodynamic limit, to a description of the surface Σ of Proposition 3.3.11 in terms of its tangent lines and planes (using the so-called Legendre transformations).

3.4.1 MICROCANONICAL ENSEMBLE. If we take into account the momenta and the kinetic energy, the microcanonical partition function is

$$\Omega_{tot}(\Lambda, n, E) = \frac{1}{n!}\int_{\mathbf{R}^{nv}} dp_1 \cdots dp_n \int_{\Lambda^n} dx_1 \cdots dx_n$$
$$\times \delta^- \left(\sum_{i=1}^n \frac{p_i^2}{2m} + U(x_1, \ldots, x_n) - E \right) \qquad (4.1)$$

Define a kinetic partition function by

$$\Omega_{kin}(n, E) = \int_{\mathbf{R}^{nv}} dp_1 \cdots dp_n \delta\left(\sum_{i=1}^{n} \frac{p_i^2}{2m} - E\right) = \frac{(2\pi m E)^{nv/2}}{E\Gamma(nv/2)} \tag{4.2}$$

then we have

$$\Omega_{tot}(\Lambda, n, E) = \int dt \int_{\mathbf{R}^{nv}} dp_1 \cdots dp_n \delta\left(\sum_{i=1}^{n} \frac{p_i^2}{2m} - t\right)$$

$$\times \frac{1}{n!} \int_{\Lambda^n} dx_1 \cdots dx_n \delta^-(U(x_1, \ldots, x_n) + t - E)$$

$$= \int dt \Omega_{kin}(n, t)\Omega(\Lambda, n, E - t) \tag{4.3}$$

Let $\Lambda \to \infty$ (Fisher), $V(\Lambda)^{-1}n \to \rho$, $V(\Lambda)^{-1}E \to \varepsilon$, then

$$\lim_{\Lambda \to \infty} V(\Lambda)^{-1} \log \Omega_{kin}(n, E) = s_{kin}(\rho, \varepsilon) = \frac{v\rho}{2}\left(\log \frac{4\pi m \varepsilon}{v\rho} + 1\right) \tag{4.4}$$

We suppose now $0 \leq \rho < \rho_{cp}$ and $\varepsilon > \varepsilon_0(\rho)$ and let

$$s_{tot}(\rho, \varepsilon) = \sup_{\varepsilon' \in (0, \varepsilon - \varepsilon_0(\rho))} [s_{kin}(\rho, \varepsilon') + s(\rho, \varepsilon - \varepsilon')] \tag{4.5}$$

Theorem 3.3.12 and Eq. (4.4) show that there exists a nonempty interval (t_1, t_2) such that, given $\delta > 0$, we have for large Λ and all $t \in (t_1, t_2)$

$$\Omega_{kin}(n, t)\Omega(\Lambda, n, E - t) > \exp[V(\Lambda)(s_{tot}(\rho, \varepsilon) - \delta)]$$

Inserting this in (4.3) yields

$$\liminf_{\Lambda \to \infty} V(\Lambda)^{-1} \log \Omega_{tot}(\Lambda, n, E) \geq s_{tot}(\rho, \varepsilon) \tag{4.6}$$

We show that on the other hand, given $\delta > 0$, we have for large Λ and all t

$$\Omega_{kin}(n, t)\Omega(\Lambda, n, E - t) < \exp[V(\Lambda)(s_{tot}(\rho, \varepsilon) + \delta)] \tag{4.7}$$

The left-hand side vanishes for $t < 0$ and, by the stability condition, for large t; if (4.7) were not satisfied on the remaining compact interval we could, as in Section 3.3.13, choose a sequence (Λ_i, n_i, E_i) violating the thermodynamic limit [Theorem 3.3.12,(a), (b), or (c)]. From the inequality (4.7) and the fact that its left-hand side vanishes outside of a fixed compact interval we get

$$\limsup_{\Lambda \to \infty} V(\Lambda)^{-1} \log \Omega_{tot}(\Lambda, n, E) \leq s_{tot}(\rho, \varepsilon) \tag{4.8}$$

3.4.2 PROPOSITION. *Let* $\Lambda \to \infty$ *in the sense of Fisher and* $V(\Lambda)^{-1}n \to \rho$, $V(\Lambda)^{-1}E \to \varepsilon$ *where* $0 \leq \rho < \rho_{cp}$ *and* $\varepsilon_0(\rho) < \varepsilon$; *then, if* Ω_{tot} *is given*

by (4.1),

$$\lim_{\Lambda \to \infty} V(\Lambda)^{-1} \log \Omega_{tot}(\Lambda, n, E) = s_{tot}(\rho, \varepsilon) \tag{4.9}$$

where s_{tot} is given by Eq. (4.5).

The cases where $(\rho, \varepsilon) \notin \Theta$ are treated analogously. According to (4.5) we see that ε is distributed between kinetic and potential (=configurational) energy so as to maximize the sum of the kinetic and the configurational entropies. The (specific) configurational energy is contained in the interval $[\varepsilon_0(\rho), \varepsilon_1(\rho)]$

3.4.3 CANONICAL ENSEMBLE. The canonical partition function is

$$\left(\frac{2\pi m}{\beta}\right)^{nv/2} Q(\Lambda, n, \beta) \tag{4.10}$$

where Q is the configurational partition function:

$$Q(\Lambda, n, \beta) = \frac{1}{n!} \int_{\Lambda^n} dx_1 \cdots dx_n \exp(-\beta U(x_1, \ldots, x_n)) \tag{4.11}$$

If $0 \le \rho < \rho_{cp}, \beta > 0$, we define

$$f(\rho, \beta) = \inf_{\varepsilon > \varepsilon_0(\rho)} (\varepsilon - \beta^{-1} s(\rho, \varepsilon)) \tag{4.12}$$

Observe that the infimum is reached at $\varepsilon = \varepsilon^*$ if the slope of a tangent to the curve $\varepsilon \to s(\rho, \varepsilon)$ has the value β for $\varepsilon = \varepsilon^*$, $f(\rho, \beta)$ is thus the abscissa of intersection of this tangent with the ε axis (see Fig. 9). It follows from (4.12) that $f(\rho, \beta)$ is a convex function of ρ; $\beta f(\rho, \beta)$ is a concave function of β; or $f(\rho, \beta)$ is a concave function of β^{-1}.

Let $\Lambda \to \infty$ (Fisher), $V(\Lambda)^{-1} n \to \rho$. Given $\delta > 0$ we can find $\tilde{\varepsilon}$ such that $\beta \tilde{\varepsilon} - s(\rho, \tilde{\varepsilon}) < \beta f(\rho, \beta) + \delta$. Let $\tilde{E} = \tilde{\varepsilon} V(\Lambda)$. For sufficiently large Λ we

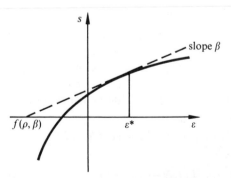

FIGURE 9 Entropy versus energy.

have

$Q(\Lambda, n, \beta)$

$$\geq \frac{1}{n!} \int_{\Lambda^n} dx_1 \cdots dx_n \delta^-(U(x_1, \ldots, x_n) - \tilde{E}) \exp(-\beta U(x_1, \ldots, x_n))$$

$$\geq \exp(-\beta \tilde{E})\Omega(\Lambda, n, \tilde{E}) \geq \exp[V(\Lambda)(-\beta \tilde{\varepsilon} + s(\beta, \tilde{\varepsilon}) - \delta)]$$

$$\geq \exp[V(\Lambda)(-\beta f(\rho, \beta) - 2\delta)]$$

so that

$$\liminf_{\Lambda \to \infty} V(\Lambda)^{-1} \log Q(\Lambda, n, \beta) \geq -\beta f(\rho, \beta) \tag{4.13}$$

An estimate in the opposite direction is easily obtained by writing

$$Q = \sum_1^n Q_i + \tilde{Q}$$

$$Q_i = \frac{1}{n!} \int_{\Lambda^n} dx_1 \cdots dx_n \delta^{E_{i+1} - E_i}(U(x_1, \ldots, x_n) - E_i) \exp(-\beta U(x_1, \ldots, x_n))$$

where $E_i = V(\Lambda)\varepsilon_i$, $-\rho B > \varepsilon_1 < \cdots < \varepsilon_{n+1}$, and $\beta \varepsilon_i - s(\rho, \varepsilon_{i+1}) > \beta f(\rho, \beta) - \delta$. The last term

$$\tilde{Q} \leq \frac{V(\Lambda)^n}{n!} \exp(-\beta E_{n+1}) \tag{4.14}$$

is negligible if ε_{n+1} is chosen sufficiently large. Altogether

$$\limsup_{\Lambda \to \infty} V(\Lambda)^{-1} \log Q(\Lambda, n, \beta) \leq -\beta f(\rho, \beta) \tag{4.15}$$

Analogous results are obtained when $\lim V(\Lambda)^{-1} n \geq \rho_{cp}$, yielding the following theorem.

3.4.4 THEOREM. Let $Q(\Lambda, n, \beta)$ be defined by (4.11). A function $f(\rho, \beta)$ is defined for $0 \leq \rho < \rho_{cp}$ and $\beta > 0$ by

$$f(\rho, \beta) = \inf_{\varepsilon > \varepsilon_0(\rho)} (\varepsilon - \beta^{-1} s(\rho, \varepsilon)) \tag{4.16}$$

$f(\rho, \beta)$ is a concave function of β^{-1} and a convex function of ρ, $\rho^{-1} f(\rho, \beta)$ is a convex and decreasing function of ρ^{-1}.

Let $\Lambda \to \infty$ in the sense of Fisher and $V(\Lambda)^{-1} n \to \rho$.

(a) If $0 \leq \rho < \rho_{cp}$ then $\lim_{\Lambda \to \infty} V(\Lambda)^{-1} \log Q(\Lambda, n, \beta) = -\beta f(\rho, \beta)$.

(b) If $\rho = \rho_{cp}$ then $\limsup_{\Lambda \to \infty} V(\Lambda)^{-1} \log Q(\Lambda, n, \beta) \leq -\beta f^*(\beta)$ where $f^*(\beta) = \lim_{\rho \to \rho_{cp}, \rho < \rho_{cp}} f(\rho, \beta)$.

(c) If $\rho > \rho_{cp}$ then $\lim_{\Lambda \to \infty} V(\Lambda)^{-1} \log Q(\Lambda, n, \beta) = -\infty$.

The function f is known as specific free energy (free energy per unit volume). The free energy per particle $\rho^{-1}f(\rho, \beta)$ is a decreasing function of ρ^{-1} because $Q(\Lambda, n, \beta)$ is, for fixed n, β, an increasing function of Λ.

Formula (4.16) establishes a relation between the parameters of the canonical and microcanonical ensemble, (ρ, β) corresponding to (ρ, ε^*) where ε^* is the abscissa of the point of contact (if it is unique) of the tangent with slope β to the curve $\varepsilon \to s(\rho, \varepsilon)$ (see Fig. 9). Assuming differentiability[18] we readily obtain the standard thermodynamic relations

$$f(\rho, \beta) = \varepsilon^* - \beta^{-1}s(\rho, \varepsilon^*) \qquad \beta = \frac{\partial s(\rho, \varepsilon^*)}{\delta\varepsilon^*} \qquad s(\rho, \varepsilon^*) = \beta^2 \frac{\partial f(\rho, \beta)}{\partial \beta}$$

They express here the equivalence of ensembles.

3.4.5 GRAND CANONICAL ENSEMBLE. The grand partition function is

$$\Xi(\Lambda, z, \beta) = \sum_{n=0}^{\infty} z^n Q(\Lambda, n, \beta) \tag{4.17}$$

where $Q(\Lambda, 0, \beta) = 1$ and $z > 0$ is the activity. We define

$$p(z, \beta) = \sup_{0 \le \rho < \rho_{cp}} (\beta^{-1}\rho \log z - f(\rho, \beta)) \tag{4.18}$$

The supremum is reached at $\rho = \rho^*$ if the slope of a tangent to the curve $\rho \to f(\rho, \beta)$ has the value $\beta^{-1} \log z$ for $\rho = \rho^*$; $-\beta p(z, \beta)$ is the ordinate of the intersection of this tangent with the f axis (see Fig. 10). It follows from (4.18) that $\beta p(z, \beta)$ is a convex function of $\log z$ and β.

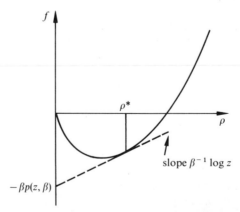

FIGURE 10 Free energy versus density.

[18] Because of convexity, derivatives actually exist almost everywhere.

Let $\Lambda \to \infty$ (Fisher). Given $\delta > 0$ we can find $\tilde{\rho}$, $0 \leq \tilde{\rho} < \rho_{cp}$, such that $\beta^{-1}\tilde{\rho} \log z - f(\tilde{\rho}, \beta) \geq p(z, \beta) - \beta^{-1}\delta$ and, choosing \tilde{n} such that $V(\Lambda)^{-1}\tilde{n} \to \tilde{\rho}$, we have for sufficiently large Λ

$$\Xi(\Lambda, z, \beta) \geq z^{\tilde{n}}Q(\Lambda, \tilde{n}, \beta)$$

$$\exp[V(\Lambda)(\tilde{\rho} \log z - \beta f(\tilde{\rho}, \beta) - \delta)] \geq \exp[V(\Lambda)(\beta p(z, \beta)) - 2\delta)]$$

So that

$$\liminf_{\Lambda \to \infty} V(\Lambda)^{-1} \log \Xi(\Lambda, z, \beta) \geq \beta p(z, \beta) \tag{4.19}$$

We have by the stability condition

$$z^n Q(\Lambda, n, \beta) \leq \frac{1}{n!}[zV(\Lambda) e^{\beta B}]^n \leq \left[z\frac{V(\Lambda)}{n} e^{\beta B + 1}\right]^n$$

Defining $v_0 = \frac{1}{2}z^{-1} e^{-\beta B - 1}$, we have

$$\sum_{n > v_0^{-1}V(\Lambda)} z^n Q(\Lambda, n, \beta) \leq \sum_{n > v_0^{-1}V(\Lambda)} (\tfrac{1}{2})^n \tag{4.20}$$

On the other hand, given $\delta > 0$, we have for $V(\Lambda)^{-1}n$ in the compact interval $[0, v_0^{-1}]$ and for sufficiently large Λ

$$z^n Q(\Lambda, n, \beta) \leq \exp[V(\Lambda)(\rho \log z - \beta f(V(\Lambda)^{-1}n, \beta) + \delta)]$$

$$\leq \exp[V(\Lambda)(\beta p(z, \beta) + \delta)]$$

so that

$$\Xi(\Lambda, z, \beta) \leq v_0^{-1}V(\Lambda) \exp[V(\Lambda)(\beta p(z, \beta) + \delta)] + \sum_{n > v_0^{-1}V(\Lambda)} (\tfrac{1}{2})^n$$

hence

$$\limsup_{\Lambda \to \infty} V(\Lambda)^{-1} \log \Xi(\Lambda, z, \beta) \leq \beta p(z, \beta) \tag{4.21}$$

3.4.6 THEOREM. *Let $\Xi(\Lambda, z, \beta)$ be defined by* (4.17). *A function $p(z, \beta)$ is defined for $z > 0$, $\beta > 0$ by*

$$p(z, \beta) = \sup_{0 \leq \rho \leq \rho_{cp}} (\beta^{-1}\rho \log z - f(\rho, \beta)) \tag{4.22}$$

$\beta p(z, \beta)$ is a convex function of $\log z$ and β, $p(z, \beta) \geq 0$, and $p(z, \beta)$ is an increasing function of z.

Let $\Lambda \to \infty$ in the sense of Fisher, then

$$\lim_{\Lambda \to \infty} V(\Lambda)^{-1} \log \Xi(\Lambda, z, \beta) = \beta p(z, \beta) \tag{4.23}$$

The function p is the pressure; we have $p \geq 0$ because $\Xi \geq 1$ and $p(z, \beta)$ is an increasing function of z because $\Xi(\Lambda, z, \beta)$ is. Assuming differentiability,

(4.22) yields the following thermodynamic relations

$$\beta p(z, \beta) = \rho^* \log z - \beta f(\rho^*, \beta) \qquad \beta^{-1} \log z = \frac{\partial f(\rho^*, \beta)}{\partial \rho^*} \qquad \rho^* = \frac{\partial(\beta p(z, \beta))}{\partial(\log z)}$$

3.4.7 EXTENSIONS. The results obtained in Sections 3.3 and 3.4 can certainly be generalized. We mention the following points.

(a) The notion of temperedness is somewhat unnatural. In particular we shall use a different notion in the next chapter to study the thermodynamic limit for pair interactions at small activity.

(b) We can hope to replace convergence of regions to infinity in the sense of Fisher by a more general notation. It can be proved that convergence in the sense of Van Hove is sufficient for the canonical ensemble if pair potentials of the Lennard–Jones type (see Section 3.2.10) are used. See also Exercise 3D and Theorem 4.3.1.

(c) The results proved for continuous systems hold, with suitable modifications, for lattice systems also.

3.4.8 CONTINUITY OF THE PRESSURE. The thermodynamic stability conditions of Theorem 3.4.4 imply that, at fixed β, the pressure p is a decreasing function of the specific volume ρ^{-1}:

$$\frac{\partial p}{\partial(\rho^{-1})} = -\frac{\partial^2}{\partial(\rho^{-1})^2}[\rho^{-1} f(\rho, \beta)] \le 0 \qquad (4.24)$$

Experimentally, it appears that p is, furthermore, a continuous function of ρ^{-1}; no (jump) discontinuities appear. In the case of a pair interaction with positive or hard core pair potential, the continuity of p results from the inequality (4.28) below, but a general proof is lacking.[19]

3.4.9 PROPOSITION.[20] Let the pair potential Φ be positive ($\Phi \ge 0$) or have a hard core and satisfy (2.14). Define $\Phi_+(x) = \max(\Phi(x), 0)$ and

$$D = e^{\beta B} \int dx(1 - \exp[-\beta\Phi_+(x)]) \qquad (4.25)$$

Also, let

$$p_\Lambda(z) = \beta^{-1} V(\Lambda)^{-1} \log \Xi(\Lambda, z, \beta) \qquad (4.26)$$

$$\rho_\Lambda(z) = z \frac{d}{dz} V(\Lambda)^{-1} \log \Xi(\Lambda, z, \beta) \qquad (4.27)$$

[19] Besides the cases covered by Proposition 3.4.9, it can be proved that p is continuous if Φ is bounded above (see Ruelle [1]) or if Φ is of Lennard–Jones type (see R. L. Dobrushin and R. A. Minlos, Existence and Continuity of Pressure in Classic Statistical Physics, *Teorija Verojatn. i ee Prim.* **12**, 596–618 (1967)).

[20] Due to Penrose (unpublished) and Ginibre [2].

Then

$$0 \le \frac{dp_\Lambda}{d\rho_\Lambda} = \frac{dp_\Lambda(z)/dz}{d\rho_\Lambda(z)/dz} \le \beta^{-1}(1 + Dz) \tag{4.28}$$

Define

$$P_n = \Xi^{-1} z^n \int_{\Lambda^n} dx_1 \cdots dx_n \exp[-\beta U(x_1, \ldots, x_n)] \tag{4.29}$$

We apply the Schwarz inequality to

$$P_{n+1} = \Xi^{-1} z^{n+1} \int_{\Lambda^n} dx_1 \cdots dx_n \exp[-\beta U(x_1, \ldots, x_n)]$$

$$\times \int_\Lambda dx_{n+1} \exp\left[-\beta \sum_{i=1}^{n} \Phi(x_{n+1} - x_i)\right]$$

obtaining

$$(P_{n+1})^2 \le P_n \Xi^{-1} z^{n+2} \int_{\Lambda^{n+2}} dx_1 \ldots dx_{n+2} \exp[-\beta U(x_1, \ldots, x_{n+1})]$$

$$\times \exp\left[-\beta \sum_{i=1}^{n} \Phi(x_{n+2} - x_i)\right] \tag{4.30}$$

We now write

$$\exp\left[-\beta \sum_{i=1}^{n} \Phi(x_{n+2} - x_i)\right] = \exp\left[-\beta \sum_{i=1}^{n+1} \Phi(x_{n+2} - x_i)\right]$$

$$+ [1 - \exp(-\beta\Phi(x_{n+2} - x_{n+1}))] \times \exp\left[-\beta \sum_{i=1}^{n} \Phi(x_{n+1} - x_i)\right]$$

$$\le \exp\left[-\beta \sum_{i=1}^{n+1} \Phi(x_{n+2} - x_i)\right] + e^{\beta B} \times [1 - \exp(-\beta\Phi_+(x_{n+2} - x_{n+1}))] \tag{4.31}$$

Inserting (4.31) into (4.30) yields

$$(P_{n+1})^2 \le P_n(P_{n+2} + DzP_{n+1}) \tag{4.32}$$

A probability measure on the integers greater than or equal to zero is defined so that the probability of occurrence of n is $(1/n!)P_n$. If $\langle \cdot \rangle$ is the corresponding expectation value, the Schwarz inequality and (4.32) yield

$$[\langle n \rangle (1 + Dz)]^2 = \left[\sum_{n=0}^{\infty} \frac{1}{n!} P_n \left(\frac{P_{n+1}}{P_n} + nDz\right)\right]^2$$

$$\le \sum_{n=0}^{\infty} \frac{1}{n!} P_n \left(\frac{P_{n+1}}{P_n} + nDz\right)^2$$

$$\leq \sum_{n=0}^{\infty} \frac{1}{n!} (P_{n+2} + DzP_{n+1} + 2P_{n+1}nDz + P_n n^2 D^2 z^2)$$

$$= \langle n(n-1) + nDz + 2n(n-1)Dz + n^2 D^2 z^2 \rangle$$

$$= \langle n^2 \rangle (1 + Dz)^2 - \langle n \rangle (1 + Dz)$$

Therefore

$$\langle n \rangle^2 \leq \langle n^2 \rangle - \frac{\langle n \rangle}{1 + Dz} \tag{4.33}$$

and we obtain

$$\frac{dp_\Lambda(z)/dz}{d\rho_\Lambda(z)/dz} = \beta^{-1} \frac{\langle n \rangle}{\langle n^2 \rangle - \langle n \rangle^2} \leq \beta^{-1}(1 + Dz) \tag{4.34}$$

which concludes the proof.

3.5 THERMODYNAMIC LIMIT FOR THE QUANTUM ENSEMBLES

The first problem that we encounter if we attempt to study the statistical mechanics of quantum systems is to define the Hamiltonian of a system in a box Λ as an unbounded self-adjoint operator. To begin with, the Hamiltonian is given as the formal expression

$$H_n = T + U \tag{5.1}$$

$$T = \sum_{i=1}^{n} \left(-\frac{1}{2m} \Delta_i \right) \tag{5.2}$$

$$U(x_1, \ldots, x_n) = \sum_{k \geq 2} \sum_{1 \leq i_1 < \cdots < i_k \leq n} \Phi^k(x_{i_1}, \ldots, x_{i_k}) \tag{5.3}$$

where Δ_i is the Laplace operator for the coordinate x_i of the ith particle, and Φ^k is a k-body potential in the sense of Section 3.1. The operator T is naturally defined as an (unbounded) self-adjoint operator in $L^2(\mathbf{R}^{nv})$ (use Fourier transformation); we call $\mathscr{D}(T)$ its domain. On the other hand, U is defined on $\mathscr{D}(U) = \{\varphi \in L^2(\mathbf{R}^{nv}) : U\varphi \in L^2(\mathbf{R}^{nv})\}$. We mention one important case where $\mathscr{D}(T) \subset \mathscr{D}(U)$.

3.5.1 THEOREM (T. Kato).[21] We may write $\Phi^k(x_1, \ldots, x_k) = \hat{\Phi}^k(x_2 - x_1, \ldots, x_k - x_1)$. Suppose that we have

$$\hat{\Phi}^k = \hat{\Phi}^{k,p} + \hat{\Phi}^{k,\infty} \tag{5.4}$$

[21] See Kato [1], [2] and Nelson [2].

where

$$\hat{\Phi}^{k,p} \in L^p(\mathbf{R}^{(k-1)\nu}) \qquad \hat{\Phi}^{k,\infty} \in L^\infty(\mathbf{R}^{(k-1)\nu}) \qquad (5.5)$$

and $p \geq 2$, $p > \frac{1}{2}(k-1)\nu$. Then the following results hold.

(a) $\mathscr{D}(T) \subset \mathscr{D}(U)$.

(b) *Given $a > 0$ there exists $b > 0$ such that for all $\varphi \in \mathscr{D}(T)$:*

$$\|U\varphi\| \leq a\|T\varphi\| + b\|\varphi\| \qquad (5.6)$$

(c) *$T + U$ is self-adjoint and bounded below on $\mathscr{D}(T)$.*

This theorem holds for several species of particles. Notice that for pair interactions and $\nu \leq 3$ the assumptions of the theorem reduce to $\Phi \in L^2(\mathbf{R}^\nu) + L^\infty(\mathbf{R}^\nu)$, which is in particular satisfied for Coulomb interactions between point charges.

If the assumptions of Theorem 3.5.1 are not satisfied, the domain $\mathscr{D}(T) \cap \mathscr{D}(U)$ need not be dense in $L^2(\mathbf{R}^{n\nu})$. This is not too annoying as we expect to define the Hamiltonian for systems with hard cores only on a subspace of $L^2(\mathbf{R}^{n\nu})$. The conclusion in Theorem 3.5.1 that $T + U$ is bounded below is important and this property should be required to hold in all cases since we shall, for the canonical ensemble, have to take the trace of $e^{-\beta H_n}$. This gives a hint for the definition of the Hamiltonian in the cases of interest for statistical mechanics: assume that the symmetric operator $T + U$ defined on $\mathscr{D}(T) \cap \mathscr{D}(U) \cap L^2(\Lambda^n)$ is bounded below and take its Friedrichs extension, which is a self-adjoint operator in the closure of $\mathscr{D}(T) \cap \mathscr{D}(U) \cap L^2(\Lambda^n)$.

We first describe the Friedrichs extension, then discuss its use in the present situation.

3.5.2 THEOREM (*Friedrichs*).[22] *Let A be a symmetric[23] operator, defined on a domain \mathscr{D} dense in the Hilbert space \mathscr{H}. We assume that A is bounded below: $A \geq \alpha\mathbf{1}$. We may then introduce on \mathscr{D} the scalar product*

$$[\psi, \varphi] = (\psi, [A + (1 - \alpha)\mathbf{1}]\varphi) \qquad (5.7)$$

and complete \mathscr{D} to a Hilbert space \mathscr{L} with respect to (5.7). The following properties hold.

(a) *The canonical mapping $\mathscr{L} \to \mathscr{H}$ is one-to-one into.*

(b) *There is one and only one self-adjoint extension \tilde{A} of A such that its domain $\tilde{\mathscr{D}} \subset \mathscr{L}$; \tilde{A} has the same lower bound as A and is called the Friedrichs extension of A.*

Let us remark that even in the situation of Theorem 3.4.1 where $T + U$ is self-adjoint in $L^2(\mathbf{R}^{n\nu})$, we have still to define the Hamiltonian as

[22] See, e.g., Riesz and Nagy [1].
[23] That is, if $\varphi, \psi \in \mathscr{D}$, $(\varphi, A\psi) = (A\varphi, \psi)$.

self-adjoint operator for bounded regions Λ, and that this requires the prescription of some sort of boundary conditions. A choice of boundary conditions for the differential operator $T + U$ amounts to the choice of a self-adjoint extension from the domain $\mathscr{D}(T) \cap \mathscr{D}(U) \cap L^2(\Lambda^n)$. In particular the Friedrichs extension corresponds to the "vanishing of the wave function on the boundary" ("infinitely repulsive wall") condition, but is more general. To see this notice first that a "wave function" φ is an element of $\tilde{\mathscr{D}}$; therefore the norm (5.7) of φ is finite. Let us assume that U is bounded below (this will usually be the case); then $(\varphi, T\varphi)$ is finite and therefore the first-order derivatives of φ (in the sense of distributions) are again square integrable. This implies that φ has some smoothness properties and, since φ vanishes outside of Λ, the wave function must in some sense tend to zero at the boundary of Λ. As an example the reader can check that if $U = 0$ (free particles) and if Λ is a cube, the Friedrichs extension yields the familiar eigenfunctions[24] vanishing at the boundary of Λ.

We shall denote by $H_n(\Lambda)$ (resp. H_n) the self-adjoint operator obtained by applying the Friedrichs extension to $T + U$ defined on $\mathscr{D}(T) \cap \mathscr{D}(U) \cap L^2(\Lambda^n)$ [resp. $\mathscr{D}(T) \cap \mathscr{D}(U)$], assuming that $T + U$ is bounded below.[25] In cases of physical interest the domain of $H_n(\Lambda)$ will be dense in $L^2(\Lambda^n)$ except if we treat hard core systems. In the case of spherical hard cores of diameter $R > 0$ the domain of $H_n(\Lambda)$ would be dense in $L^2(\Lambda^n \setminus S_R^n)$. The fact that the hard cores collide elastically is expressed by the vanishing of the wave function at the boundary of S_R^n and results again from the use of the Friedrichs extension.

3.5.3 THE STABILITY CONDITION. We have requested that $H_n(\Lambda)$ be bounded below; we impose now a stronger condition, analogous to the stability condition for classical systems:

$$H_n \geq -nB\mathbf{1} \tag{5.8}$$

for some $B \geq 0$ and all n. If this condition is violated the binding energy per particle in the ground state of an n particle system diverges for large n or is not defined. In the case of B.–E. or F.–D. statistics, H_n [resp. $H_n(\Lambda)$] is of course restricted to $L_\varepsilon^2(\mathbf{R}^{nv})$ [resp. $L_\varepsilon^2(\Lambda^n)$].

Since $T \geq 0$, the classical stability condition

$$U(x_1, \ldots, x_n) \geq -nB \tag{5.9}$$

[24] It suffices to show that these eigenfunctions can be approximated in the metric $\|\varphi\|' = (\varphi, [T + \mathbf{1}]\varphi)^{1/2}$ by twice differentiable functions with support in Λ.

[25] The reader should keep in mind that $H_n(\Lambda)$ is just one possible definition of the Hamiltonian in Λ; this definition does not distinguish between two regions Λ which differ by a set of Lebesgue measure zero.

implies (5.8), which in turn implies

$$H_n(\Lambda) \geq -nB1 \tag{5.10}$$

A system for which (5.10) does not hold is catastrophic because the grand partition function[26]

$$\Xi = \sum_n e^{n\beta\mu} \operatorname{Tr} \exp[-\beta H_n(\Lambda)]$$

diverges for all values of the chemical potential μ.

It is remarkable that, usually, if (5.9) is not satisfied, then (5.10) is not satisfied either. We proved catastrophic behavior in the classical case by constructing configurations of n particles in a bounded region Λ with a potential energy less than or equal to $-Cn^2$ for large n[27] (see Section 3.2.2). We can approximate these configurations by quantum wave functions $\varphi_n \in L_\varepsilon^2(\Lambda^n)$ such that for large n[28]

$$(\varphi_n, U\varphi_n) \leq -Cn^2 \tag{5.11}$$

and

$$(\varphi_n, T\varphi_n) \leq \begin{cases} C'n & \text{(bosons)} \\ C''n^{1+2/\nu} & \text{(fermions)} \end{cases} \tag{5.12}$$

The inequalities (5.11), (5.12) are in contradiction with (5.10) for bosons and, if $\nu \geq 3$, also for fermions.[29] This argument holds for all upper semi-continuous potentials which violate (5.9); for instance, the potential of Fig. 4 ($\nu = 3$) remains catastrophic in quantum statistical mechanics for bosons and even fermions.

There is however one case of great physical interest where classical stability is violated but (5.8) holds: systems of point electric charges (with Coulomb interactions) if the particles of one sign of charge are fermions. A finite number q of species of fermions is allowed, i.e., the Hamiltonian acts on a subspace $\mathscr{L}_{n,q}$ of $L^2(\mathbf{R}^{n\nu})$ which consists of functions φ such that their n arguments may be grouped in q families and φ is antisymmetric with respect to permutations within each family.

3.5.4 THEOREM. (Dyson–Lenard).[30] Let

$$H_n(y_1, \ldots, y_p) = -\sum_{i=1}^{n} \frac{1}{2m} \Delta_i + \left[\sum_{1 \leq i < j \leq n} \frac{1}{|x_j - x_i|} \right.$$
$$\left. + \sum_{1 \leq i < j \leq p} \frac{1}{|y_j - y_i|} - \sum_{i=1}^{n} \sum_{j=1}^{p} \frac{1}{|y_j - x_i|} \right] \tag{5.13}$$

[26] We consider only the case of B.–E. and F.–D. statistics.
[27,28] Or at least for n belonging to an infinite subset of the positive integers.
[29] The Pauli principle is thus not sufficient to prevent collapse in this case.
[30] See Dyson and Lenard [1], Dyson [1], [2].

act on the space $\mathscr{L}_{n,q}$ *(consisting of functions of the variables* x_1, \ldots, x_n; Δ_i *is the Laplace operator with respect to* x_i*). There exists a constant* $B > 0$ *independent of* $q, n, p, y_1, \ldots, y_p$ *such that*

$$H_n(y_1, \ldots, y_p) \geq -nq^{2/3}B\mathbf{1} \tag{5.14}$$

This beautiful theorem, which is also extremely hard to prove, implies stability in the sense of (5.8) if the particles of one sign of the charge are fermions; the particles of the other sign may be fermions or bosons or may be considered as classical particles. If the particles of both signs are bosons, we have a catastrophic behavior: there exists $D > 0$ and n-particle wave functions φ_n such that for large n

$$(\varphi_n, H_n\varphi_n) < -n^{7/5}D \tag{5.15}$$

where

$$H_n = -\sum_{i=1}^{n} \frac{1}{2m}\Delta_i + \sum_{1 \leq i < j \leq n} \frac{\varepsilon_i\varepsilon_j}{|x_j - x_i|} \tag{5.16}$$

and $\varepsilon_i = \pm 1$ is the charge of the ith particle.

The study of the thermodynamic limit for quantum systems of charged particles presents special difficulties due to the "long range" character of the Coulomb interaction (which is far from tempered). Lieb and Lebowitz, using the theorem of Dyson–Lenard, have recently overcome these difficulties.

We turn to the problem of proving the existence of the thermodynamic limit for other quantum systems. For simplicity we shall make exactly the same assumptions as in the classical case, namely that the interaction $(\Phi^k)_{k \geq 2}$ is tempered (see Definition 3.1.1) and stable (see Definition 3.2.1). We shall restrict our attention to B.–E. and F.–D. statistics. The following version of the minimax principle will play an essential role.

3.5.5 PROPOSITION. *Let* A *be a symmetric operator, defined on a domain* \mathscr{D} *dense in the Hilbert space* \mathscr{H}*. For every finite-dimensional subspace* M *in the domain of* A *let*

$$\lambda(M) = \sup_{\varphi \in M, \|\varphi\| = 1} (\varphi, A\varphi) \tag{5.17}$$

For every integer $m \geq 1$*, let*

$$\lambda_m = \inf_{\dim M = m} \lambda(M) \tag{5.18}$$

Let $\lambda_1 \neq -\infty$*, then* $\lambda_m \to \infty$ *if and only if the spectrum of the Friedrichs extension* \tilde{A} *of* A *consists of discrete eigenvalues with finite multiplicity; these eigenvalues in increasing order and repeated according to multiplicity are then the* λ_m*.*

These results follow from Proposition 2.5.1 if we notice that every finite-dimensional subspace \tilde{M} in the domain $\tilde{\mathscr{D}}$ of \tilde{A} can be approximated in the metric (5.7) by a subspace M in the domain \mathscr{D} of A (see Theorem 3.5.2), and that the numbers λ_m are therefore the same whether they are defined for A or for \tilde{A}.

It is easy to compute the spectrum of $H_n(\Lambda_0)$ if Λ_0 is a cube and the interaction $(\Phi^k)_{k \geq 0}$ vanishes. We find that this spectrum consists of discrete eigenvalues with finite multiplicity. This determines the numbers $\lambda_m{}^0$ computed according to (5.18) from the operator T on the domain $\mathscr{D}(T) \cap L_\varepsilon^2(\Lambda_0{}^n)$. If we replace Λ_0 by a bounded Lebesgue measurable $\Lambda \subset \Lambda_0$ we find numbers $\lambda'_m \geq \lambda_m{}^0$. Finally if we introduce a stable inter-action and compute the numbers λ_m from the operator $T + U$ on the domain $\mathscr{D}(T) \cap \mathscr{D}(U) \cap L_\varepsilon^2(\Lambda^n)$ we have

$$\lambda_m \geq \lambda'_m - nB \geq \lambda_m{}^0 - nB \tag{5.19}$$

According to Proposition 3.5.5 we have proved the following result.

3.5.6 PROPOSITION. *The spectrum of the operator $H_n(\Lambda)$ consists of discrete eigenvalues with finite multiplicity.*

Generalizing the above argument we obtain the following.

3.5.7 PROPOSITION. *The mth eigenvalue of $H_n(\Lambda)$ decreases if either Λ increases or $U(x_1, \ldots, x_n)$ decreases.*

The quantum microcanonical partition function is

$$\Omega(\Lambda, n, E) = \operatorname{Tr} \delta^-(H_n(\Lambda) - E) \tag{5.20}$$

i.e., the number of eigenvalues of $H_n(\Lambda)$ (with multiplicity) below E. The quantum entropy is

$$S(\Lambda, n, E) = \log \Omega(\Lambda, n, E) \tag{5.21}$$

It is an increasing function of Λ (by Proposition 3.5.7) and of E. As in the classical case we invert $S(\Lambda, n, E)$ with respect to E to obtain $E(\Lambda, n, S)$ which is an increasing function of S satisfying

$$E(\Lambda, n, S) \geq -nB \tag{5.22}$$

The quantity $E(\Lambda, n, S)$ satisfies the same inequalities, given by Propositions 3.3.1 and 3.3.2, as in the classical case. This is seen by a simple adaptation of the proofs, using Proposition 3.5.5. As an example we prove Proposition 3.3.2,(a). Consider two regions Λ_1, Λ_2 with distance $r \geq R_0$. There are $\Omega(\Lambda_1, n_1, E_1)$ eigenvalues of $H_{n_1}(\Lambda_1)$ below E_1. By Proposition 3.5.5, if $\varepsilon > 0$ we can thus find a subspace M_1 of $\mathscr{D}(T) \cap \mathscr{D}(U) \cap L_\varepsilon^2(\Lambda_1{}^n)$ such that $\dim M_1 = \Omega(\Lambda_1, n_1, E_1)$ and

$$\sup_{\varphi_1 \in M_1, \|\varphi_1\| = 1} (\varphi_1, [T + U]\varphi_1) \leq E_1 + \varepsilon \tag{5.23}$$

Similarly we can find a subspace M_2 of $\mathcal{D}(T) \cap \mathcal{D}(U) \cap L_\varepsilon^2(\Lambda_2^{n_2})$ such that $\dim M_2 = \Omega(\Lambda_2, n_2, E_2)$ and

$$\sup_{\varphi_2 \in M_2, \|\varphi_2\| = 1} (\varphi_2, [T + U]\varphi_2) \leq E_2 + \varepsilon \tag{5.24}$$

Let $\varphi_1 \bigotimes_\varepsilon \varphi_2$ be obtained by symmetrizing ($\varepsilon = +1$) or antisymmetrizing ($\varepsilon = -1$) the product $\varphi_1 \cdot \varphi_2$ with $\varphi_1 \in M_1$, $\varphi_2 \in M_2$. The space $M \subset L_\varepsilon^2((\Lambda_1 \cup \Lambda_2)^{n_1 + n_2})$ generated by the $\varphi_1 \bigotimes_\varepsilon \varphi_2$ has dimension $\Omega(\Lambda_1, n_1, E_1) \cdot \Omega(\Lambda_2, n_2, E_2)$. Using temperedness we check that if $\varphi \in M$ and $\|\varphi\| = 1$, then

$$(\varphi, [T + U]\varphi) \leq (E_1 + \varepsilon) + (E_2 + \varepsilon) + An_1 n_2 r^{-\lambda}$$

Therefore by Proposition 3.5.5

$$\Omega(\Lambda_1 \cup \Lambda_2, n_1 + n_2, E_1 + E_2 + 2\varepsilon + An_1 n_2 r^{-\lambda}) \geq$$
$$\geq \Omega(\Lambda_1, n_1, E_1) \cdot \Omega(\Lambda_2, n_2, E_2) \tag{5.25}$$

Letting ε go to zero and taking the logarithm yields

$$S(\Lambda_1 \cup \Lambda_2, n_1 + n_2, E_1 + E_2 + An_1 n_2 r^{-\lambda}) \geq S(\Lambda_1, n_1, E_1) + S(\Lambda_2, n_2, E_2)$$

from which the quantum version of Proposition 3.3.2,(a) follows. The other inequalities are proved similarly.

The study of the thermodynamic limit for classical systems in Sections 3.3 and 3.4 was based essentially upon general inequalities which, as we have just seen, are also satisfied in the quantum case. The transition from the classical (configurational) ensembles to the quantum ensembles is made by replacing $U(x_1, \ldots, x_n)$ by $H_n(\Lambda)$ and $(n!)^{-1} \int dx_1 \cdots dx_n$ by Tr. There are a few places however where specific estimates were used in the classical case which have to be modified in the quantum situation. We first state the quantum equivalent of the theorems proved in Sections 3.3 and 3.4 for classical systems and then list the modifications to be brought to the proofs.

3.5.8 THEOREM. *Let $H_n(\Lambda)$ be the Friedrichs extension of $T + U$ defined on $\mathcal{D}(T) \cap \mathcal{D}(U) \cap L^2(\Lambda^n)$ where U is computed from a tempered and stable interaction. Let $S(\Lambda, n, E)$ be defined by (5.20) and (5.21). Let*

$$Q(\Lambda, n, \beta) = \mathrm{Tr}[-\beta H_n(\Lambda)] \tag{5.26}$$

where $\beta > 0$ is the inverse temperature. Let

$$\Xi(\Lambda, z, \beta) = \sum_{n=0}^{\infty} z^n Q(\Lambda, n, \beta) \tag{5.27}$$

where $Q(\Lambda, 0, \beta) = 1$ and $z = e^{\beta\mu} > 0$. Theorems 3.3.12, 3.4.4, and 3.4.6 remain true with these new definitions, except that in 3.4.6 the function $p(z, \beta)$

may be defined only for $z < z_0(\beta)$ *where the region* $\{\log z, \beta) : 0 < z < z_0(\beta)\}$
is convex, and

$$\lim_{\Lambda \to \infty} V(\Lambda)^{-1} \log \Xi(\Lambda, z, \beta) = +\infty \qquad (5.28)$$

if $z > z_0(\beta)$.

The following modifications have to be brought in the proofs. The bound (3.19) does not hold in the quantum situation, but computation of $E(\Lambda_0, 1, \sigma)$ yields another bound which serves the same purposes; the situation is similar in (3.32). The second inequality (3.43) is false in the quantum case, but no essential use is made of it. The estimates (4.14) and (4.20) present a more serious problem; we replace them by the use of the following results.

3.5.9 PROPOSITION. *If* $\Lambda \to \infty$ *(Fisher),* $V(\Lambda)^{-1}n \to \rho$ *and* $V(\Lambda)^{-1}E \to \infty$, *then the points* $(V(\Lambda)^{-1}E, V(\Lambda)^{-1}S(\Lambda, n, E))$ *go and stay below any tangent to the graph of* $\varepsilon \to s(\rho, \varepsilon)$.

This is seen by considering ε as a function of s and applying Proposition 3.3.2,(a).

3.5.10 PROPOSITION. *If* $\Lambda \to \infty$ *(Fisher) and* $V(\Lambda)^{-1}n \to \infty$, *then the points* $(V(\Lambda)^{-1}n, V(\Lambda)^{-1} \log Q(\Lambda, n, \beta))$ *go and stay below any tangent to the graph of* $\rho \to -\beta f(\rho, \beta)$.

This follows from the inequality

$$\mathrm{Tr}\, \exp[-\beta H_{n_1 + n_2}(\Lambda_1 \cup \Lambda_2)]$$

$$\geq \mathrm{Tr}\, \exp[-\beta H_{n_1}(\Lambda_1)] \cdot \mathrm{Tr}\, \exp[-\beta H_{n_2}(\Lambda_2)] \exp(-\beta A n_1 n_2 r^{-\lambda})$$

which results from (5.25) and holds if $r \geq R_0$ is the distance of Λ_1, Λ_2.

In using Propositions 3.5.9 and 3.5.10 we must keep in mind the possibility that the slope of the tangents to the graph of $\varepsilon \to s(\rho, \varepsilon)$ may be bounded below by $\beta_0 > 0$ or that the slope of the tangents to the graph of $-\beta f(\rho, \beta)$ may be bounded below by $-\log z_0$. The former possibility does not occur, as we see by comparison with the case of a "free" system [vanishing interaction $(\Phi^k)_{k \geq 2}$]. The latter possibility does occur for free bosons, justifying the restriction $z < z_0(\beta)$ in Theorem 3.5.8.

3.5.11 GROUND STATE. Let $E(\Lambda, n)$ be the lowest eigenvalue of $H_n(\Lambda)$, i.e., the *ground state energy* of n particles in Λ. Using the notation ρ_{cp}[31] and $\varepsilon_0(\cdot)$ with the same meaning as in the quantum version of Theorem 3.3.12, we have the following result:

[31] Remember that for quantum systems we may have $\rho_{cp} < +\infty$ even if the potential energy U does not take the value $+\infty$.

Let $\Lambda \to \infty$ in the sense of Fisher and $V(\Lambda)^{-1}n \to \rho$

(a) *If $0 \le \rho < \rho_{cp}$, then $V(\Lambda)^{-1}E(\Lambda, n) \to \varepsilon_0(\rho)$.*
(b) *If $\rho = \rho_{cp}$, then $\liminf V(\Lambda)^{-1}E(\Lambda, n) \ge \varepsilon^*$ where*

$$\varepsilon^* = \lim_{\rho \to \rho_{cp}, \rho < \rho_{cp}} \varepsilon_0(\rho)$$

(c) *If $\rho > \rho_{cp}$, then $V(\Lambda)^{-1}E(\Lambda, n) \to +\infty$.*

This is a corollary of Theorem 3.3.12 obtained by noticing that $S(\Lambda, n, E) \ge 0$ if $E > E(\Lambda, n)$ and $S(\Lambda, n, E) = -\infty$ if $E < E(\Lambda, n)$.

BIBLIOGRAPHICAL NOTE

The first proof of the existence of the thermodynamic limit was published by Van Hove [1] for the canonical ensemble in 1949. Unaware of this work, Yang and Lee [1] derived a proof for the grand canonical ensemble (hard core systems) in 1952. A large part of the scientific public questioned that the existence of the thermodynamic limit required any proof (Van Hove, private communication); anyway further progress was halted until 1962–63. At that time van Kampen (unpublished) pointed to a serious difficulty in Van Hove's proof; independently Ruelle [1] published a new proof, and Fisher [1] obtained similar results which were published only later after incorporation of an idea of Wils [1] which permits the treatment of tempered interactions. Further results have been obtained by various authors, notably Griffiths [1], [3] and Dobrushin [1]. The important notion of stable interaction may be traced back to a 1939 paper by Onsager [1].

Sections 3.3–3.5 present a new organization of material contained mostly in Fisher [1], Griffiths [3], and Ruelle [2]. See also R. A. Minlos and A. Ya. Povzner, On the Thermodynamic Limit of the Entropy, *Trud. Moskovsk. Mat. Obshch.* **17**, 243–272 (1967).

EXERCISES

3.A. Show that the Lennard–Jones potential for $v = 3$

$$\Phi(x) = 4\varepsilon \left[\left(\frac{a}{|x|} \right)^{12} - \left(\frac{a}{|x|} \right)^{6} \right]$$

(with $\varepsilon, a > 0$) is stable and tempered (use 3.2.10).

3.B. Show that the Morse potential for $v = 3$

$$\Phi(x) = \varepsilon \left[\exp\left(-\frac{2(|x| - R_0)}{a} \right) - 2 \exp\left(-\frac{|x| - R_0}{a} \right) \right]$$

(with $\varepsilon, a, R_0 > 0$) is stable and tempered if $\exp(R_0/a) \geq 16$ and catastrophic if $\exp(R_0/a) < 16$ (use 3.2.4 and 3.2.6).

3.C. Prove the lower bound for the electrostatic interaction energy between charge distributions mentioned in 3.2.11.

3.D. If temperedness is replaced by the stronger condition (1.11), the proofs in Section 3.3 simplify quite a bit: instead of having to place the regions at a variable distance R_N we can use R_0 everywhere. Show that we can also replace Fisher convergence by the following weaker conditions:

(a) $\Lambda \to \infty$ in the sense of Van Hove, and
(b) there exists $C > 0$ and for each Λ a parallelepiped $\Lambda(a)$ such that $\Lambda \subset \Lambda(a)$, $V(a)^{-1}V(\Lambda) \geq C$, and $a \to \infty$.

[The main difference appears in the proof of (3.34), where Λ is placed in $\Lambda(a)$ instead of $\Lambda_{N'}$; an estimate of $E(\Lambda(a), n_1, S_1)$ is obtained by filling a cube $\Lambda_{N'}$ with translates of $\Lambda(a)$.]

3.E. If the stability condition is satisfied, the classical grand partition function (4.17) is an entire function $\Xi(z)$ of the activity considered as a complex variable. Show that this entire function is of order at most 1, and is of order 0 if (2.27) holds. In the latter case Hadamard's factorization yields

$$\Xi(z) = \prod_i \left(1 - \frac{z}{\zeta_i}\right)$$

where the ζ_i are the zeros of Ξ. In particular in the presence of hard cores Ξ is a polynomial. For quantum systems (B.–E. or F.–D. statistics) (2.27) implies again that Ξ is an entire function of order 0, but the grand partition function for a free boson system ($U = 0$) is not an entire function of z.

3.F. If (2.27) holds, then in Theorem 3.5.8 we have $z_0(\beta) = +\infty$.

3.G.[32] Let D be a simply connected domain of the complex plane C, symmetric with respect to the real axis and having a nonvoid intersection with the positive real axis. Prove that if the grand partition function $\Xi(z)$ has no zeros in D (for sufficiently large Λ), then $p(z, \beta)$ extends from the positive real axis to an analytic function of z in D. (Use Vitali's theorem on bounded sequences of analytic functions.)

3.H. Consider a classical continuous system with spherical hard cores of diameter R [excluded region given by (2.12) of Chapter 1]. Let \mathscr{B}_0 be the space of sequences $(\Phi^k)_{k \geq 1}$ where $\Phi^1 \in \mathbf{R}$ and $(\Phi^k)_{k \geq 2}$ is an interaction such that $\Phi^k = 0$ for large k and Φ^k is continuous with compact support (finite

[32] The solution of this exercise is given in Section 5.1.

range) for all k. If $\Phi = (\Phi^k)_{k \geq 1} \in \mathscr{B}_0$, let

$$\|\Phi\| = \sup_n \ \sup_{(x_1,\ldots,x_n)\in \mathbf{R}^{n\nu}\backslash S_{R^n}} \left[\Phi^1 + \sum_{k=2}^{n} \ \sum_{1 < i_2 < \cdots < i_k < n} \right.$$

$$\left. \times \ \frac{|\Phi^k(x_{i_1}, x_{i_2}, \ldots, x_{i_k})|}{k} \right]$$

$$U_\Phi(x_1,\ldots,x_n) = \sum_{k=1}^{n} \ \sum_{1 \leq i_1 < \cdots < i_k \leq n} \Phi^k(x_{i_1},\ldots,x_{i_k})$$

$$Z_\Lambda(\Phi) = 1 + \sum_{n=1}^{\infty} \frac{1}{n!} \int_{\Lambda^n} dx_1 \cdots dx_n \exp[-U_\Phi(x_1,\ldots,x_n)]$$

$$P_\Lambda(\Phi) = V(\Lambda)^{-1} \log Z_\Lambda(\Phi)$$

Show that $\lim_{\Lambda \to \infty} P_\Lambda(\Phi)$ exists when $\Lambda \to \infty$ in the sense of Van Hove (we may use Theorem 3.4.6 for cubes and proceed along the lines of Section 2.3). Let \mathscr{B} be the Banach space obtained by completing \mathscr{B}_0 with respect to the norm $\|\cdot\|$; then Theorem 2.3.3 holds with these new definitions, except for the replacement of (3.9) by

$$|P(\Phi) - P(\Psi)| \leq \rho_{cp}\|\Phi - \Psi\|$$

where ρ_{cp} is the "close packing" density for hard spheres of diameter R.

CHAPTER 4

LOW DENSITY EXPANSIONS AND CORRELATION FUNCTIONS

A rather detailed analysis of dilute systems is possible, using series expansions of quantities of interest in powers of the activity or the density. In this chapter we prove the convergence of these expansions from a study of the correlation functions. The correlation functions provide a description of the state of an infinite system, and their investigation is in itself of great interest.

4.1 DEFINITIONS

When the activity z or the density ρ of a system of particles tends to zero, the behavior of the system approaches that of a system without interactions. It is therefore natural to try to obtain expansions of quantities of interest in powers of z or ρ. Such expansions are indeed well known, notably the *virial expansion* (pressure p in powers of ρ) and the *Mayer expansions* (p and ρ in powers of z). Rules have been devised for the systematic calculation of their coefficients, labeling the contributions by means of various diagrams.[1] Until recently, however, the convergence of these expansions remained unproved.

4.1.1 CORRELATION FUNCTIONS. It will be convenient to use throughout this chapter the notation $(x\|)_n$ for (x_1, \ldots, x_n) where $x_1, \ldots, x_n \in \mathbf{R}^\nu$. We recall that the (configurational) grand canonical ensemble is defined by a measure on $\sum_{n=0}^{\infty} \Lambda^n$, which on Λ^n reduces to

$$\frac{z^n}{n!} \exp[-\beta U(x)_n] \, dx_1 \cdots dx_n \tag{1.1}$$

[1] See for instance Uhlenbeck and Ford [1].

The total mass of this measure is the grand partition function

$$\Xi(\Lambda, z, \beta) = \sum_{n=0}^{\infty} \frac{z^n}{n!} \int_{\Lambda^n} dx_1 \cdots dx_n \exp[-\beta U(x)_n] \qquad (1.2)$$

The m-point *correlation function* $\rho_\Lambda(x)_m = \rho_\Lambda(x_1, \ldots, x_m)$ is defined as the probability density (with respect to Lebesgue measure) for finding m different particles at prescribed positions $x_1, \ldots, x_m \in \Lambda$. Using (1.1) we find

$$\rho_\Lambda(x)_m = \Xi(\Lambda, z, \beta)^{-1} \sum_{n=0}^{\infty} \frac{z^{m+n}}{n!} \int_{\Lambda^n} dx_{m+1} \cdots dx_{m+n} \exp[-\beta U(x)_{m+n}] \qquad (1.3)$$

We expect that the correlation function $\rho_\Lambda(x)_m$ has a thermodynamic limit $\rho(x)_m$ (infinite volume correlation function) when $\Lambda \to \infty$ in a suitable manner.

We shall in the rest of this chapter restrict ourselves mostly to pair interactions. We shall assume that the pair potential Φ satisfies the stability condition 3.2.1. Instead of temperedness we shall need the following weak decrease property of Φ at infinity.

4.1.2 DEFINITION. *We say that the pair potential Φ is regular if it is bounded below and satisfies*

$$C(\beta) = \int dx\, |e^{-\beta\Phi(x)} - 1| < +\infty \qquad (1.4)$$

for some $\beta > 0$ and therefore for all $\beta > 0$.

If Φ is stable, then it is automatically bounded below by $-2B$.[2] If Φ is bounded below, (1.4) is equivalent to requiring that Φ be absolutely integrable outside of a set of finite Lebesgue measure [take the set $\{x : \Phi(x) \geq 1\}$]; this shows that (1.4) holds for all $\beta > 0$ if it holds for some $\beta > 0$.

4.2 THE METHOD OF INTEGRAL EQUATIONS

The correlation functions satisfy various systems of linear integral equations, in particular the Kirkwood–Salsburg and the Mayer–Montroll equations. These integral equations are a powerful tool for the study of the correlation functions at small activity.

4.2.1 THE KIRKWOOD–SALSBURG EQUATIONS. Let χ_Λ be the characteristic function of the bounded Lebesgue measurable set $\Lambda \subset \mathbf{R}^\nu$. We write

$$\chi_\Lambda(x)_n = \prod_{i=1}^{n} \chi_\Lambda(x_i) \qquad (2.1)$$

[2] Because $\Phi(x) = U(0, x) \geq -2B$.

We assume that the pair potential Φ is stable and regular and let

$$U(x)_n = \sum_{1 \le i < j \le n} \Phi(x_j - x_i) \tag{2.2}$$

$$\psi(x)_n = \exp[-\beta U(x)_n] \tag{2.3}$$

where $\beta > 0$; (1.2) and (1.3) may then be rewritten as

$$\Xi(\Lambda, z, \beta) = \sum_{n=0}^{\infty} \frac{z^n}{n!} \int d(x)_n \chi_\Lambda(x)_n \psi(x)_n \tag{2.4}$$

$$\rho_\Lambda(x)_m = \Xi(\Lambda, z, \beta)^{-1} \sum_{n=0}^{\infty} \frac{z^{m+n}}{n!} \int dx_{m+1} \cdots dx_{m+n} \chi_\Lambda(x)_{m+n} \psi(x)_{m+n} \tag{2.5}$$

We introduce the notation

$$W^1(x)_m = \sum_{i=2}^{m} \Phi(x_i - x_1) \tag{2.6}$$

$$(x)'_{m-1} = (x_2, \ldots, x_m) \tag{2.7}$$

then

$$\psi(x)_{m+n} = \exp[-\beta W^1(x)_{m+n}] \psi(x)'_{m+n-1}$$
$$= \exp[-\beta W^1(x)_m]$$
$$\times \left[\prod_{j=1}^{n} (1 + (\exp[-\beta \Phi(x_{m+j} - x_1)] - 1)) \right] \psi(x)'_{m+n-1} \tag{2.8}$$

Writing $x_{m+j} = y_j$ and defining

$$K(x_1, (y)_n) = \prod_{j=1}^{n} (\exp[-\beta \Phi(y_j - x_1)] - 1) \tag{2.9}$$

we obtain upon inserting (2.8) in (2.5):

$$\rho_\Lambda(x)_m = \chi_\Lambda(x)_m z \exp[-\beta W^1(x)_m] \Xi(\Lambda, z, \beta)^{-1} M(x)_m \tag{2.10}$$

where

$M(x)_m$

$$= \sum_{n=0}^{\infty} \frac{z^{m+n-1}}{n!} \int dy_1 \cdots dy_n \sum_{s=0}^{n} \frac{n!}{s!(n-s)!} K(x_1, (y)_s) \chi_\Lambda(x)'_{m+n-1} \psi(x)'_{m+n-1}$$

$$= \sum_{n=0}^{\infty} \sum_{s=0}^{n} \frac{z^{m+n-1}}{s!(n-s)!} \int dy_1 \cdots dy_s K(x_1, (y)_s)$$

$$\times \int dy_{s+1} \cdots dy_n \chi_\Lambda(x)'_{m+n-1} \psi(x)'_{m+n-1}$$

$$= \sum_{n=0}^{\infty} \frac{z^{m+n-1}}{n!} \int dy_1 \cdots dy_n \chi_\Lambda(x)'_{m+n-1} \psi(x)'_{m+n-1}$$

$$+ \sum_{s=1}^{\infty} \frac{1}{s!} \int dy_1 \cdots dy_s K(x_1, (y)_s) \sum_{n=s}^{\infty} \frac{z^{m+n-1}}{(n-s)!}$$

$$\times \int dy_{s+1} \cdots dy_n \chi_\Lambda(x)'_{m+n-1} \psi(x)'_{m+n-1} \qquad (2.11)$$

and the interchanges of summations and integrations are justified by stability. If we introduce in (2.11) the definition (2.5) of $\rho_\Lambda(x)'_{m+n-1}$ we obtain

$$\rho_\Lambda(x_1) = \chi_\Lambda(x_1) z \left[1 + \sum \frac{1}{n!} \int dy_1 \cdots dy_n K(x_1, (y)_n) \rho_\Lambda(y)_n \right] \qquad (2.12)$$

$$\rho_\Lambda(x)_m = \chi_\Lambda(x)_m z \, \exp[-\beta W^1(x)_m]$$

$$\times \left[\rho_\Lambda(x)'_{m-1} + \sum_{n=1}^{\infty} \frac{1}{n!} \int dy_1 \cdots dy_n K(x_1, (y)_n) \rho_\Lambda((x)'_{m-1}, (y)_n) \right] \qquad (2.13)$$

where (2.13) holds for $m > 1$.

The integral equations (2.12), (2.13) are the Kirkwood–Salsburg equations.[3] They form a linear inhomogeneous system for the correlation functions $\rho_\Lambda(x)_n$, $n \geq 1$.

4.2.2 THE BANACH SPACES E_ξ. Let $\xi > 0$ and let E_ξ be the space of sequences $\boldsymbol{\varphi} = (\varphi(x)_n)_{n \geq 1}$, where $\varphi(x)_n$ is a complex Lebesgue measurable function on $\mathbf{R}^{n\nu}$, such that

$$\|\boldsymbol{\varphi}\|_\xi = \sup_{n \geq 1} (\xi^{-n} \operatorname*{ess\,sup}_{(x)_n \in \mathbf{R}^{n\nu}} |\varphi(x)_n|) < +\infty \qquad (2.14)$$

Because of stability and (2.5) we have

$$0 \leq \rho_\Lambda(x)_n \leq \Xi(\Lambda, z, \beta)^{-1} \exp(zV(\Lambda) e^{\beta B})(z \, e^{\beta B})^n \qquad (2.15)$$

so that $\boldsymbol{\rho}_\Lambda = (\rho_\Lambda(x)_n)_{n \geq 1} \in E_\xi$ if $\xi \geq z \, e^{\beta B}$. We shall now transform the Kirkwood–Salsburg equations into an equation for $\boldsymbol{\rho}_\Lambda$ in E_ξ.

We may define a linear operator χ_Λ in E_ξ by

$$\chi_\Lambda \boldsymbol{\varphi} = (\chi_\Lambda(x)_n \varphi(x)_n)_{n \geq 1} \qquad (2.16)$$

χ_Λ is idempotent[4] and of norm 1.[5] Also let K be defined by

$$(K\varphi)(x_1) = \sum_{n=1}^{\infty} \frac{1}{n!} \int d(y)_n K(x_1, (y)_n) \varphi(y)_n \qquad (2.17)$$

$$(K\varphi)(x)_m = \exp[-\beta W^1(x)_m]$$

[3] See, e.g., Hill [1], p. 251. These equations actually go back to Mayer [1] [see Eq. (54')].
[4] That is, $\chi_\Lambda \cdot \chi_\Lambda = \chi_\Lambda$.
[5] That is, $\sup_{\|\varphi\|_\xi = 1} \|\chi_\Lambda \varphi\|_\xi = 1$.

$$\times \left[\varphi(x)'_{m-1} + \sum_{n=1}^{\infty} \frac{1}{n!} \int d(y)_n K(x_1,(y)_n) \varphi((x)'_{m-1},(y)_n) \right] \quad (2.18)$$

where (2.18) holds for $m > 1$. It follows from the stability condition that Φ is bounded below by $-2B$ so, using also the regularity condition 4.1.2, we get

$$|(K\varphi)(x)_m| \leq \|\boldsymbol{\varphi}\|_{\xi} \exp[\xi C(\beta)] (e^{2\beta B}\xi)^{m-1} \quad (2.19)$$

Hence, K maps E_{ξ} in $E_{e^{2\beta B}\xi}$. With these notations and results we see that it is possible to rewrite (2.12) and (2.13) as one equation in $E = \bigcup_{\xi>0} E_{\xi}$:

$$\boldsymbol{\rho}_{\Lambda} = z\chi_{\Lambda}\boldsymbol{\alpha} + z\chi_{\Lambda}K\boldsymbol{\rho}_{\Lambda} \quad (2.20)$$

where $\boldsymbol{\alpha}$ is such that $\alpha(x_1) = 1$ and $\alpha(x)_n = 0$ for $n > 1$.

It is possible to transform (2.20) into an equation which holds in one of the E_{ξ} (rather than $E = \bigcup_{\xi>0} E_{\xi}$). For this we have, in (2.19), to get rid of the factor $(e^{2\beta B})^{m-1}$, which comes from $\exp[-\beta W^1(x)_m]$ in (2.18). The variable x_1 plays a special role in Eq. (2.18), but we can, by permutation of variables, consider other equations where this role is played by x_i, $i = 2,\ldots,m$, $W^1(x)_m$ being replaced by $W^i(x)_m$. Notice now that

$$\sum_{i=1}^{m} W^i(x)_m = 2U(x)_m \geq -2mB \quad (2.21)$$

This implies that for each $(x)_m \in \mathbf{R}^{mv}$ we can choose i such that

$$W^i(x)_m \geq -2B \quad (2.22)$$

In view of (2.22) we can construct an operator Π on E_{ξ} which for each $(x)_m = (x_1,\ldots,x_m)$ replaces $\varphi(x)_m$ by $\varphi(x_{i_1},\ldots,x_{i_m})$ where the permutation $(1,\ldots,m) \to (i_1,\ldots,i_m)$ is chosen such that $W^{i_1}(x)_m \geq -2B$. Since $\rho_{\Lambda}(x)_m$ is symmetric in its m arguments we may replace (2.20) by

$$\boldsymbol{\rho}_{\Lambda} = z\chi_{\Lambda}\boldsymbol{\alpha} + z\chi_{\Lambda}\Pi K\boldsymbol{\rho}_{\Lambda} \quad (2.23)$$

and (2.19) is replaced by

$$|(\Pi K\varphi)(x)_m| \leq \|\boldsymbol{\varphi}\|_{\xi} e^{2\beta B} \exp[\xi C(\beta)]\xi^{m-1} \quad (2.24)$$

so that ΠK maps E_{ξ} into itself with norm

$$\|\Pi K\|_{\xi} = \sup_{\|\varphi\|_{\xi}=1} \|\Pi K\varphi\|_{\xi} \leq e^{2\beta B\xi^{-1}} \exp[\xi C(\beta)] \quad (2.25)$$

Besides (2.23) let us consider also the equation

$$\boldsymbol{\rho} = z\boldsymbol{\alpha} + z\Pi K\boldsymbol{\rho} \quad (2.26)$$

In both (2.23) and (2.26) we allow z to be complex; in view of (2.24) the kernels of these equations have norms smaller than

$$|z| e^{2\beta B\xi^{-1}} \exp[\xi C(\beta)] \quad (2.27)$$

Therefore (2.23) and (2.26) have unique solutions in E_ξ^6:

$$\boldsymbol{\rho}_\Lambda = (1 - z\chi_\Lambda \Pi K)^{-1} z\chi_\Lambda \alpha \qquad (2.28)$$

$$\boldsymbol{\rho} = (1 - z\Pi K)^{-1} z\alpha \qquad (2.29)$$

as soon as the expression (2.27) is smaller than one, i.e.,

$$|z| < e^{-2\beta B} \xi \exp[-\xi C(\beta)] \qquad (2.30)$$

which is possible [take $\xi = C(\beta)^{-1}$] provided

$$|z| < e^{-2\beta B - 1} C(\beta)^{-1} \qquad (2.31)$$

This situation is depicted in Fig. 11. Because of (2.15) and of the uniqueness of the solution (2.28), this solution extends the correlation functions to the complex values of z for which (2.31) holds.

For real $z > 0$, (2.4) and (2.5) yield

$$\int dx_1 \rho_\Lambda(x_1) = \Xi(\Lambda, z, \beta)^{-1} \sum_{n=1}^{\infty} \frac{z^n}{(n-1)!} \int d(x)_n \chi_\Lambda(x)_n \psi(x)_n$$

$$= z \frac{d}{dz} \log \Xi(\Lambda, z, \beta) \qquad (2.32)$$

Because of (2.28), the left-hand side continues to an analytic function of z in the circle (2.31). This shows that $\Xi(\Lambda, z, \beta)$, which is an analytic function of z, has no zero in (2.31). The definition (2.5) of correlation functions continues therefore to make sense for z in (2.31) and, by analytic continuation,[7] this definition is equivalent to (2.28).

4.2.3 THEOREM.[8] *Let Φ be a stable, regular pair potential, and let z be a complex number satisfying*

$$|z| < e^{-2\beta B - 1} C(\beta)^{-1} \qquad (2.33)$$

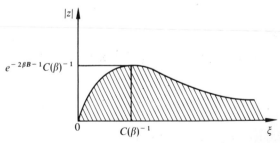

FIGURE 11 The inequality $|z| < e^{-2\beta B} \xi \exp[-\xi C(\beta)]$ holds in the shaded area.

[6] The inverses in (2.28) and (2.29) can be expressed as Liouville–Neumann series.

[7] Notice that (2.28) and (2.29) define $\boldsymbol{\rho}_\Lambda, \boldsymbol{\rho}$ as vector-valued analytic functions (values in E_ξ) because the Liouville–Neumann series give a convergent expansion of $\boldsymbol{\rho}_\Lambda$ in powers of z.

[8] See Ruelle [3].

Then the grand partition function (2.4) has no zero in (2.33). If the correlation functions $\rho_\Lambda(x)_n$ are defined by (2.5) for z in (2.33), there exist "infinite volume correlation functions" $\rho(x)_n$ and a function $\varepsilon(\cdot)$ positive and decreasing, such that

$$\lim_{\lambda \to \infty} \varepsilon(\lambda) = 0 \qquad (2.34)$$

and

$$|\rho_\Lambda(x)_n - \rho(x)_n| \le \xi^n \varepsilon(\lambda) \qquad (2.35)$$

if ξ, z satisfy (2.30) and λ is the minimum distance from $x_1, \ldots, x_n \in \Lambda$ to the boundary of Λ.

The sequence $\boldsymbol{\rho}_\Lambda$ of the $\rho_\Lambda(x)_n$ and the sequence $\boldsymbol{\rho}$ of the $\rho(x)_n$ are the only solutions of (2.23) and (2.26), respectively, in E_ξ when ξ satisfies (2.30) [in particular if $\xi = C(\beta)^{-1}$], $\boldsymbol{\rho}_\Lambda$ and $\boldsymbol{\rho}$ depend analytically on z.

It remains only to prove (2.35). This is obtained by a term by term comparison of the power series in z giving $\boldsymbol{\rho}_\Lambda$ and $\boldsymbol{\rho}$, as we now show.

If $\Lambda' \subset \Lambda''$ we have

$$0 \le \chi_{\Lambda''}(y)_n - \chi_{\Lambda'}(y)_n \le \sum_{j=1}^{n} (1 - \chi_{\Lambda'}(y_j)) \qquad (2.36)$$

Let $\Lambda \subset \Lambda'$ be a set of points which have a distance greater than δ to the boundary of Λ'. Using Eqs. (2.9), (2.36), and (1.4), we obtain

$$\operatorname*{ess\,sup}_{(x)_m} |\chi_\Lambda(x)_m \int d(y)_n K(x_1, (y)_n) [\chi_{\Lambda''}((x)'_{m-1}, (y)_n) - \chi_{\Lambda'}((x)'_{m-1}, (y)_n)]$$

$$\times \; \varphi((x)'_{m-1}, (y)_n)|$$

$$\le \|\varphi\|_\xi \cdot \xi^{m+n-1} \chi_\Lambda(x)_m \int d(y)_n |K(x_1, (y)_n)| [\chi_{\Lambda''}((x)'_{m-1}, (y)_n)$$
$$- \chi_{\Lambda'}((x)'_{m-1}, (y)_n)]$$

$$\le n \|\varphi\|_\xi \cdot \xi^{m+n-1} C(\beta)^{n-1} \int dy \, |\exp[-\beta\Phi(y - x_1)] - 1| \chi_\Lambda(x)_m [1 - \chi_{\Lambda'}(y)]$$

$$\le n \|\varphi\|_\xi \cdot \xi^{m+n-1} C(\beta)^{n-1} C^\delta(\beta) \qquad (2.37)$$

where we have written

$$C^\delta(\beta) = \int_{|x| > \delta} dx \, |e^{-\beta\Phi(x)} - 1| \qquad (2.38)$$

Therefore, by the definition (2.17), (2.18) of K and the definition of Π we have

$$\operatorname*{ess\,sup}_{(x)_m} |[(\chi_\Lambda \Pi K \chi_{\Lambda''} - \chi_\Lambda \Pi K \chi_{\Lambda'}) \varphi(x)_m| \le e^{2\beta B} \|\varphi\|_\xi \xi^m \exp[\xi C(\beta)] \cdot C^\delta(\beta)$$

or, using the operator norm,

$$\|\chi_\Lambda \Pi K \chi_{\Lambda''} - \chi_\Lambda \Pi K \chi_{\Lambda'}\|_\xi \le \eta(\delta) \qquad (2.39)$$

$$\eta(\delta) = e^{2\beta B} \exp[\xi C(\beta)] \cdot C^\delta(\beta) \qquad (2.40)$$

Now let $\Lambda(\delta)$ be the set of points of Λ with distance greater than δ to the boundary of Λ; (2.39) yields

$$\|\chi_{\Lambda(j\delta)} \Pi K \chi_{\Lambda(k\delta)} - \chi_{\Lambda(j\delta)} \Pi K \chi_{\Lambda((j-1)\delta)}\|_\xi \le \eta(\delta) \qquad \text{for} \quad k \le j - 1$$

$$\|\chi_{\Lambda(j\delta)} \Pi K - \chi_{\Lambda(j\delta)} \Pi K \chi_{\Lambda((j-1)\delta)}\|_\xi \le \eta(\delta)$$

from which we obtain, respectively,

$$\|\chi_{\Lambda(k\delta)} (\Pi K \chi_\Lambda)^k - \chi_{\Lambda(k\delta)} \Pi K \chi_{\Lambda((k-1)\delta)} \Pi K \cdots \chi_{\Lambda(\delta)} \Pi K \chi_\Lambda\|_\xi \le k\eta(\delta)(\|\Pi K\|_\xi)^{k-1}$$

$$\|\chi_{\Lambda(k\delta)} \Pi K \chi_{\Lambda((k-1)\delta)} \Pi K \cdots \chi_{\Lambda(\delta)} \Pi K \chi_\Lambda - \chi_{\Lambda(k\delta)} (\Pi K)^k\|_\xi \le k\eta(\delta)(\|\Pi K\|_\xi)^{k-1}$$

Therefore

$$\|\chi_{\Lambda(k\delta)} (\chi_\Lambda \Pi K)^k \chi_\Lambda - \chi_{\Lambda(k\delta)} (\Pi K)^k\|_\xi \le 2k\eta(\delta)(\|\Pi K\|_\xi)^{k-1} \qquad (2.41)$$

Replacing δ in this formula by $l\delta/k$ we obtain

$$\|\chi_{\Lambda(l\delta)} (\chi_\Lambda \Pi K)^k \chi_\Lambda - \chi_{\Lambda(l\delta)} (\Pi K)^k\|_\xi \le 2k\eta(\delta)(\|\Pi K\|_\xi)^{k-1} \qquad (2.42)$$

for $k = 1, \ldots, l$. On the other hand

$$\left\| \chi_{\Lambda(l\delta)} \left[(1 - z\chi_\Lambda \Pi K)^{-1} - \left(1 + \sum_{k=1}^{l} (z\chi_\Lambda \Pi K)^k \right) \right] \chi_\Lambda \right\|_\xi \le \frac{(\|z\Pi K\|_\xi)^{l+1}}{1 - \|z\Pi K\|_\xi}$$
$$(2.43)$$

$$\left\| \chi_{\Lambda(l\delta)} \left[(1 - z\Pi K)^{-1} - \left(1 + \sum_{k=1}^{l} (z\Pi K)^k \right) \right] \right\|_\xi \le \frac{(\|z\Pi K\|_\xi)^{l+1}}{1 - \|z\Pi K\|_\xi}$$
$$(2.44)$$

From Eqs. (2.42), (2.43), and (2.44), we find

$$\|\chi_{\Lambda(l\delta)} (1 - \chi_\Lambda \Pi K)^{-1} \chi_\Lambda - \chi_{\Lambda(l\delta)} (1 - \Pi K)^{-1}\|_\xi$$

$$\le \frac{2(\|z\Pi K\|_\xi)^{l+1}}{1 - \|z\Pi K\|_\xi} + 2\eta(\delta)|z| \sum_{k=1}^{l} k(\|z\Pi K\|_\xi)^{k-1}$$

$$\le \frac{2(\|z\Pi K\|_\xi)^{l+1}}{1 - \|z\Pi K\|_\xi} + \frac{2|z|\eta(\delta)}{(1 - \|z\Pi K\|_\xi)^2} \qquad (2.45)$$

The right-hand side of (2.45) tends to zero when l, δ tend to infinity. Writing $\lambda = l\delta$ we see that there exists a function $\varepsilon(\cdot)$ positive, decreasing,

and independent of Λ such that

$$\lim_{\lambda \to \infty} \varepsilon(\lambda) = 0 \qquad (2.46)$$

$$\|\chi_{\Lambda(\lambda)}(1 - z\chi_\Lambda \Pi K)^{-1}\chi_\Lambda - \chi_{\Lambda(\lambda)}(1 - z\Pi K)^{-1}\|_\xi \leq \frac{\varepsilon(\lambda)}{\|z\alpha\|_\xi} \qquad (2.47)$$

Making use of (2.28) and (2.29), we obtain finally

$$\|\chi_{\Lambda(\lambda)}\boldsymbol{\rho}_\Lambda - \chi_{\Lambda(\lambda)}\boldsymbol{\rho}\|_\xi \leq \varepsilon(\lambda) \qquad (2.48)$$

which is equivalent to (2.35) and concludes the proof of the theorem.

4.2.4 COMPLEMENTS TO THEOREM 4.2.3

(a) If K is a compact set in the region (2.33) and $\xi = C(\beta)^{-1}$, the function $\varepsilon'(\cdot)$ such that $\varepsilon(\cdot) = |z|\varepsilon'(\cdot)$ may be chosen independently of $|z| \in K$ as we see from (2.45).

(b) If $e^{-\beta\Phi}$ is a continuous function on \mathbf{R}^ν (as is usually the case) we may use Banach spaces of bounded continuous instead of bounded measurable functions; in particular, the $\rho(x)_n$ are continuous.

4.2.5 THE MAYER–MONTROLL EQUATIONS. Defining

$$K((x)_m, (y)_n) = \prod_{j=1}^n \left[\left(\prod_{i=1}^m \exp[-\beta\Phi(y_j - x_i)] \right) - 1 \right] \qquad (2.49)$$

we can prove

$$\rho_\Lambda(x)_m = \chi_\Lambda(x)_m z^m \exp[-\beta U(x)_m]$$
$$\times \left[1 + \sum_{n=1}^\infty \frac{1}{n!} \int d(y)_n K((x)_m, (y)_n) \rho_\Lambda(y)_n \right] \qquad (2.50)$$

These integral equations are the Mayer–Montroll equations; they form a linear inhomogeneous system for the correlation functions $\rho_\Lambda(x)_n$, $n \geq 1$. The Mayer–Montroll equations may be used instead of the Kirkwood–Salsburg equations to study the infinite volume limit of the correlation functions. Their usefulness is, however, restricted to positive and hard core potentials.[9]

4.2.6 LATTICE SYSTEMS. Let us consider a classical lattice system in the sense of Section 2.4. We assume that $\Phi \in \mathscr{B}$ and that $\Phi(X) = 0$ unless X has exactly one or two elements, and we put

$$z = \exp[-\Phi(\{x\})] \qquad (2.51)$$

$$\beta\varphi(\xi) = \Phi(\{x, y\}) \qquad \text{if} \quad x - y = \xi \neq 0 \qquad (2.52)$$

$$\varphi(0) = +\infty \qquad (2.53)$$

[9] See Penrose [1].

The condition $\Phi \in \mathscr{B}$ becomes, in terms of φ,

$$\sum_{\xi \neq 0} |\varphi(\xi)| < +\infty \tag{2.54}$$

We can now, using the pair potential φ, construct a theory completely analogous to that described above for continuous systems, with the obvious replacement of integrals over \mathbf{R}^ν by sums over \mathbf{Z}^ν. We notice that (2.54) implies the analog of stability because of (4.8) of Chapter 2, and the analog of regularity of the pair potential because

$$\sum_{\xi \in \mathbf{Z}^\nu} |e^{-\beta\varphi(\xi)} - 1| \leq 1 + \sum_{\xi \neq 0} (e^{\beta|\varphi(\xi)|} - 1)$$

$$\leq \exp\left(\beta \sum_{\xi \neq 0} |\varphi(\xi)|\right) \tag{2.55}$$

where we have used the superadditivity[10] of the function $x \to e^x - 1$ on $[0, +\infty)$. We leave to the reader the task of rephrasing Theorem 4.2.3 in the present situation.

We indicate now another treatment of lattice systems which is not restricted to pair interactions. This treatment is analogous to that given by the Kirkwood–Salsburg equations, but makes essential use of the fact that two particles cannot occupy the same lattice site.

The interaction potential Φ is a function of finite subsets of \mathbf{Z}^ν, and we assume that $\Phi(X) = 0$ if $N(X) = 1$ (i.e., if X contains just one point of \mathbf{Z}^ν); the chemical potential will appear in the activity z. We assume also that

$$D = \sum_{X \ni 0} |\Phi(X)| < +\infty \tag{2.56}$$

This condition is more stringent than the condition $\|\Phi\| < +\infty$ used in Section 2.4, but permits all pair potentials satisfying (2.54). We define

$$\Xi(\Lambda, z, \beta) = \sum_{X \subset \Lambda} z^{N(X)} e^{-\beta U(X)} \tag{2.57}$$

$$\rho_\Lambda(X) = \Xi(\Lambda, z, \beta)^{-1} \sum_{Y \subset \Lambda \setminus X} z^{N(X)+N(Y)} \exp[-\beta U(X \cup Y)] \tag{2.58}$$

If X is a finite subset of \mathbf{Z}^ν, we let X' be obtained by suppressing one point x_1 of X, e.g., the first point in lexicographic order.[11] We introduce the following definitions, where $X \cap Y = \varnothing$

$$W^1(X) = \sum_{S:x_1 \in S \subset X} \Phi(S) \tag{2.59}$$

$$W^1(X, Y) = \sum_{S:x_1 \in S \subset X} \Phi(S \cup Y) \tag{2.60}$$

[10] If $a, b \geq 0$, then $(e^a - 1)(e^b - 1) \geq 0$, hence $(e^a - 1) + (e^b - 1) \leq e^{a+b} - 1$.

[11] If $a = (a^1, \ldots, a^\nu) \in \mathbf{Z}^\nu$ and $b = (b^1, \ldots, b^\nu) \in \mathbf{Z}^\nu$ we say that a is before b in lexicographic order if $(a^1 < b^1)$ or $(a^1 = b^1$ and $a^2 < b^2)$ or \cdots or $(a^1 = b^1, a^2 = b^2, \ldots, a^{\nu-1} = b^{\nu-1}$ and $a^\nu < b^\nu)$.

$$I(X, Y) = \sum_{S:x_i \in S \subset X} \sum_{T:\varnothing \ne T \subset Y} \Phi(S \cup T) = \sum_{T:\varnothing \ne T \subset Y} W^1(X, T) \quad (2.61)$$

$$K(X, Y) = \sum_n \sum_{\{T_1, \dots, T_n\}}^* \prod_{j=1}^n (\exp[-\beta W^1(X, T_j)] - 1) \quad (2.62)$$

In (2.62), Σ^* is extended over all sets $\{T_1, \dots, T_n\}$ of nonempty subsets of Y which cover Y (i.e., $\bigcup_j T_j = Y$), and $Y \ne \varnothing$. Notice the identities

$$U(X \cup Y) = W^1(X) + I(X, Y) + U(X' \cup Y) \quad (2.63)$$

$$e^{-\beta I(X,Y)} = \prod_{T:\varnothing \ne T \subset Y} (1 + (e^{-\beta W^1(X,T)} - 1)) = 1 + \sum_{T \subset Y} K(X, T) \quad (2.64)$$

We obtain from (2.56) the inequality

$$\sum_{Y:Y \cap X = \varnothing} |W^1(X, Y)| \le D \quad (2.65)$$

From this and the superadditivity of $e^x - 1$ for $x \ge 0$ we get

$$\sum_{T:T \cap X = \varnothing} |\exp[-\beta W^1(X, T)] - 1| \le \sum_{T:T \cap X = \varnothing} (\exp[\beta |W^1(X, T)|] - 1)$$

$$\le \exp\left(\beta \sum_{T:T \cap X = \varnothing} |W^1(X, T)|\right) - 1$$

$$\le e^{\beta D} - 1 \quad (2.66)$$

and therefore

$$\sum_{Y:Y \cap X = \varnothing} |K(X, Y)| \le \sum_{Y:Y \cap X = \varnothing} \sum_n \sum_{\{T_1, \dots, T_n\}}^* \prod_{j=1}^n |\exp[-\beta W^1(X, T_j)] - 1|$$

$$\le \sum_n \frac{1}{n!} \sum_{T_1:T_1 \cap X = \varnothing} \cdots \sum_{T_n:T_n \cap X = \varnothing} \prod_{j=1}^n$$
$$\times |\exp[-\beta W^1(X, T_j)] - 1|$$

$$= \sum_n \frac{1}{n!} \left(\sum_{T:T \cap X = \varnothing} |\exp[-\beta W^1(X, T)] - 1|\right)^n$$

$$\le \sum_{n \ge 1} \frac{1}{n!} (e^{\beta D} - 1)^n = \exp(e^{\beta D} - 1) - 1 \quad (2.67)$$

Let $\chi_\Lambda(X) = 1$ if $X \subset \Lambda$, 0 otherwise. Using (2.63), (2.64) we have

$$\rho_\Lambda(X) = \Xi^{-1} \sum_{Y:Y \cap X - \varnothing} \chi_\Lambda(X \cup Y) z^{N(X \cup Y)} \exp[-\beta U(X \cup Y)]$$

$$= \chi_\Lambda(X) z \exp[-\beta W^1(X)] \Xi^{-1} \sum_{Y:Y \cap X = \varnothing} \chi_\Lambda(X' \cup Y) z^{N(X' \cup Y)}$$

$$\times \left(1 + \sum_{T \subset Y} K(X, T)\right) \exp[-\beta U(X' \cup Y)]$$

$$= \chi_\Lambda(X) z \exp[-\beta W^1(X)]$$

$$\times \left[\rho_\Lambda(X') - \rho_\Lambda(X) + \sum_{T:T \cap X = \varnothing} K(X, T)(\rho_\Lambda(X' \cup T) - \rho_\Lambda(X \cup T)) \right]$$

and therefore

$$\rho_\Lambda(X) = \chi_\Lambda(X) \frac{z \exp[-\beta W^1(X)]}{1 + z \exp[-\beta W^1(X)]}$$

$$\times \left[\rho_\Lambda(X') + \sum_{T:T \cap X = \varnothing} K(X, T)(\rho_\Lambda(X' \cup T) - \rho_\Lambda(X \cup T)) \right] \quad (2.68)$$

4.2.7 THEOREM.[12] *Let E be the Banach space of complex bounded functions of nonempty finite subsets of \mathbf{Z}^ν, with the uniform norm. If X is a finite subset of \mathbf{Z}^ν, X' is obtained by suppressing the first element of X in lexicographic order. An operator K is defined on E by*

$$(K\varphi)(X) = \frac{z \exp[-\beta W^1(X)]}{1 + z \exp[-\beta W^1(X)]}$$

$$\left[\varphi(X') + \sum_{T:T \cap X = \varnothing} K(X, T)(\varphi_\Lambda(X' \cup T) - \varphi_\Lambda(X \cup T)) \right] \quad (2.69)$$

where the first term $\varphi(X')$ in the square brackets has to be omitted if $N(X) = 1$ and $W^1(X), K(X, T)$ are defined by (2.59), (2.62). Writing

$$C = \sum_{X \ni 0} \Phi(X) \qquad D = \sum_{X \ni 0} |\Phi(X)| \qquad (2.70)$$

we have

$$\|K\| \leq \left| \frac{z\, e^{\beta(D-C)}}{1 + z\, e^{\beta(D-C)}} \right| \cdot [2 \exp(e^{\beta D} - 1) - 1] \qquad (2.71)$$

provided

$$\left| \frac{z\, e^{\beta(D-C)}}{1 + z\, e^{\beta(D-C)}} \right| \cdot [2 \exp(e^{\beta D} - 1) - 1] < 1 \qquad (2.72)$$

In the circle of the z plane where (2.72) holds, Ξ defined by (2.57) does not vanish and ρ_Λ defined by (2.58) is the only solution of the following equation in E:

$$\rho_\Lambda = \chi_\Lambda \frac{z}{1+z} \alpha + \chi_\Lambda K \rho_\Lambda \qquad (2.73)$$

where $\alpha(X) = 1$ if $N(X) = 1, 0$ otherwise. Given $\varepsilon > 0$ we can find a finite

[12] See Gallavotti and Miracle [2] and Gallavotti, Miracle, and Robinson [1].

set $\Delta \subset \mathbf{Z}^\nu$ such that

$$|\rho_\Lambda(X) - \rho(X)| < \varepsilon \qquad (2.74)$$

if $X + \Delta \subset \Lambda$, and ρ is the only solution in E of

$$\rho = \frac{z}{1+z}\alpha + K\rho \qquad (2.75)$$

The vector $\rho \in E$ depends analytically on z[13] in (2.72).

We first prove (2.71). Notice that if a is real, the curve $\{z:|(1 + az)^{-1}az| = C\}$ is a circle symmetric with respect to the real axis. It is therefore enough, using (2.67), to prove

$$\left| \frac{z \exp[-\beta W^1(X)]}{1 + z \exp[-\beta W^1(X)]} \right| \leq \left| \frac{z\, e^{\beta(D-C)}}{1 + z\, e^{\beta(D-C)}} \right| \qquad (2.76)$$

for real z where it follows from $W^1(X) \geq -(D - C)$ and the fact that the right-hand side of (2.76) is less than one by (2.72).

Equation (2.73) is just (2.68). The remaining statements are proved in the same manner as the analogous statements of Theorem 4.2.3.

4.3 MAYER SERIES AND VIRIAL EXPANSION

We place ourselves again in the situation of Theorem 4.2.3 and consider the first components $\rho_\Lambda(x_1)$, $\rho(x_1)$ of ρ_Λ and ρ. As previously

$$|z| < e^{-2\beta B - 1}C(\beta)^{-1} \qquad (3.1)$$

We notice the following facts

(a) Because of the translation invariance of Eq. (2.26), $\rho(x_1)$ is a constant ρ_1, which depends analytically on z in the circle (3.1).

(b) The relation (2.28) in E_ξ implies that $z^{-1}|\rho_\Lambda(x_1)|$ has a bound independent of x_1, Λ, and which can be chosen independent of z if $z \in K$, K a compact subset of (3.1).

(c) In view of Eq. (2.35) and 4.2.4,(a), we have

$$|\rho_\Lambda(x_1) - \rho_1| \leq |z|C(\beta)^{-1}\varepsilon'(\lambda) \qquad (3.2)$$

where $\varepsilon'(\cdot)$ may be chosen independent of z if $z \in K$.

From (a), (b), (c) it follows that if $\Lambda \to \infty$ in the sense of Van Hove, then

$$\lim_{\Lambda \to \infty} V(\Lambda)^{-1}z^{-1} \int dx_1 \rho_\Lambda(x_1) = z^{-1}\rho_1 \qquad (3.3)$$

uniformly on every compact set K contained in (3.1). By reference to (2.32)

[13] ρ is also a real analytic function of β and Φ because K is.

we see that $V(\Lambda)^{-1} \int dx_1 \rho_\Lambda(x_1)$ is the grand canonical average of $V(\Lambda)^{-1} n$, i.e., the density, and also that

$$V(\Lambda)^{-1} \log \Xi(\Lambda, z, \beta) = \int_0^z \frac{dz}{z} V(\Lambda)^{-1} \int dx_1 \rho_\Lambda(x_1) \qquad (3.4)$$

It follows that if $\Lambda \to \infty$ in the sense of Van Hove, then

$$\beta p = \lim V(\Lambda)^{-1} \log \Xi(\Lambda, z, \beta) = \int_0^z \frac{dz}{z} \rho_1 \qquad (3.5)$$

We summarize the preceding results.

4.3.1 THEOREM. *If $\Lambda \to \infty$ in the sense of Van Hove, then the following limits exist*

$$\lim_{\Lambda \to \infty} \beta^{-1} V(\Lambda)^{-1} \log \Xi(\Lambda, z, \beta) = p \qquad (3.6)$$

$$\lim_{\Lambda \to \infty} z \frac{d}{dz} V(\Lambda)^{-1} \log \Xi(\Lambda, z, \beta) = \rho_1 \qquad (3.7)$$

when z satisfies (1.1), and these limits continue analytically to (1.1), the physical pressure, and density (defined for $z > 0$). In particular the Mayer series

$$\beta p = \sum_{n=1}^{\infty} b_n z^n \qquad (3.8)$$

$$\rho_1 = \sum_{n=1}^{\infty} n b_n z^n \qquad (3.9)$$

have a radius of convergence at least $e^{-2\beta B - 1} C(\beta)^{-1}$.

This theorem extends in an obvious manner to lattice systems (see 4.2.5). Furthermore, in the situation of Theorem 4.2.7 we see that p and ρ_1 are analytic in the circle (2.72). If we adopt the point of view that a phase transition occurs when the thermodynamic functions exhibit a singularity[14] [see Section 1.1,(c')], the above results imply that no phase transition occurs for $z < e^{-2\beta B - 1} C(\beta)^{-1}$. We shall say that at small activity a system is in the gas phase.[15]

Instead of the activity, we may use the density ρ_1 as parameter: since $b_1 = 1$ we may invert (3.9) in a neighborhood of the origin and insert in (3.8); we obtain in this way the virial expansion

$$p = \sum_{n=1}^{\infty} c_n \rho_1^n \qquad (3.10)$$

[14] That p and ρ_1 are also real analytic functions of the inverse temperature β follows from the fact that the kernels of the equations considered in Section 4.2 are analytic in β.

[15] We shall check in Section 4.4 that the correlation functions satisfy a cluster property as expected for a gas.

where the coefficients c_n satisfy

$$c_n = \frac{1}{2\pi i} \int_C \left(\frac{dp}{dz}\right) \frac{dz}{n\rho_1{}^n} = \frac{\beta^{-1}}{2\pi i} \int_C \frac{dz}{nz\rho_1^{n-1}} \tag{3.11}$$

if C is a circle of radius less than $e^{-2\beta B-1}C(\beta)^{-1}$ around the origin.

In view of (2.29) and (2.25), if $\xi = C(\beta)^{-1}$ we have

$$|\rho_1 - z| \leq \xi \cdot \|\boldsymbol{\rho} - z\boldsymbol{\alpha}\|_\xi \leq |z| \sum_{l=1}^{\infty} |z|^l \|\Pi K\|_\xi^l$$

$$= \frac{|z|^2 \cdot \|\Pi K\|_\xi}{1 - |z| \cdot \|\Pi K\|_\xi} \leq \frac{|z|^2}{e^{-2\beta B-1}C(\beta)^{-1} - |z|} \tag{3.12}$$

hence

$$|\rho_1| \geq |z| - \frac{|z|^2}{e^{-2\beta B-1}C(\beta)^{-1} - |z|} \tag{3.13}$$

From (3.13) we see that $|\rho_1| \geq (3 - 2^{3/2})e^{-2\beta B-1}$ on the circle $|z| = (1 - 2^{-1/2})e^{-2\beta B-1}C(\beta)^{-1}$; this implies that

(a) the radius of convergence of the expansion of z in powers of ρ_1 is at least $(3 - 2^{3/2})e^{-2\beta B-1}C(\beta)^{-1}$, and the same is true for (3.10);

(b) the coefficients c_n given by (3.11) are bounded by

$$|c_n| \leq \frac{\beta^{-1}}{n} \left[\frac{e^{-2\beta B-1}C(\beta)}{(3 - 2^{3/2})} \right]^{n-1} \tag{3.14}$$

The results (a) and (b) can be improved if the crude estimate of the coefficient b_l implicit in (3.12) is replaced by

$$|nb_n| \leq \frac{n^{n-2}}{(n-1)!} e^{2(n-2)\beta B}C(\beta)^{n-1} \tag{3.15}$$

which result from Eqs. (4.31) and (4.34) below. We then obtain the following theorem.

4.3.2 THEOREM.[16] *The radius of convergence of the virial expansion*

$$p = \sum_{n=1}^{\infty} c_n\rho_1{}^n \tag{3.16}$$

is at least $\alpha(e^{2\beta B} + 1)^{-1}C(\beta)^{-1}$ *and*

$$|c_n| \leq \frac{\beta^{-1}}{n} [\alpha^{-1}(e^{2\beta B} + 1)C(\beta)]^{n-1} \tag{3.17}$$

where $\alpha = 0.28952$.

[16] Lebowitz and Penrose [1]. The estimates of this theorem hold also before the limit of an infinite volume is taken. For $\rho_1 < \alpha(e^{2\beta B} + 1)C(\beta)^{-1}$ not only does the virial expansion converge but its limit is the physical pressure (the series may be convergent beyond the region where it has physical significance).

Using more refined techniques, Groeneveld [2] has shown that

$$|c_n| \le \frac{\beta^{-1}}{n} [e(e^{2\beta B} + 1)]^{n-2} C(\beta)^{n-1}$$

and in the case of a positive potential

$$|c_n| \le \frac{\beta^{-1}}{n} (1 + \gamma)^{-1} \gamma^{-n+2} C(\beta)^{n-1}$$

where $\gamma = 0.27846$ is the positive root of $\gamma \cdot e^{1+\gamma} = 1$. The radius of convergence of the virial expansion is thus at least $e^{-1}(e^{2\beta B} + 1)^{-1} C(\beta)^{-1}$ and, for positive Φ, $\gamma C(\beta)^{-1}$.

4.4 THE ALGEBRAIC METHOD

In this section we consider an approach to the study of correlation functions which is parallel to that of Section 4.2 but is more suited to certain problems. The idea is to investigate directly the coefficients of the expansions of the correlation functions in powers of z.

4.4.1 DEFINITIONS. Let \mathscr{A} be the complex vector space of sequences[17] ψ:

$$\psi = (\psi(x)_n)_{n \ge 0} \tag{4.1}$$

such that, for each n, $\psi(x)_n$ is a bounded complex Lebesgue measurable function[18] on $\mathbf{R}^{n\nu}$. Notice that the 0th component ψ_0 of ψ is a complex number. It will be convenient to represent by a single letter a finite sequence (x_1, \dots, x_n) of vectors in \mathbf{R}^ν: $X = (x_1, \dots, x_n) = (x)_n$, and we shall write

$$\psi(X) = \psi(x_1, \dots, x_n) = \psi(x)_n \tag{4.2}$$

Let now $\psi_1, \psi_2 \in \mathscr{A}$; we write

$$\psi(X) = \sum_{Y \subset X} \psi_1(Y)\psi_2(X \setminus Y) \tag{4.3}$$

where the summation extends over all subsequences Y of X and $X \setminus Y$ is the subsequence of X obtained by striking out the elements of Y in X. For instance

$$\psi_0 = \psi_{10}\psi_{20} \qquad \psi(x_1) = \psi_{10}\psi_2(x_1) + \psi_1(x_1)\psi_{20}$$

$$\psi(x_1, x_2) = \psi_{10}\psi_2(x_1, x_2) + \psi_1(x_1)\psi_2(x_2) + \psi_1(x_2)\psi_2(x_1) + \psi_1(x_1, x_2)\psi_{20}$$

Equation (4.3) defines a product

$$\psi = \psi_1 * \psi_2 \tag{4.4}$$

[17] The spaces E_ξ of Section 4.2 consist of sequences (φ_n) with $n \ge 1$ instead of $n \ge 0$ here.
[18] We do not distinguish between functions which differ only on a set of Lebesgue measure zero.

in \mathscr{A}. It is easily seen that, with respect to this product, \mathscr{A} is a commutative algebra with unit element $\mathbf{1}$ [$\mathbf{1}$ is defined by $\mathbf{1}_0 = 1$, $\mathbf{1}(x)_n = 0$ for $n > 0$].

We shall call \mathscr{A}_+ the subspace of \mathscr{A} formed by the elements ψ such that $\psi_0 = 0$.[19]

The power series expansion of the exponential yields a well-defined mapping Γ of \mathscr{A}_+ onto $\mathbf{1} + \mathscr{A}_+$:

$$\Gamma\varphi = \mathbf{1} + \varphi + \frac{\varphi * \varphi}{2} + \frac{\varphi * \varphi * \varphi}{3!} + \cdots \qquad \varphi \in \mathscr{A}_+ \qquad (4.5)$$

Γ has an inverse Γ^{-1} (corresponding to the logarithm):

$$\Gamma^{-1}(\mathbf{1} + \varphi') = \varphi' - \frac{\varphi' * \varphi'}{2} + \frac{\varphi' * \varphi' * \varphi'}{3} + \cdots \qquad \varphi' \in \mathscr{A}_+ \qquad (4.6)$$

We see readily that $(\Gamma\varphi)(x)_n$ is the sum of the products

$$\varphi(X_1) \cdots \varphi(X_r) \qquad (4.7)$$

corresponding to all partitions of $(x)_n$ into subsequences $X_1, \cdots, X_r (1 \le r \le n)$. If $\varphi \in \mathscr{A}_+$ and $\psi = \Gamma\varphi$, the first few component equations of (4.5) are thus

$$\psi_0 = 1 \qquad \psi(x_1) = \varphi(x_1)$$

$$\psi(x_1, x_2) = \varphi(x_1, x_2) + \varphi(x_1)\varphi(x_2)$$

$$\psi(x_1, x_2, x_3) = \varphi(x_1, x_2, x_3) + \varphi(x_1, x_2)\varphi(x_3) + \varphi(x_1, x_3)\varphi(x_2)$$
$$+ \varphi(x_2, x_3)\varphi(x_1) + \varphi(x_1)\varphi(x_2)\varphi(x_3)$$

4.4.2 BOLTZMANN FACTORS AND URSELL FUNCTIONS. Let $\psi \in \mathbf{1} + \mathscr{A}_+$ be defined by

$$\psi_0 = 1 \qquad \psi(x)_n = \exp[-\beta U(x)_n] \qquad (4.8)$$

where $U(x)_n$ is given by (2.2).

Define also $\varphi_0 = 0$, $\varphi(x)_1 = 0$, and

$$\varphi(x)_n = \sum_\gamma \prod_{i<j}{}^\gamma (\exp[-\beta\Phi(x_j - x_i)] - 1) \qquad n > 1 \qquad (4.9)$$

where the sum extends over all connected graphs γ with vertices $1, \ldots, n$ (see Fig. 12) and the product over all pairs (i, j) such that the points i and j are joined by a line in γ.

The functions $\psi(x)_n$ and $\varphi(x)_n$ are known, respectively, as *Boltzmann factors* and *Ursell functions*. If we write

$$\psi(x)_n = \prod_{1 \le i < j \le n} (1 + (\exp[-\beta\Phi(x_j - x_i)] - 1))$$

[19] \mathscr{A}_+ is an ideal of \mathscr{A}.

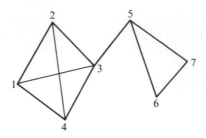

FIGURE 12 A connected graph.

and develop the product in the right-hand side we obtain, using (4.7),

$$\psi = \Gamma \varphi \qquad (4.10)$$

The form (4.10) of the relation between Boltzmann factors and Ursell functions will allow us to use the algebraic properties of the exponential rather than the graph-theoretical considerations which are usual in the study of the Ursell functions.

4.4.3 THE CLUSTER PROPERTY. Let $\psi \in 1 + \mathscr{A}_+$ and let the functions[20] $\psi(x)_n$ be continuous and invariant under translations:

$$\psi(x_1 + a, \ldots, x_n + a) = \psi(x_1, \ldots, x_n)$$

Let X_1 be a subsequence of $(x)_n$ and $X_2 = (x)_n \backslash X_1$. It may occur that as the minimum of the distances $|x_j - x_i|$ for $x_i \in X_1$ and $x_j \in X_2$ goes to infinity,

$$|\psi(x)_n - \psi(X_1)\psi(X_2)| \to 0 \qquad (4.11)$$

If this is true for all n and every subsequence X_1 of $(x)_n$, we say that the functions $\psi(x)_n$ have the *cluster property*. This name corresponds to the fact that if the arguments of $\psi(x)_n$ are separated into two distant "clusters" X_1, X_2, the function $\psi(x)_n$ factorizes into the product $\psi(X_1)\psi(X_2)$. For instance, if the pair potential Φ is continuous and tends to zero at infinity, the Boltzmann factors have the cluster property.

Let $\varphi = \Gamma^{-1}\psi$; the functions $\psi(x)_n$ have the cluster property if and only if the $\varphi(x)_n$, considered as functions of the differences $x_{i+1} - x_i$, tend to zero at infinity.

This result follows easily from (4.7) and illustrates the interest of introducing the mapping Γ. Of course, various cluster properties may be defined where, instead of continuous functions tending to zero at infinity, we consider functions which are integrable or have some other kind of decrease at infinity.

[20] Here the $\psi(x)_n$ are not necessarily Boltzmann factors.

4.4.4 FURTHER DEFINITIONS. Let $\chi(\cdot)$ be a Lebesgue integrable function on \mathbf{R}^ν. To every $\psi \in \mathcal{A}$ we associate the formal power series

$$\langle \chi, \psi \rangle(z) = \psi_0 + \sum_{n=1}^{\infty} \frac{z^n}{n!} \int dx_1 \cdots dx_n \chi(x_1) \cdots \chi(x_n) \psi(x)_n \qquad (4.12)$$

If $\psi_1, \psi_2 \in \mathcal{A}$, the following identity between formal power series holds

$$\langle \chi, \psi_1 * \psi_2 \rangle(z) = \langle \chi, \psi_1 \rangle(z) \cdot \langle \chi, \psi_2 \rangle(z) \qquad (4.13)$$

as we check readily.[21] If $\varphi \in \mathcal{A}_+$, we deduce from (4.13), and the fact that Γ is the exponential, the formula

$$\langle \chi, \Gamma\varphi \rangle(z) = \exp[\langle \chi, \varphi \rangle(z)] \qquad (4.14)$$

We introduce now an operation $D_x : \psi \to D_x \psi$ by

$$D_x \psi(y)_n = \psi(x, y_1, \ldots, y_n) \qquad (4.15)$$

More generally if $X = (x_1, \ldots, x_m)$ we shall write

$$D_X \psi = D_{x_1} \cdots D_{x_m} \psi \qquad (4.16)$$

It is clear that D_x and D_X are linear; D_x furthermore satisfies the identity

$$D_x(\psi_1 * \psi_2) = (D_x\psi_1) * \psi_2 + \psi_1 * (D_x\psi_2) \qquad (4.17)$$

which is characteristic of a *derivation*. From the rule of differentiation of an exponential we obtain the important formula

$$D_x(\Gamma\varphi) = D_x\varphi * \Gamma\varphi \qquad \varphi \in \mathcal{A}_+ \qquad (4.18)$$

4.4.5 EXPRESSIONS FOR Ξ AND THE CORRELATION FUNCTIONS. We concentrate now on the case where ψ is the sequence (4.8) of the Boltzmann factors. Let χ_Λ be the characteristic function of the bounded Lebesgue measurable set $\Lambda \subset \mathbf{R}^\nu$. We may rewrite the definition of the grand partition function as

$$\Xi(\Lambda, z, \beta) = \exp[\beta V(\Lambda) p(\Lambda, z, \beta)] = \langle \chi_\Lambda, \psi \rangle(z) \qquad (4.19)$$

Because of the assumed stability of the pair potential Φ, (4.19) is an entire series. Since $\Xi(\Lambda, 0, \beta) = 1$, the logarithm of Ξ is analytic in a neighborhood of the origin and given by the series

$$\beta V(\Lambda) p(\Lambda, z, \beta) = \langle \chi_\Lambda, \varphi \rangle(z) \qquad (4.20)$$

where we have used (4.14) and (4.10), and φ is the sequence of the Ursell functions.

[21] The mapping $\psi \to \langle \chi, \psi \rangle(z)$ is thus a homomorphism of \mathcal{A} in the algebra of formal power series.

If the complex activity z is in a suitable neighborhood of the origin in the complex plane we may define $\psi_\Lambda{}^z \in 1 + \mathscr{A}_+$ by

$$\psi_\Lambda{}^z(X) = \Xi^{-1} \langle \chi_\Lambda, D_X\psi \rangle(z) \tag{4.21}$$

Actually, since Ξ does not vanish for z real and greater than zero, (4.21) is defined for the physical values of z and by comparison with (2.5) we obtain

$$\rho_\Lambda(x)_m = z^m \chi_\Lambda(x_1) \cdots \chi_\Lambda(x_m)\psi_\Lambda{}^z(x)_m \qquad m > 0 \tag{4.22}$$

Let $X = (x_1, \ldots, x_m)$, $m > 0$; we define $\tilde{\varphi}_X \in \mathscr{A}$ by

$$\tilde{\varphi}_X = \psi^{-1} * D_X\psi \tag{4.23}$$

Notice that if $m = 1$, i.e., $X = (x_1)$, we obtain, using $\psi = \Gamma\varphi$ and (4.18)

$$\tilde{\varphi}_{(x_1)} = \psi^{-1} * D_{x_1}\varphi * \psi = D_{x_1}\varphi \tag{4.24}$$

We have the following identity between power series

$$
\begin{aligned}
\psi_\Lambda{}^z(X) &= \Xi^{-1} \langle \chi_\Lambda, \psi * (\psi^{-1} * D_X\psi) \rangle(z) \\
&= \Xi^{-1} \langle \chi_\Lambda, \psi \rangle(z) \cdot \langle \chi_\Lambda, \psi^{-1} * D_X\psi \rangle(z) \\
&= \langle \chi_\Lambda, \tilde{\varphi}_X \rangle(z)
\end{aligned}
\tag{4.25}
$$

which yields an expansion of $\psi_\Lambda{}^z(X)$ in powers of the activity.

4.4.6 CONVERGENCE OF THE ACTIVITY EXPANSIONS. If we use (4.22) and insert the expansion (4.25) in the Kirkwood–Salsburg equations we obtain a set of relations between the functions $\tilde{\varphi}_X(Y)$. We shall now indicate a direct derivation of such relations and the estimates to which they lead.

Since $X = (x_1, \ldots, x_m)$, we may write $D_X\psi = D_{x_k}D_{X_k}\psi$, where $X_k = X \setminus (x_k)$. Therefore (4.23) becomes $\tilde{\varphi}_X = \psi^{-1} * D_{x_k}D_{X_k}\psi$. Using the definitions (2.6), (2.9), and (4.3) we have thus

$$
\begin{aligned}
\tilde{\varphi}_X(Y) &= \sum_{Z \subset Y} \psi^{-1}(Y \setminus Z) D_{x_k}D_{X_k}\psi(Z) \\
&= \sum_{Z \subset Y} \psi^{-1}(Y \setminus Z) \exp[-\beta W^k(X \cup Z)]\psi(X_k \cup Z) \\
&= \exp[-\beta W^k(X)] \sum_{Z \subset Y} \psi^{-1}(Y \setminus Z) \sum_{S \subset Z} K(x_k, S)\psi(X_k \cup Z) \\
&= \exp[-\beta W^k(X)] \sum_{S \subset Y} K(x_k, S) \sum_{Z:S \subset Z \subset Y} \psi^{-1}(Y \setminus Z)\psi(X_k \cup Z)
\end{aligned}
$$

Writing $T = Z \setminus S$ we obtain

$$\tilde{\varphi}_X(Y) = \exp[-\beta W^k(X)] \sum_{S \subset Y} K(x_k, S) \sum_{T \subset Y \setminus S} \psi^{-1}(Y \setminus (S \cup T))\psi(X_k \cup S \cup T)$$

hence finally the recurrence relation[22]

$$\tilde{\varphi}_X(Y) = \exp[-\beta W^k(X)] \sum_{S \subset Y} K(x_k, S) \tilde{\varphi}_{X_k \cup S}(Y \setminus S) \qquad (4.26)$$

If $\xi > 0$, we obtain from (4.26) by induction on $m + n$ the majorization[23]

$$\sup_{(x)_m} \int d(y)_n |\tilde{\varphi}_{(x)_m}(y)_n| \leq n! \xi^{m-1} \{e^{2\beta B} \xi^{-1} \exp[C(\beta)\xi]\}^{m+n-1} \qquad (4.27)$$

This estimate is of the same type as those of Section 4.2 and implies [taking $\xi = C(\beta)^{-1}$] that the series (4.25) converges for

$$|z| < e^{-2\beta B - 1} C(\beta)^{-1} \qquad (4.28)$$

We can somewhat improve the bound (4.27). For instance if $m = 1$ we have $U(x)_m = 0$ and therefore in view of (4.24),

$$\int dx_2 \cdots dx_n |\varphi(x_1, x_2, \ldots, x_n)| \leq (n-1)! \, e^{-2\beta B} \{e^{2\beta B} \xi^{-1} \exp[C(\beta)\xi]\}^{n-1} \qquad (4.29)$$

By a more careful majorization we obtain from (4.26), instead of (4.27) and (4.29), the estimates[24]

$$\sup_{(x)_m} \int d(y)_n |\tilde{\varphi}_{(x)_m}(y)_n| \leq m(m+n)^{n-1} e^{2(m+n-1)\beta B} C(\beta)^n \qquad (4.30)$$

$$\int dx_2 \cdots dx_n |\varphi(x_1, x_2, \ldots, x_n)| \leq n^{n-2} e^{2(n-2)\beta B} C(\beta)^{n-1} \qquad (4.31)$$

4.4.7 CLUSTER PROPERTY OF THE CORRELATION FUNCTIONS. We know by Theorem 4.2.3 that $\rho_\Lambda(x)_m$ tends to a limit $\rho(x)_m$ when $\Lambda \to \infty$; therefore, ψ_Λ^z tends to a limit ψ^z [see (4.22)]. The activity expansion of ψ^z is obtained by letting $\Lambda \to \infty$ in (4.25); we therefore get

$$\psi^z(X) = \sum_{n=0}^{\infty} \frac{z^n}{n!} \int d(y)_n \tilde{\varphi}_X(y)_n \qquad (4.32)$$

[22] Note that for $m = 1$, $n = 0$, $\tilde{\varphi}_{(x)_m}(y)_n = 1$.

[23] The simple calculation is left to the reader; notice that we can replace the right-hand side of (4.26) by its geometric means over k and that the first factor yields then

$$\exp\left[-\beta \frac{1}{m} \sum W^k(X)\right] = \exp\left[-\frac{2}{m} \beta U(X)\right] \leq e^{2\beta B}$$

[24] See Penrose [1]. The inequalities obtained by Penrose differ from (4.30) and (4.31) in having the absolute value signs outside rather than inside the integral. It may be noticed that (4.30) does not improve the estimate (4.28) for the circle of convergence of the activity expansion of ψ_Λ^z.

which converges again in (4.28). Naturally

$$\rho(x)_m = z^m \psi^z(x)_m \tag{4.33}$$

In particular, for $m = 1$, we get

$$\rho_1 = \sum_{n=1}^{\infty} \frac{z^n}{(n-1)!} \int dx_2 \cdots dx_n \varphi(x_1, \ldots, x_n) \tag{4.34}$$

The $\rho(x)_m$ with $\rho_0 = 1$ form an element ρ of $\mathbf{1} + \mathcal{A}_+$; let

$$\rho = \Gamma\omega \qquad \psi^z = \Gamma\varphi^z \tag{4.35}$$

Then (4.33) yields

$$\omega(x)_m = z^m \varphi^z(x)_m \tag{4.36}$$

The functions $\omega(x)_m$ are the *cluster functions* of the correlation functions. According to 4.4.3 the correlation functions $\rho(x)_m$ have a cluster property if the cluster functions $\omega(x)_m$ have a property of decrease at infinity as functions of the differences $x_{i+1} - x_i$. We show that such a property holds, the "decrease at infinity" being replaced by absolute integrability.

4.4.8 THEOREM.[25] *We assume the interaction to be given by a stable regular pair potential* Φ. *The cluster functions* $\omega(x)_m$ *of the correlation functions (defined by* $\rho = \Gamma\omega$) *are absolutely integrable as functions of the differences* $x_{i+1} - x_i$ *when* $|z| < e^{-2\beta B - 1} C(\beta)^{-1}$:

$$\int dx_2 \cdots dx_m |\omega(x)_m| \leq \frac{(m-1)! |z| e^{-2\beta B} [|z| e^{2\beta B + 1} C(\beta)]^{m-1}}{[1 - |z| e^{2\beta B + 1} C(\beta)]^m} \tag{4.37}$$

Before proving (4.37) we express $\varphi^z(x)_m$ in terms of the Ursell functions $\varphi(x)_m$. If we remember the definition (4.16) of D_X and formula (4.18) we obtain

$$D_{(x)_m} \Gamma\varphi = \Gamma\varphi * \sum D_{X_1}\varphi * \cdots * D_{X_r}\varphi \tag{4.38}$$

where the summation extends over all partitions of $(x)_m$ into subsequences $X_1, \ldots, X_r (1 \leq r \leq m)$. Inserting (4.38) into the definition (4.23) of $\tilde{\varphi}_X$ we get

$$\tilde{\varphi}_{(x)_m} = (\Gamma\varphi)^{-1} * D_{(x)_m} \Gamma\varphi = \sum D_{X_1}\varphi * \cdots * D_{X_r}\varphi$$

and inserting this into (4.32) yields

$$\psi^z(x)_m = \sum \prod_{j=1}^{r} \left(\sum_{n=0}^{\infty} \frac{z^n}{n!} \int d(y)_n D_{X_j} \varphi(y)_n \right)$$

[25] See Ruelle [4].

Comparison with (4.35) gives finally

$$\varphi^z(X) = \sum_{n=0}^{\infty} \frac{z^n}{n!} \int d(y)_n D_X \varphi(y)_n \qquad (4.39)$$

Rewriting (4.29) with $\xi = C(\beta)^{-1}$ we have

$$\int dx_2 \cdots dx_m |\varphi(x)_m| \leq (m-1)! \, e^{-2\beta B} [e^{2\beta B + 1} C(\beta)]^{m-1} \qquad (4.40)$$

From (4.39) and (4.40) we obtain the integrability of $|\varphi^z(x)_m|$:

$$\int dx_2 \cdots dx_m |\varphi^z(x)_m| \leq e^{-2\beta B} \sum_{n=0}^{\infty} |z|^n \frac{(m+n-1)!}{n!} [e^{2\beta B + 1} C(\beta)]^{m+n-1}$$

$$= \frac{(m-1)! \, e^{-2\beta B} [e^{2\beta B + 1} C(\beta)]^{m-1}}{[1 - |z| \, e^{2\beta B + 1} C(\beta)]^m} \qquad (4.41)$$

where we have used the formula

$$\sum_{n=0}^{\infty} \frac{(m+n-1)!}{n!} t^n = \frac{(m-1)!}{(1-t)^m} \qquad |t| < 1$$

Finally, (4.37) results from (4.41) and (4.36).

4.5 POSITIVE POTENTIALS

In this section we assume that the pair potential Φ is positive ($\Phi \geq 0$). A number of simplifications occur then in the formulas of the previous sections and new results can be proved. We notice that we have now $B = 0$ and

$$C(\beta) = \int dx(1 - e^{-\beta\Phi(x)}) \qquad (5.1)$$

The activity expansions converge at least in the circle

$$|z| < e^{-1} C(\beta)^{-1} \qquad (5.2)$$

4.5.1 LIEB'S INEQUALITIES.[26] In the statistical mechanics of systems with pair interactions, products of the form $\Pi \exp[-\beta\Phi(x_j - x_l)]$ occur continually. If we write $e^{-\beta\Phi(x)} = (1 + (e^{-\beta\Phi(x)} - 1)) = 1 + f(x)$ we are led to products $\Pi_{k=1}^{N}(1 + f_k)$ where $-1 \leq f_k \leq 0$ (because $\Phi \geq 0$) and the following result is relevant.

[26] See Lieb [1] and also Penrose [2], where Lieb's results are extended to hard core potentials.

4.5.2 PROPOSITION.[27] *Let* $-1 \leq f_k \leq 0$ *for* $k = 1, \ldots, N$ *and write*

$$\prod_{k=1}^{N} (1 + f_k t) = \sum_{k=0}^{N} A_k t^k \qquad (5.3)$$

The following inequalities hold

$$\prod_{k=1}^{N} (1 + f_k) \lessgtr \sum_{k=0}^{L} A_k \qquad (5.4)$$

where \leq *holds for even L and* \geq *for odd L.*

We have indeed, with $0 < \theta < 1$,

$$\prod_{k=1}^{N} (1 + f_k t) = \sum_{k=0}^{L} A_k t^k + \frac{t^{L+1}}{(L+1)!} F^{(L+1)}(\theta t) \qquad (5.5)$$

where

$$F^{(L+1)}(t) = \frac{d^{L+1}}{dt^{L+1}} \prod_{k=1}^{N} (1 + f_k t) \qquad (5.6)$$

If we take $0 \leq t \leq 1$, the right-hand side of (5.6) is a sum of products containing $L + 1$ factors $f_k \leq 0$ and $N - (L + 1)$ factors $1 + f_k t \geq 0$, hence $F^{(L+1)}(\theta)$ is greater than or equal to zero if $L + 1$ is even, less than or equal to zero if $L + 1$ is odd. Inserting this into (5.5) with $t = 1$, we obtain (5.4).

As an application of Proposition 4.5.2 we insert (5.4) into (2.8) and follow the derivation of the Kirkwood–Salsburg equations; we obtain in this way (for all $z > 0$)

$$\rho_\Lambda(x_1) \lessgtr \chi_\Lambda(x_1) z \left[1 + \sum_{n=1}^{L} \int d(y)_n K(x_1, (y)_n) \rho_\Lambda(y)_n \right] \qquad (5.7)$$

$$\rho_\Lambda(x)_m \lessgtr \chi_\Lambda(x)_m z \exp[-\beta W^1(x)_m]$$
$$\times \left[\rho_\Lambda(x)'_{m-1} + \sum_{n=1}^{L} \frac{1}{n!} \int d(y)_n K(x_1, (y)_n) \rho_\Lambda((x)'_{m-1}, (y)_n) \right] \qquad (5.8)$$

If $0 < z < e^{-1} C(\beta)^{-1}$ we can let $\Lambda \to \infty$ in (5.7) and (5.8); we get in particular

$$\rho_1 \lessgtr z \left[1 + \sum_{n=1}^{L} \int dy_1 \cdots dy_n K(x_1, (y)_n) \rho(y)_n \right] \qquad (5.9)$$

Let us expand the correlation functions $\rho_\Lambda(x)_m$ in powers of z:

$$\rho_\Lambda(x)_m = z^m \sum_{n=0}^{\infty} z^n \rho_\Lambda{}^n(x)_m \qquad (5.10)$$

[27] For a generalization of this result see Polya and Szegö [1], Vol. II, Part 5, *163*, pp. 67 and 255.

In view of (4.22) and (4.25), the coefficients are

$$\rho_\Lambda{}^n(x)_m = \chi_\Lambda(x)_m \frac{1}{n!} \int d(y)_n \chi_\Lambda(y)_n \tilde{\varphi}_{(x)_m}(y)_n \qquad (5.11)$$

By inserting (5.10) in (5.7) and (5.8) we can see[28] that, for all $z > 0$,

$$\rho_\Lambda(x)_m \lesseqgtr z^m \sum_{n=0}^{L} z^n \rho_\Lambda{}^n(x)_m \qquad (5.12)$$

i.e., the right-hand side is an upper bound to $\rho_\Lambda(x)_m$ for even L and a lower bound for odd L: the activity expansion of the correlation functions has the *alternating bound property*.

Taking $m = 1$ in (5.12) and using Eqs. (3.4), (3.6), and (3.7), we see that the Mayer series have the alternating bound property.

We mention without proof[29] that the activity expansions of $\rho(x)_m/\rho_1$ and z/ρ_1 also have the alternating bound property.

Formula (5.12) implies in particular that the function $\rho_\Lambda{}^n(x)_m$ defined by (5.11) has the sign of $(-1)^n$; a stronger result actually holds:

$$(-1)^n \tilde{\varphi}_{(x)_m}(y)_n \geq 0 \qquad (5.13)$$

which is readily obtained from (4.26) by induction on $m + n$. In particular, by (4.24), the Ursell functions have alternating signs[30]:

$$(-1)^{m-1} \varphi(x)_m \geq 0 \qquad (5.14)$$

We now collect a few results concerning the Mayer expansions.

4.5.3 THEOREM.[31] *If* $\Phi \geq 0$, *the Mayer series*

$$\beta p = \sum_{n=1}^{\infty} b_n z^n \qquad \rho_1 = \sum_{n=1}^{\infty} n b_n z^n \qquad (5.15)$$

have the alternating bound property; their coefficients satisfy $(-1)^{n-1} b_n \geq 0$
and

$$\frac{1}{n} \leq \frac{|b_n|}{C(\beta)^{n-1}} \leq \frac{n^{n-2}}{n!} \qquad (5.16)$$

The radius of convergence \mathcal{R} *of the series* (5.15) *satisfies*

$$e^{-1} C(\beta)^{-1} \leq \mathcal{R} \leq C(\beta)^{-1} \qquad (5.17)$$

and the analytic functions βp *and* ρ_1 *defined by* (5.15) *have a singularity at* $z = -\mathcal{R}$.

[28] See Penrose [2].
[29] See Penrose [2].
[30] This result was first derived by Groeneveld [1].
[31] This theorem is due to Groeneveld [1] except for the alternating bound property.

The alternating bound property of (5.15) with $(-1)^{n-1} b_n \geq 0$ has been established above. The upper bound in (5.16) is a special case of (3.15). Omitting some terms in the right-hand side of (4.26) (they all have the same sign) we get

$$|\tilde{\varphi}_{(x_1)}(y)_n| \geq \sum_{j=1}^{n} (1 - \exp[-\beta\Phi(y_j - x_1)])|\tilde{\varphi}_{(y_j)}(y_1, \ldots, \hat{y}_j, \ldots, y_n)|$$

hence

$$\int dx_2 \cdots dx_{n+1} |\varphi(x)_{n+1}| = \int d(y)_n |\tilde{\varphi}_{(x_1)}(y)_n|$$

$$\geq nC(\beta) \int d(y)_{n-1} |\tilde{\varphi}_{(x_1)}(y)_{n-1}|$$

and therefore by Eqs. (3.9) and (4.34)

$$|nb_n| = \frac{1}{(n-1)!} \int dx_2 \cdots dx_n |\varphi(x_2, \ldots, x_n)| \geq \frac{1}{(n-1)!} \cdot (n-1)! C(\beta)^{n-1}$$

which proves the lower bound in (5.16). Finally (5.17) follows from (5.16), and the fact that the coefficients b_n have alternating signs implies that βp and ρ_1 are singular at $z = -\mathcal{R}$.

From (5.9) we get in particular

$$z[1 - C(\beta)\rho_1] \leq \rho_1 \leq z \tag{5.18}$$

and the first inequality gives

$$\rho_1 \geq \frac{z}{1 + C(\beta)z} \tag{5.19}$$

From the fact, discussed in Section 4.3, that no phase transition occurs for $z \in (0, e^{-1} C(\beta)^{-1})$ and from (5.19) it follows that no phase transition occurs (the system remains a gas) for

$$0 < \rho_1 < (1 + e)^{-1} C(\beta)^{-1} \tag{5.20}$$

For particles with hard cores of diameter R and no further interaction, $C(\beta)$ is the volume of a sphere of radius R (excluded volume), so (5.20) implies thus that for sufficiently large dimension v the gas phase extends up to densities higher than the density R^{-v} of a simple cubic lattice.[32] We may thus construct large cubic boxes Λ such that the configuration space of n particles in Λ contains several connected components around simple cubic lattice configurations and one main connected component which is, as far as we can see, a perfectly normal gas.

[32] Because, for large v, $(1 + e)^{-1} [(\pi^{v/2}/\Gamma(v/2 + 1)) \cdot R^v]^{-1} > R^{-v}$.

4.6 QUANTUM SYSTEMS

As has been shown by Ginibre the methods developed in Sections 4.2 and 4.4 for classical systems can be extended to quantum systems.[33] We shall in this section describe without proofs the main features of this extension which is based upon the use of the Wiener integral.

A trajectory ω in \mathbf{R}^l is a function from a real interval $[0, \beta]$ to \mathbf{R}^l. The conditional Wiener measure P_{xy} is a measure on the trajectories[34] such that $\omega(0) = x$, $\omega(\beta) = y$. If the functional f on such trajectories is of the form

$$f(\omega) = F[\omega(t_1), \ldots, \omega(t_n)] \tag{6.1}$$

i.e., if the functional depends only on the points on the trajectory at times t_i with $0 < t_1 < \cdots < t_n < \beta$, we put

$$
P_{xy}(f) = \int dx_1 \cdots dx_n F(x_1, \ldots, x_n) \frac{1}{(\pi t_1)^{l/2}} \exp\left[-\frac{(x_1 - x)^2}{t_1} \right]
$$

$$
\times \frac{1}{[\pi(t_2 - t_1)]^{l/2}} \exp\left[-\frac{(x_2 - x_1)^2}{t_2 - t_1} \right] \times \cdots
$$

$$
\times \frac{1}{[\pi(\beta - t_n)]^{l/2}} \exp\left[-\frac{(y - x_n)^2}{\beta - t_n} \right] \tag{6.2}
$$

By the usual methods of integration theory we can extend this definition to a wide class of functionals f. We can prove that the measure P_{xy} is carried by the set of continuous trajectories from x to y (i.e., the complementary set has measure 0).

We shall use the notation

$$P_{xy}(f) = \int P_{xy}(d\omega) f(\omega) \tag{6.3}$$

and denote by χ_Λ the characteristic function of the set of trajectories which remain in Λ, where Λ is a bounded open connected region in \mathbf{R}^l. Given $\varphi \in L^2(\Lambda)$, let

$$Q\varphi(x) = \int dy \int P_{xy}(d\omega) \exp\left\{ -\int_0^\beta U[\omega(t)]\, dt \right\} \chi_\Lambda(\omega)\varphi(y) \tag{6.4}$$

Under suitable assumptions on the real function U on \mathbf{R}^l, the formula (6.4) defines a bounded operator Q on $L^2(\Lambda)$, and Q is expressed by Kac's formula

$$Q = \exp[-\beta(-\tfrac{1}{4}\Lambda + U)_\Lambda] \tag{6.5}$$

[33] See Ginibre [1].
[34] If we extend \mathbf{R}^l to a compact set $\dot{\mathbf{R}}^l$ by adding a point at infinity and if we introduce the topology of pointwise convergence on the functions ω, the set of all trajectories becomes a compact set Ω (by Tychonov's theorem) and the Wiener measure is a measure on Ω. For a good short introduction to Wiener integrals, see Nelson [2].

where Δ is the Laplace operator in \mathbf{R}^ν and $(-\frac{1}{4}\Delta + U)_\Lambda$ is a self-adjoint extension of the symmetric operator $-\frac{1}{4}\Delta + U$ defined on twice differentiable functions with support in Λ.

We are primarily interested in the case $\mathbf{R}^l = \mathbf{R}^{n\nu}$ where a trajectory $(\omega)_n$ corresponds to the trajectories $\omega_1, \cdots, \omega_n$ of n particles in \mathbf{R}^ν. Suppose that these particles interact through a pair potential Φ and put

$$U(x)_n = \sum_{1 \le i < j \le n} \Phi(x_j - x_i) \tag{6.6}$$

If Λ is a bounded open connected region in \mathbf{R}^ν, we define the Hamiltonian $H_n(\Lambda)$ of n particles in Λ as[35]

$$H_n(\Lambda) = \left(\sum_{i=1}^n -\frac{1}{4}\Delta_i + U \right)_{\Lambda^n} \tag{6.7}$$

The kernel of $\exp[-\beta H_n(\Lambda)]$ is then given by (6.5):

$$\exp[-\beta H_n(\Lambda)]((x)_n, (y)_n)$$

$$= \int P_{(x)_n(y)_n}(d(\omega)_n)\chi_\Lambda(\omega)_n \exp\left\{ -\int_0^\beta dt \sum_{i<j} \Phi(\omega_j(t) - \omega_i(t)) \right\} \tag{6.8}$$

The formula (6.8) gives us access to the quantum equivalent of the correlation functions, namely the *reduced density matrices*. Using for simplicity Maxwell–Boltzmann (M.–B.) statistics, we define the m-particle reduced density matrix as[36]

$$\rho_\Lambda((x)_m; (y)_m) = \Xi^{-1} \sum_{n=0}^\infty \frac{z^{m+n}}{n!} \int d(u)_n Q_{m+n}((x)_m, (u)_n; (y)_m, (u)_n) \tag{6.9}$$

$$\Xi = \sum_{n=0}^\infty \frac{z^n}{n!} \int d(u)_n Q_n((u)_n; (u)_n) \tag{6.10}$$

where we have written $Q_n = \exp[-\beta H_n(\Lambda)]$. Using (6.8) we may rewrite (6.9) and (6.10) as

$$\rho_\Lambda((x)_m; (y)_m) = \Xi^{-1} \sum_{n=0}^\infty \frac{z^{m+n}}{n!} \int d(u)_n \int P_{(x)_m(u)_n;(y)_m(u)_n}(d(\omega)_{m+n})$$

$$\times \chi_\Lambda(\omega)_{m+n} \exp[-U(\omega)_{m+n}] \tag{6.11}$$

$$\Xi = \sum_{n=0}^\infty \frac{z^n}{n!} \int d(u)_n \int P_{(u)_n(u)_n}(d(\omega)_n)\chi_\Lambda(\omega)_n \exp[-U(\omega)_n] \tag{6.12}$$

$$U(\omega)_n = \sum_{1 \le i < j \le n} \Phi(\omega_j - \omega_i), \qquad \Phi(\omega) = \int_0^\beta dt\Phi(\omega(t)) \tag{6.13}$$

[35] We put the mass of the particles equal to 2 to simplify the formulas.
[36] We could rewrite (6.9) and (6.10) using traces; in particular, Ξ is the grand partition function.

Finally (6.11) may be put into the form

$$\rho_\Lambda((x)_m;(y)_n) = \int P_{(x)_m(y)_n}(d(\omega)_m)\rho_\Lambda(\omega)_m \qquad (6.14)$$

where we have written

$$\rho_\Lambda(\omega)_m = \Xi^{-1}\sum_{n=0}^\infty \frac{z^{m+n}}{n!}\int d(u)_n \int P_{(u)_n(u)_n}(d(\omega)_n)\chi_\Lambda(\omega)_{m+n}\exp[-U(\omega)_{m+n}] \qquad (6.15)$$

We notice now that the above formulas are in close correspondence with the formulas defining Ξ and the correlation functions for classical systems. Actually, (6.12) and (6.15) are obtained from (2.4) and (2.5) by replacing points $x \in \mathbf{R}^v$ by trajectories ω, Lebesgue integrals $\int dx$ by Wiener integrals $\int du \int P_{uu}(d\omega)$, and $\beta\Phi(x)$ by $\Phi(\omega)$ given by (6.13). This suggests that we study the functionals $\rho_\Lambda(\omega)_m$ by the same techniques (integral equations or algebraic method) as were used for the correlation functions, and then obtain the reduced density matrices by the integration (6.14).

Before stating the results obtained in this way we indicate the conditions on the pair potential: Φ is continuous (except perhaps at the origin if $v \geq 2^{37}$), stable, and absolutely integrable. A hard core is also allowed: $\Phi(x) = \infty$ for $|x| \leq R$. In this case Φ should be continuous and integrable outside of the hard core. *Under these conditions, if z is sufficiently small and if Λ is a sphere centered at the origin with radius tending to ∞, $\rho_\Lambda((x)_m;(y)_m)$ tends to a limit $\rho((x)_m;(y)_m)$ in the sense of the strong topology of the operators on $L^2(\mathbf{R}^{mv})$. The Mayer series and the virial expansion have a nonvanishing radius of convergence.*

The above results are valid for Maxwell–Boltzmann statistics. For B.–E. or F.–D. statistics, Eqs. (6.9) and (6.10) are replaced by

$$\rho_\Lambda((x)_m;(y)_m) = \Xi^{-1}\sum_{n=0}^\infty \frac{z^{m+n}}{n!}\int d(u)_n \sum_\pi \varepsilon^\pi Q_{m+n}[(x)_m,(u_n);\pi((y)_m,(u)_n)] \qquad (6.16)$$

$$\Xi = \sum_{n=0}^\infty \frac{z^n}{n!}\int d(u)_n \sum_\pi \varepsilon^\pi Q_n((u)_n;\pi(u_n)) \qquad (6.17)$$

where π is a permutation of $m + n$ variables in (6.16), n variables in (6.17), the summation is over the symmetric group; and $\varepsilon^\pi = 1$ for bosons, ε^π is the signature of the permutation π for fermions.

Understandably, the introduction of the new statistics brings in complications. However, *the results stated above remain true for B.–E. and F.–D. statistics provided either $\Phi \geq 0$ or Φ has a hard core and condition (2.14) of Chapter. 3 is satisfied.*[38] For the proofs we must refer the reader to the original papers.

[37] It is sufficient to require that Φ be continuous except on a set of (Newtonian) capacity 0; a point has capacity 0 in \mathbf{R}^v if $v \geq 2$.

[38] It can also be proved that a certain cluster property holds for the reduced density matrices.

4.7 CONSTRUCTION OF A HILBERT SPACE

The sequence of the correlation functions possesses a remarkable positivity property which we discuss briefly in this section. The approach to statistical mechanics implicit in this discussion will be developed in Chapters 6 and 7.

Let \mathscr{B} be the space of sequences $\psi = (\psi(x)_n)_{n \geq 0}$ where $\psi(x)_n$ is a bounded complex Borel[39] function with compact support in \mathbf{R}^{nv} and $\psi(x)_n = 0$ for large n (the $n = 0$ component of ψ is a complex number ψ_0).

Let $(\rho(x)_n)_{n \geq 0}$ be the sequence of correlation functions defined by (1.3) for a bounded region Λ or a suitable thermodynamic limit of these; we put $\rho_0 = 1$. Remember that $\rho(x)_n$ is the probability density for finding n *different* particles at prescribed positions x_1, \ldots, x_n. Let $\hat{\rho}(x)_n$ be the probability density[40] for finding n particles, *not necessarily different*, at x_1, \ldots, x_n. Clearly $\hat{\rho}(x)_n$ is equal to $\rho(x)_n$ plus terms containing a product of $\delta(x_j - x_i)$ multiplied by some $\rho(x_{i_1}, \ldots, x_{i_m})$, $m < n$.[41] Explicitly, we readily check that

$$\int d(x)_m \hat{\rho}(x)_m \varphi(x)_m$$

$$= \Xi^{-1} \sum_{s=0}^{\infty} \frac{z^s}{s!} \int_{\Lambda^s} d(x)_s \exp[-\beta U(x)_s] \sum_{i_1=1}^{s} \cdots \sum_{i_m=1}^{s} \varphi(x_{i_1}, \ldots, x_{i_m}) \qquad (7.1)$$

We introduce a bilinear form[42] on \mathscr{B} by

$$(\psi, \varphi) = \sum_{m,n} \int d(x)_m \int d(y)_n \varphi(x)_m \psi(y)_n^* \hat{\rho}(x_1, \ldots, x_m, y_1, \ldots, y_n) \qquad (7.2)$$

Using (7.1) we obtain

$$(\psi, \varphi) = \Xi^{-1} \sum_{s=0}^{\infty} \frac{z^s}{s!} \int_{\Lambda^s} d(x)_s \exp[-\beta U(x)_s]$$

$$\times \left[\sum_m \sum_{i_1=1}^{s} \cdots \sum_{i_m=1}^{s} \varphi(x_{i_1}, \ldots, x_{i_m}) \right] \left[\sum_n \sum_{j_1=1}^{s} \cdots \sum_{j_n=1}^{s} \psi(x_{j_1}, \ldots, x_{j_n}) \right]^*$$

$$(7.3)$$

and it follows that $(\varphi, \varphi) \geq 0$: *the form* (7.2) *is positive semidefinite*.[43]

[39] The Borel functions include the continuous functions and their pointwise limits, in particular the characteristic functions of open sets.

[40] This "probability density" is singular since it contains δ measures.

[41] For example, $\hat{\rho}(x_1, x_2) = \rho(x_1, x_2) + \delta(x_1 - x_2)\rho(x_1)$.

[42] Actually, a *sesquilinear* form: linear in the second argument and antilinear in the first.

[43] This result remains true for an infinite system provided a suitable thermodynamic limit exists. Notice also that the grand canonical average could be replaced by a more general measure on configurations.

To exploit this property we introduce some further notation. If $\varphi \in \mathcal{B}$ we define $\varphi^* \in \mathcal{B}$ by[44]

$$\varphi^*(x)_m = \varphi(x_m, \ldots, x_1)^* \tag{7.4}$$

and if $a \in \mathbf{R}^\nu$ we define $\tau_a \varphi \in \mathcal{B}$ by

$$\tau_a\varphi(x)_m = \varphi(x_1 - a, \ldots, x_m - a) \tag{7.5}$$

If $\varphi, \psi \in \mathcal{B}$ we define a product $\varphi \circledast \psi$ by

$$(\varphi \circledast \psi)(x)_m = \sum_{i=0}^{m} \varphi(x_1, \ldots, x_i)\psi(x_{i+1}, \ldots, x_m) \tag{7.6}$$

Let us write

$$\hat{\rho}(\varphi) = \sum_{n=0}^{\infty} \int d(x)_n \hat{\rho}(x)_n \varphi(x)_n \tag{7.7}$$

then (7.2) becomes

$$(\psi, \varphi) = \hat{\rho}(\psi^* \circledast \varphi) \tag{7.8}$$

We suppose from now on that the $\rho(x)_n$ are infinite volume correlation functions and invariant under translations[45]:

$$\rho(x_1 + a, \ldots, x_n + a) = \rho(x_1, \ldots, x_n) \tag{7.9}$$

This can be expressed by

$$\hat{\rho}(\tau_a\varphi) = \hat{\rho}(\varphi) \tag{7.10}$$

The properties of the $\hat{\rho}(x)_n$ expressed by $(\varphi, \varphi) \geq 0$ and (7.10) are strongly reminiscent of the positivity and invariance properties of the vacuum expectation values of products of fields in relativistic quantum field theory. The *Wightman construction*[46] based upon these properties may also be performed here; we first sketch the argument and then state the result.

If we take the quotient of \mathcal{B} by the subspace $\mathcal{N} = \{\varphi : (\varphi, \varphi) = 0\}$, the scalar product (7.8) becomes positive definite on \mathcal{B}/\mathcal{N}, which may be completed to a Hilbert space \mathfrak{H}. We let $\varphi \to \alpha(\varphi)$ be the canonical mapping from \mathcal{B} to \mathfrak{H}. Let $\Omega = \alpha(\mathbf{1}) \in \mathfrak{H}$ where $\mathbf{1}$ is the identity of \mathcal{B} for the \circledast product.[47]

If $\varphi \in \mathcal{B}$, a linear operator $A(\varphi)$ is defined on the set $\mathcal{D} = \alpha(\mathcal{B})$ (which is dense in \mathfrak{H}) by

$$A(\varphi)\alpha(\psi) = \alpha(\varphi \circledast \psi) \tag{7.11}$$

[44] That is, $\varphi(x)_m$ is replaced by its complex conjugate and the order of the arguments is reversed. Actually this reversal will play no role in the commutative situation discussed here.
[45] This applies in particular to the $\rho(x)_n$ obtained in Theorem 4.2.3.
[46] The Wightman [1] construction is in fact a variant of the Gel'fand–Segal construction which we shall consider in A.3.5.
[47] That is, $\mathbf{1}_0 = 1$, $\mathbf{1}(x)_n = 0$ for $n > 0$.

[Notice that if $\alpha(\psi) = 0$, then $\alpha(\varphi \circledast \psi) = 0$ since for all $\omega \in \mathscr{B}$:

$$|(\alpha(\omega), \alpha(\varphi \circledast \psi))| = |\hat{\rho}_1(\omega^* \circledast \varphi \circledast \psi)| = |(\varphi^* \circledast \omega, \psi)|$$
$$\leq (\varphi^* \circledast \omega, \varphi^* \circledast \omega)^{1/2}(\psi, \psi)^{1/2} = 0.]$$

A unitary representation U of the additive group \mathbf{R}^ν in \mathfrak{H} is defined by

$$U(a)\alpha(\psi) = \alpha(\tau_a\psi) \tag{7.12}$$

because $(\tau_a\psi, \tau_a\psi) = \hat{\rho}(\tau_a(\psi^* \circledast \psi)) = \hat{\rho}(\psi^* \circledast \psi) = (\psi, \psi)$.
 The following theorem is now easily proved.

4.7.1 THEOREM. *There exist* (a) *a Hilbert space* \mathfrak{H}, (b) *a linear manifold* \mathscr{D} *dense in* \mathfrak{H}, (c) *a vector* $\Omega \in \mathscr{D}$ *such that* $\|\Omega\| = 1$, (d) *a linear mapping* A *from the elements* φ *of* \mathscr{B} *to commuting linear operators* $A(\varphi): \mathscr{D} \to \mathscr{D}$ *such that* $A(\varphi \circledast \psi) = A(\varphi)A(\psi)$, *and* (e) *a unitary representation* U *of* \mathbf{R}^ν *in* \mathfrak{H} *such that*

$$U(a)\Omega = \Omega \qquad U(a)A(\varphi)U(-a) = A(\tau_a\varphi) \tag{7.13}$$

The following identity is satisfied

$$(\Omega, A(\varphi)\Omega) = \hat{\rho}(\varphi) \tag{7.14}$$

and determines $(\mathfrak{H}, \Omega, A, U)$ *up to unitary equivalence*[48] *provided we assume that* $\mathscr{D} = \{A(\varphi)\Omega: \varphi \in \mathscr{B}\}$.
 In field theory, Ω would be called the *vacuum* and (7.14) expresses an ensemble average[49] as a vacuum expectation value. If $\psi \in \mathscr{B}$ is such that $\psi(x)_n = 0$ for $n \neq 1$ we may write the mapping $\psi \to A(\psi)$ as

$$A(\psi) = \int dx_1 A(x_1)\psi(x_1) \tag{7.15}$$

More generally, we can write

$$A(\varphi) = \varphi_0 1 + \sum_{n=1}^{\infty} \int dx_1 \cdots dx_n \varphi(x_1, \ldots, x_n)A(x_1) \cdots A(x_n) \tag{7.16}$$

and

$$\hat{\rho}(x)_n = (\Omega, A(x_1) \cdots A(x_n)\Omega) \tag{7.17}$$

The "field" $A(x)$ is an operator-valued measure. To understand its physical meaning let χ be such that $\chi(x)_m = 0$ for $m \neq 1$ and $\chi(x)_1$ is the characteristic function of a set M. Insertion in (7.1) shows that $(\Omega, A(\chi)^n\Omega)$ is the nth moment of the number of particles in M; therefore, $A(\chi)$ may be interpreted as the

[48] That is, if $\mathfrak{H}', \Omega', A', U'$ satisfy the theorem, there exists an isometry V of \mathfrak{H} onto \mathfrak{H}' such that $V\Omega = \Omega'$, $VAV^{-1} = A'$, $VUV^{-1} = U'$.
[49] More precisely the infinite volume limit of an ensemble average.

operator number of particles in M. It is then tempting, if f_m is a bounded continuous function such that $f_m(m) = 1$ and $f_m(n) = 0$ for integer $n \neq m$, to write $(\Omega, f_m(A(\chi))\Omega)$ for the probability of finding exactly m particles in M. Unfortunately $A(\chi)$ has not been defined as self-adjoint operator and $f_m(A(\chi))$ does not in general make sense. It is however sufficient to know that $A(\chi)$ is essentially self-adjoint, i.e., has exactly one self-adjoint extension. The following result is thus relevant.

4.7.2 THEOREM (Nelson).[50] *Let A be a symmetric operator with domain $\mathscr{D} \subset \mathfrak{H}$. We say that Ψ is an analytic vector for A if Ψ is in the domain of A^n for all integers $n \geq 0$ and the following power series in z has nonvanishing convergence radius*

$$\sum_{n=0}^{\infty} \frac{z^n}{n!} \| A^n \Psi \| \qquad (7.18)$$

If the set of analytic vectors is dense in \mathfrak{H}, A is essentially self-adjoint.
 If the series

$$\sum_{n=0}^{\infty} \frac{z^n}{n!} (\Omega, A(\chi)^n\Omega) = \sum_{n=0}^{\infty} \frac{z^n}{n!} \int d(x)_n \hat{\rho}(x)_n \chi(x_1) \cdots \chi(x_n) \qquad (7.19)$$

has a nonvanishing radius of convergence, the same is true of

$$\sum_{n=0}^{\infty} \frac{z^n}{n!} (\Psi, A(\chi)^n\Psi) = \sum_{n=0}^{\infty} \frac{z^n}{n!} (\alpha(\psi^*)\Psi, A(\chi)^n\Omega)$$

for all $\Psi = \alpha(\psi) \in \mathscr{D}$ (use Schwarz's inequality), and it follows readily that Ψ is an analytic vector. It can be shown that (7.19) converges when there exists a constant $\xi > 0$ such that for all n

$$\sup_{(x)_n} \rho(x)_n \leq \xi^n \qquad (7.20)$$

Under this condition $A(\chi)$ is thus essentially self-adjoint.[51] Notice that (7.20) is satisfied by the correlation functions obtained in Theorem 4.2.3 (see also Exercise 4.D).
 There are cases where $A(\chi)$ is not essentially self-adjoint and the probability for finding exactly m particles in M is not determined by the correlation functions; the description provided by correlation functions is then incomplete (see Exercise 4.E). Notice also that the existence of the correlation functions implies the existence of all moments of the distribution of the number of particles[52] in M, and this is a severe restriction on the probabilistic situation

[50] See Nelson [1], Lemma 5.1. For an application in field theory analogous to that made here, see Borchers and Zimmermann [1].
 [51] If (7.20) holds, all $A(\varphi)$ with $\varphi = \varphi^*$ are essentially self-adjoint and their spectral projections commute.
 [52] In the field theoretical description this corresponds to the fact that Ω is in the domain of $A(\chi)^n$ for all n.

described. In view of these remarks it appears interesting to describe the state of an infinite system by an expectation value functional defined on bounded quantities, rather than on the unbounded field A. This amounts to replacing the correlation functions by a state on a suitable C^*-algebra as discussed in Chapter 7.

We mention that the Wightman construction may be performed with the reduced density matrices instead of the correlation functions and that the "field operators" thus obtained yield a representation of the canonical commutation relations (for bosons), or anticommutation relations (for fermions). This again will be discussed in Chapter 7.

BIBLIOGRAPHICAL NOTE

The first nontrivial result on the convergence of the activity expansions was published by Groeneveld [1] in 1962 and deals with positive pair potentials (our Theorem 3.1.3). Results valid for more general pair potentials by Ruelle [3] and Penrose [1] followed. Actually, Penrose had been in possession of his main results[53] for several years. Thinking, with some appearance of reason, that there would be little interest among physicists in bounds on radii of convergence, he refrained from publishing until Groeneveld's results became known.

We have followed Ruelle [3] and Gallavotti–Miracle [2] in Section 4.2, Lebowitz and Penrose [1] (in part) in Section 4.3, Ruelle [4] in Section 4.4, Lieb [1], Penrose [2], and Groeneveld [1] in Section 4.5, Ginibre [1] in Section 4.6. The techniques of axiomatic quantum field theory applied to statistical mechanics in Section 4.7 are standard.

EXERCISES

4.A. Check that for a pair interaction

$$\int d(x)_{m+n} \exp[-\beta U(x)_{m+n}]$$

$$= \int d(x)_{m+n} \exp\left[-\beta \frac{2mn}{(m+n)(m+n-1)} U(x)_{m+n}\right]$$

$$\times \prod \left[\exp\left(-\frac{\beta}{(m+n)!} U(x_{i_1}, \ldots, x_{i_m})\right)\right.$$

$$\left. \times \exp\left(-\frac{\beta}{(m+n)!} U(x_{i_{m+1}}, \ldots, x_{i_{m+n}})\right)\right]$$

[53] Penrose had obtained bounds on the radius of convergence of the Mayer series which were independent of the size of the box Λ containing the system. He used the Mayer–Montroll equations and his results were valid for hard core potentials (unpublished).

where the product is over all permutations (i_1, \ldots, i_{m+n}) of $(1, \ldots, m+n)$. Applying stability and the Hölder inequality, this is

$$\leq \exp\left[\frac{2mn}{m+n-1}\beta B\right] \cdot \left[\int d(x)_m \exp[-\beta U(x)_m]\right]$$

$$\times \left[\int d(x)_n \exp[-\beta U(x)_n]\right]$$

From this check that the pressure satisfies, as a function of the activity, the following inequalities, valid for all $z_1 > 0$, $z_2 > 0$,

$$p(z_1 + z_2) \leq p(z_1 e^{2\beta B}) + p(z_2)$$

$$p(z_1 + z_2) \leq p(z_1 e^{\beta B}) + p(z_2 e^{\beta B})$$

4.B. The coefficients of the Mayer series are given [see Eqs. (3.9), (4.34), and (4.9)] by

$$b_n = \frac{1}{n!}\int dx_2 \cdots dx_n \varphi(x_1, \ldots, x_n)$$

$$= \frac{1}{n!}\int dx_2 \ldots dx_n \sum_\gamma \prod_{i<j}^\gamma (\exp[-\beta\Phi(x_j - x_i)] - 1)$$

We may think of estimating $|b_n|$ by

$$|b_n| \leq \frac{1}{n!}\sum_\gamma \int dx_2 \cdots dx_n \prod_{i<j}^\gamma |\exp[-\beta\Phi(x_j - x_i)] - 1| \qquad (*)$$

Show that this amounts to replacing the potential Φ by $\tilde{\Phi} = -(1/\beta) \times \log(|e^{-\beta\Phi} - 1| + 1)$ which is catastrophic if $\Phi \neq 0$, and that therefore the estimate $(*)$ does not permit us to conclude that the Mayer series has a nonvanishing convergence radius.

4.C. Suppose that Φ has finite range: $\Phi(x) = 0$ if $|x| \geq L$. Show that $\varphi(x)_m = 0$ if $|x_2 - x_1| \geq (m-1)L$ and that the cluster function $\omega(x)_2 = \rho(x)_2 - \rho_1^2$ satisfies

$$\int_{|x_2 - x_1| > mL} dx_2 |\omega(x_1, x_2)| \leq |z| e^{-2\beta B} \sum_{n=m}^\infty (n+1)t^{n+1}$$

$$= |z| e^{2\beta B}[m(1-t)^{-1} + (1-t)^{-2}]t^{m+1}$$

where $t = |z| e^{2\beta B + 1}C(\beta) < 1$. This integral thus tends exponentially to zero as $\lambda = mL \to \infty$. Use this result to prove the analyticity of the Fourier transform of $\omega(x)_2$ (considered as a function of $x_2 - x_1$).[54]

[54] For another result on the asymptotic behavior of $\omega(x)_2$ see Groeneveld [2], Chapter III, Theorem IV.

4.D. If the interaction $(\Phi^k)_{k \geq 2}$ is positive (i.e., $\Phi^k \geq 0$ for all k) show that $\rho_\Lambda(x)_m$ defined by (1.3) satisfies

$$\rho_\Lambda(x)_m \leq z^m \exp[-\beta U(x)_m] \leq z^m$$

If Φ is a pair potential with hard core satisfying Eq. (2.14) of Chapter 3 show that

$$\rho_\Lambda(x)_m \leq z^m e^{2m\beta B} \exp[-\beta U(x)_m]$$

$$\rho_\Lambda(x)_m \leq z^m e^{2m\beta B}$$

4.E. Let ρ_α be the sequence of infinite volume correlation functions corresponding to $z = \alpha$, $\Phi = 0$; show that $\rho_\alpha(x)_n = \alpha^n$. We define ρ as a superposition[55] of ρ_α using a measure $\mu \geq 0$ of total mass 1 on the positive real axis:

$$\rho(x)_n = \int d\mu(\alpha)\rho_\alpha(x)_n = \int \alpha^n \, d\mu(\alpha)$$

The $\rho(x)_n$ are thus just the moments of μ. From the fact that there exist different measures $\mu, \mu' \geq 0$ with the same moments of all orders,[56] deduce that the description provided by the correlation functions $\rho(x)_n$ is incomplete.

4.F. Let Φ be a stable regular pair potential and $\varphi \in \mathscr{A}_+$ be the sequence of the Ursell functions. If $|z| < e^{-2\beta B - 1}C(\beta)^{-1}$, $\varphi^z \in \mathscr{A}_+$ is defined by (4.35) or (4.39); verify that for sufficiently small z_0, z we have

$$\varphi^{z_0 + z}(X) = \sum_{n=0}^{\infty} \frac{z^n}{n!} \int d(y)_n D_X \varphi^{z_0}(y)_n$$

4.G. Let Φ be a stable regular pair potential and $|z| < e^{-2\beta B - 1}C(\beta)^{-1}$. Show that the infinite volume correlation functions $\rho(x)_m$ satisfy the Mayer–Montroll equations (2.50) with the factor $\chi_\Lambda(x)_m$ removed in the right-hand side

$$\rho(x)_m = z^m \exp[-\beta U(x)_m] \cdot \sigma(x)_m$$

$$\sigma(x)_m = 1 + \sum_{n=1}^{\infty} \frac{1}{n!} \int d(y)_n K((x)_m, (y)_n)\rho(y)_n$$

Show that $\sigma(x)_m$ is continuous on \mathbf{R}^{nv}. Show that if Φ has finite range and $e^{-\beta\Phi}$ is differentiable of class C^k, then $\sigma(x)_m$ and $\rho(x)_m$ are differentiable of class C^k. (This result extends to certain potentials with infinite range.)

[55] We do not claim that the $\rho(x)_m$ are the correlation functions corresponding to any pair potential Φ, but they have the probabilistic interpretation of correlation functions.

[56] See, e.g., Shohat and Tamarkin [1].

4.H. With the assumptions and notations of 4.G, show that for sufficiently small $|z|$,

$$\sigma(x)_m = \exp\left[\sum_n \frac{1}{n!} \int d(y)_n K((x)_m, (y)_n)\omega(y)_n\right]$$

where the $\omega(y)_n$ are the cluster functions of the infinite volume correlation functions.

CHAPTER 5

THE PROBLEM OF PHASE TRANSITIONS

In this chapter we shall adopt the traditional point of view that a phase transition manifests itself as a singularity of a thermodynamic function, which is otherwise real analytic. From this point of view a theory of phase transitions would consist in proving that the thermodynamic functions are piecewise analytic and in discussing the nature of the singularities which occur. While no theory of phase transitions in this sense exists at present, some analyticity properties of thermodynamic functions are known. In particular we have seen in Chapter 4 that, for classical and quantum systems interacting through a suitable pair potential, no phase transition occurs at small density. In the present chapter we shall discuss various other results, pertaining mostly to classical lattice systems.

5.1 THE THEOREM OF LEE AND YANG

In 1952, Yang and Lee [1] proposed a theory of phase transitions based on the idea that certain regions of the complex activity plane may remain free of zeros of the grand partition function when $\Lambda \to \infty$ (see Exercise 3.G). We have seen in Theorems 4.2.3 and 4.2.7 that this situation can arise. Lee and Yang [1] have actually exhibited a class of lattice systems for which, as we shall now see, all the zeros lie on the circle $z = 1$. We establish first the relevant mathematical results without reference to physics.

5.1.1 PROPOSITION. *Let* $(A_{ij})_{i \neq j}$ *be a family of real numbers such that* $-1 \leq A_{ij} \leq 1, A_{ij} = A_{ji}$ *for* $i = 1, \ldots, n; j = 1, \ldots, n$. *We define a polynomial* \mathscr{P}_n *in n variables by*

$$\mathscr{P}_n(z_1, \ldots, z_n) = \sum_S z^S \left(\prod_{i \in S} \prod_{j \in S'} A_{ij} \right) \tag{1.1}$$

where the summation is over all subsets $S = \{i_1, \ldots, i_s\}$ *of* $\{1, \ldots, n\}$,[1]
$z^S = z_{i_1} \cdots z_{i_s}$ *and* $S' = \{j_1, \ldots, j_{n-s}\}$ *is the complement of* S *in* $\{1, \ldots, n\}$.
Then $\mathscr{P}_n(z_1, \ldots, z_n) = 0$ *and* $|z_1| \geq 1, \ldots, |z_{n-1}| \geq 1$ *imply* $|z_n| \leq 1$.

We shall assume that the A_{ij}'s are different from 0 and ± 1. If the proposition is proved in that case it holds in general by continuity.

We have

$$\mathscr{P}_1(z_1) = 1 + z_1 \tag{1.2}$$

Clearly $\mathscr{P}_1(z_1) = 0$ implies $|z_1| = 1$. This proves the proposition for $n = 1$. We proceed now by induction on n for $n \geq 2$.

We assume that $|z_1| \geq 1, \ldots, |z_{n-2}| \geq 1$. Let us define $\Delta_n = \{1, \ldots, n\}$; then we may write

$$\mathscr{P}(z_1, \ldots, z_n) = \sum_{S \subset \Delta_n} z^S \left(\prod_{i \in S} \prod_{j \in \Delta_n \setminus S} A_{ij} \right)$$

$$= A + B z_{n-1} + C z_n + D z_{n-1} z_n \tag{1.3}$$

where

$$D = \sum_{S \subset \Delta_{n-2}} z^S \left(\prod_{i \in S \cup \{n-1, n\}} \prod_{j \in \Delta_{n-2} \setminus S} A_{ij} \right)$$

Defining $\zeta_i = (A_{n-1,i} A_{n,i})^{-1} z_i$, we get

$$D = \left[\prod_{i=1}^{n-2} A_{n-1,i} A_{n,i} \right] \mathscr{P}_{n-2}(\zeta_1, \ldots, \zeta_{n-2})$$

and since $|\zeta_1| > 1, \ldots, |\zeta_{n-2}| > 1$, the induction assumption shows that $D \neq 0$.

The equation $\mathscr{P}(z_1, \ldots, z_n) = 0$ determines a fractional linear transformation $z_n \to z_{n-1}$:

$$z_{n-1} = -\frac{A + C z_n}{B + D z_n} \tag{1.4}$$

which maps ∞ to $z_{n-1} = -C/D$ so that

$$0 = C + D z_{n-1} \equiv \sum_{S \subset \Delta_{n-1}} z^S \left(\prod_{i \in S \cup \{n\}} \prod_{j \in \Delta_{n-1} \setminus S} A_{ij} \right) \tag{1.5}$$

If we define $\zeta_i = (A_{ni})^{-1} z_i$, (1.5) becomes

$$\left[\prod_{i=1}^{n-1} A_{ni} \right] \mathscr{P}_{n-1}(\zeta_1, \ldots, \zeta_{n-1}) - 0 \tag{1.6}$$

and since $|\zeta_1| > 1, \ldots, |\zeta_{n-2}| > 1$, we have by the induction assumption $|z_{n-1}| < |\zeta_{n-1}| \leq 1$.

[1] When S or S' is the empty set, the corresponding coefficient in (1.1) is 1.

Suppose now that the proposition to be proved does not hold; then there exist z'_{n-1}, z'_n such that $\mathscr{P}_n(z_1, \ldots, z_{n-2}, z'_{n-1}, z'_n) = 0$ and $|z'_{n-1}| \geq 1, |z'_n| > 1$. The transformation (1.4) maps z'_n, ∞ to z'_{n-1}, z_{n-1} such that $|z'_{n-1}| \geq 1$, $|z_{n-1}| < 1$; therefore we can find z''_n such that $|z''_n| > 1$ and its image z''_{n-1} satisfies $|z''_{n-1}| = 1$. Repeating this argument with the index $n-1$ replaced by $i = 1, \ldots, n-2$, we obtain $\tilde{z}_1, \ldots, \tilde{z}_n$ such that

$$\mathscr{P}_n(\tilde{z}_1, \ldots, \tilde{z}_{n-1}, \tilde{z}_n) = 0 \tag{1.7}$$

with $|\tilde{z}_1| = 1, \ldots, |\tilde{z}_{n-1}| = 1, |\tilde{z}_n| > 1$.

From $A_{ij} = A_{ji}$ we obtain the identity

$$\mathscr{P}_n(z_1^{-1}, \ldots, z_n^{-1}) = z_1^{-1} \cdots z_n^{-1} \mathscr{P}_n(z_1, \ldots, z_n) \tag{1.8}$$

and using $\tilde{z}_1^* = \tilde{z}_1^{-1}, \ldots, \tilde{z}_{n-1}^* = \tilde{z}_{n-1}^{-1}$, we get

$$\mathscr{P}_n(\tilde{z}_1, \ldots, \tilde{z}_{n-1}, (\tilde{z}_n^*)^{-1}) = \mathscr{P}_n((\tilde{z}_1^*)^{-1}, \ldots, (\tilde{z}_n^*)^{-1})$$

$$= (\tilde{z}_1^*)^{-1} \cdots (\tilde{z}_n^*)^{-1} \cdot \mathscr{P}_n(\tilde{z}_1^*, \ldots, \tilde{z}_n^*)$$

$$= (\tilde{z}_1^*)^{-1} \cdots (\tilde{z}_n^*)^{-1} \cdot \mathscr{P}_n(\tilde{z}_1, \ldots, \tilde{z}_n)^* = 0$$

Comparison with (1.7) yields[2] $\tilde{z}_n = (\tilde{z}_n^*)^{-1}$ in contradiction with $|\tilde{z}_n| > 1$. This concludes the proof.

5.1.2 THEOREM (Lee–Yang).[3] *With the notation and assumptions of Proposition 5.1.1 we define a polynomial \mathscr{P}^n of degree n in z by*

$$\mathscr{P}^n(z) = \mathscr{P}_n(z, \ldots, z) = \sum_S z^{N(S)} \left(\prod_{i \in S} \prod_{j \in S'} A_{ij} \right) \tag{1.9}$$

where $N(S)$ is the number of elements in S. Then the zeros of \mathscr{P}^n all lie on the circle $\{z : |z| = 1\}$.

This results immediately from Proposition 5.1.1.

Consider now a lattice gas with pair interaction so that the potential energy of a configuration $X = \{x_1, \ldots, x_m\}$ of occupied sites is given by

$$U(X) = N(X)\Phi^1 + \sum_{\{x,y\} \subset X} \Phi^2(\{x, y\}) \tag{1.10}$$

In Section 2.4.3 we have associated with this lattice gas a spin system such that, if the spins at x_1, \ldots, x_m are "up" and the others "down" in the region Λ, the corresponding potential energy is

$$U'_\Lambda(X) = N(X)(\Phi^1 + C_{\Phi^2}) - \tfrac{1}{2} \sum_{x \in X} \sum_{y \in \Lambda \setminus X} \Phi^2(\{x, y\}) \tag{1.11}$$

[2] The coefficient of z_n in \mathscr{P}_n is the right-hand side of (1.5) or the left-hand side of (1.6) and is not equal to zero if $|z_1| \geq 1, \ldots, |z_{n-1}| \geq 1$ by the induction assumption.

[3] See Lee and Yang [1].

where

$$C_{\Phi^2} = \tfrac{1}{2} \sum_{x \neq 0} \Phi^2(\{0, x\}) \tag{1.12}$$

We have shown that[4]

$$P(\Phi) = \lim_{\Lambda \to \infty} N(\Lambda)^{-1} \log \sum_{X \subset \Lambda} \exp\left[-N(X)\Phi^1 - \sum_{\{x,y\} \subset X} \Phi^2(\{x, y\}) \right]$$

$$= \lim_{\Lambda \to \infty} N(\Lambda)^{-1} \log \sum_{X \subset \Lambda} \exp\left[-N(X)(\Phi^1 + C_{\Phi^2}) \right.$$

$$\left. + \tfrac{1}{2} \sum_{x \in X} \sum_{y \in \Lambda \setminus X} \Phi^2(\{x, y\}) \right] \tag{1.13}$$

Let us put $z = \exp[-(\Phi^1 + C_{\Phi^2})]$, $A_{xy} = \exp[\tfrac{1}{2}\Phi^2(\{x, y\})]$; then

$$\sum_{X \subset \Lambda} \exp\left[-N(X)(\Phi^1 + C_{\Phi^2}) + \tfrac{1}{2} \sum_{x \in X} \sum_{y \in \Lambda \setminus X} \Phi^2(\{x, y\}) \right]$$

$$= \sum_{X \subset \Lambda} z^{N(X)} \left(\prod_{x \in X} \prod_{y \in \Lambda \setminus X} A_{xy} \right) = \mathscr{P}^{N(\Lambda)}(z)$$

so that

$$P(\Phi) = \lim_{\Lambda \to \infty} N(\Lambda)^{-1} \log \mathscr{P}^{N(\Lambda)}(z) \qquad z > 0 \tag{1.14}$$

If $\Phi^2 \leq 0$, then $0 < A_{xy} \leq 1$ and Theorem 5.1.2 applies, showing that the zeros of $\mathscr{P}^{N(\Lambda)}(z)$ lie on the circle $|z| = 1$. Therefore, if $|z| < 1$, $[\mathscr{P}^{N(\Lambda)}(z)]^{1/N(\Lambda)}$ is analytic (take the determination which is real for $z > 0$) and satisfies

$$[\mathscr{P}^{N(\Lambda)}(z)]^{1/N(\Lambda)} \leq [\mathscr{P}^{N(\Lambda)}(1)]^{1/N(\Lambda)} \tag{1.15}$$

By reference to their logarithms, we see that the functions $[\mathscr{P}^{N(\Lambda)}(z)]^{1/N(\Lambda)}$ are uniformly bounded for $|z| < 1$ and converge if $z > 0$ [use (1.15) and (1.14)]. The theorem of Vitali implies then that they converge uniformly on any disk $\{z : |z| \leq \alpha\}$ with $\alpha < 1$. Let $\lambda(z)$ be their limit for $z < 1$; λ is analytic and has no zero (if it had one, a little circle around it should, for large Λ, also contain a zero of $[\mathscr{P}^{N(\Lambda)}(z)]^{1/N(\Lambda)}$). This shows that $P(\Phi)$ may be continued from the interval $0 < z < 1$ to a function $p(z) = \log \lambda(z)$ analytic if $|z| < 1$. From the identity

$$\mathscr{P}^{N(\Lambda)}(z^{-1}) - z^{-N(\Lambda)}\mathscr{P}^{N(\Lambda)}(z) \tag{1.16}$$

which follows from (1.8) we see that $P(\Phi)$ may also be continued from the interval $1 < z < +\infty$ to $p(z^{-1}) + \log z$. We may summarize these results as follows.

[4] Assuming $\|\Phi^2\| = \tfrac{1}{2}\sum_{x \neq 0}|\Phi^2(\{0, x\})| < +\infty$.

5.1.3 THEOREM. *If a lattice gas has a pair interaction (1.10) with fixed negative pair potential $\Phi^2 \leq 0$, $\|\Phi^2\| < +\infty$, and if we put*

$$z = \exp\left[-\Phi^1 - \tfrac{1}{2} \sum_{x \neq 0} \Phi^2(\{0, x\}) \right] \tag{1.17}$$

then the thermodynamic function[5] defined by

$$P(\Phi) \quad \lim_{\Lambda \to \infty} N(\Lambda)^{-1} \log \sum_{X \subset \Lambda} \exp[-U(X)] \tag{1.18}$$

extends from the interval $0 < z < 1$ to a function $p(z)$ analytic in $\{z : |z| < 1\}$, and from the interval $1 < z < +\infty$ to $p(z^{-1}) + \log z$. Therefore a phase transition can occur at most for $z = 1$.[6]

We shall in the following sections exhibit cases where a phase transition does actually take place at $z = 1$ and other cases where no phase transition occurs.

5.2 ABSENCE OF PHASE TRANSITIONS AT HIGH TEMPERATURE

In this section we show under rather mild conditions on the potential that classical lattice systems have no phase transition at high temperature β^{-1}.[7]

5.2.1 THEOREM. *Let the fixed potential Φ satisfy $\Phi(X) = 0$ if $N(X) = 1$ or $N(X) > k$ and $\|\Phi\| < +\infty$. We define*

$$p_\Phi(z, \beta) = \beta^{-1} \lim_{\Lambda \to \infty} N(\Lambda)^{-1} \log \sum_{X \subset \Lambda} z^{N(X)} \exp[-\beta U(X)] \tag{2.1}$$

Then, if β is sufficiently small, $p_\Phi(z, \beta)$ is a real analytic function of $z \geq 0$.[8]

From Theorem 4.2.7 and the same arguments as in Section 4.3 we find that $p_\Phi(z, \beta)$ is analytic in the circle

$$\left| \frac{z\, e^{\beta(D-C)}}{1 + z\, e^{\beta(D-C)}} \right| \cdot [2\exp(e^{\beta D} - 1) - 1] < 1 \tag{2.2}$$

of the complex activity plane.

We apply now the symmetry property 2.4.4. Putting $\Phi'(X) = \mathscr{L}\Phi(X)$ if $N(X) > 1$, $\Phi(X) = 0$ for $N(X) = 1$, Eq. (4.32) of Chapter 2 becomes

$$p_{\Phi'}(z^{-1} \exp(\beta C), \beta) = p_\Phi(z, \beta) - \beta^{-1} \log z + C_\Phi \tag{2.3}$$

[5] Notice that a factor β has been absorbed in Φ^1, Φ^2 and that z is the activity up to the fixed factor $\exp[-\tfrac{1}{2}\Sigma_{x \neq 0}\Phi^2(\{0, x\})]$.

[6] We disregard here the dependence of p on β.

[7] This result was first obtained by Dobrushin (private communication) for pair interactions; then generalized by Gallavotti, Miracle, and Robinson [1].

[8] It is also real analytic in β and Φ.

Let C', D' be the quantities corresponding to C, D for Φ' [see Eq. (2.70) of Chapter 4], then

$$
\begin{aligned}
C' = \sum_{X \ni 0} \Phi'(X) &= \sum_{X \ni 0, N(X) \geq 2} (-1)^{N(X)} \sum_{Y \supset X} \Phi(Y) \\
&= \sum_{Y \ni 0} \Phi(Y) \sum_{X : Y \supset X \ni 0, N(X) \geq 2} (-1)^{N(X)} = \sum_{Y \ni 0} \Phi(Y) = C \qquad (2.4)
\end{aligned}
$$

It follows thus from (2.2) and (2.3) that $p_\Phi(z, \beta)$ is analytic, except for a cut due to $\log z$, in the region

$$
\left| \frac{z^{-1} e^{\beta D'}}{1 + z^{-1} e^{\beta D'}} \right| \cdot [2 \exp(e^{\beta D'} - 1) - 1] < 1 \qquad (2.5)
$$

For small β it is clear that the regions (2.2) and (2.5) cover the positive z axis (see Fig. 13), proving the theorem.

5.3 EXISTENCE OF A FIRST–ORDER PHASE TRANSITION AT LOW TEMPERATURE

We consider in this section a lattice gas with pair interaction. Theorem 5.1.3 shows that, if the pair potential Φ^2 is negative, a phase transition can occur only at $z = 1$ [where z is given by (1.17)]. We shall now prove that for a suitable class of pair potentials a phase transition does occur at $z = 1$ if the temperature β^{-1} is low enough. This phase transition is of "first order", i.e., the density ρ_1 exhibits a jump discontinuity as a function of the activity.[9]

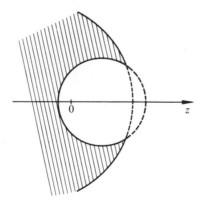

FIGURE 13 $p_\Phi(z, \beta)$ is analytic outside of the shaded area.

[9] The idea of the proof goes back to Peierls [1], and the argument was then made rigorous for nearest neighbor interactions by Griffiths [2] and Dobrushin [2]. We follow here Ginibre, Grossmann, and Ruelle [1]. See also R. L. Dobrushin, "Existence of Phase Transitions in Models of a Lattice Gas," in *Proceedings of the Fifth Berkeley Symposium on Mathematical Statistics and Probability (1965–66)* (University of California Press, Berkeley, California), Vol. 3, pp. 73–87.

5.3.1 THEOREM. *We consider a lattice system with pair interaction, i.e.*

$$U(X) = N(X)\Phi^1 + \sum_{\{x,y\} \subset X} \Phi^2(\{x, y\}) \tag{3.1}$$

and we write

$$z = \exp\left[-\beta\Phi^1 - \tfrac{1}{2}\beta \sum_{x \neq 0} \Phi^2(\{0, x\}) \right] \tag{3.2}$$

$$p(z, \beta) = \beta^{-1} \lim_{\Lambda \to \infty} N(\Lambda)^{-1} \log \sum_{X \subset \Lambda} \exp[-\beta U(X)] \tag{3.3}$$

Let Φ_1, \ldots, Φ_ν be the values of Φ^2 for nearest neighbors in the directions of the ν coordinate axes[10] and let

$$D_i = \tfrac{1}{2} \sum_x{}' |x^i| \cdot |\Phi(x)| \tag{3.4}$$

where Σ' extends over all $x \in \mathbf{Z}^\nu$ except 0 and the 2ν nearest neighbors of 0. If we have $\nu \geq 2$ and

$$\Phi_i + D_i < 0 \tag{3.5}$$

for $i = 1, \ldots, \nu$, then a first-order phase transition occurs at $z = 1$ for sufficiently small β^{-1}.

Let us define

$$\mathscr{Q}(\Lambda, n) = \sum_{S \subset \Lambda, N(S) = n} \exp\left[\frac{\beta}{2} \sum_{x \in S} \sum_{y \in \Lambda \setminus S} \Phi^2(\{x, y\}) \right] \tag{3.6}$$

The quantity \mathscr{Q} has the structure of a canonical partition function, computed for a spin system with interaction potential $-\tfrac{1}{2}\Phi^2(\{x, y\})$ between the sites x, y if they have different spins, 0 otherwise. The corresponding grand partition function is, in the notation of Section 5.1

$$\mathscr{P}^{N(\Lambda)}(z) = \sum_{n=0}^{N(\Lambda)} z^n \mathscr{Q}(\Lambda, n) \tag{3.7}$$

5.3.2 PROPOSITION. *Let the cubic "box" Λ tend to infinity.*

(a) *If $N(\Lambda)^{-1} n \to \rho$, then the limit*

$$g(\rho) = \beta^{-1} \lim N(\Lambda)^{-1} \log \mathscr{Q}(\Lambda, n) \tag{3.8}$$

exists. It is a positive concave continuous function of ρ on $[0, 1]$ (see Fig. 14), and it satisfies

$$g(1 - \rho) = g(\rho) \tag{3.9}$$

[10] That is, $\Phi_i = \Phi(e_i)$ where $e_i^i = 1$, $e_i^j = 0$ for $i \neq j$.

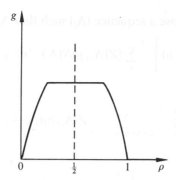

FIGURE 14 Free energy versus density in the presence of a first-order phase transition.

(b) *If $z > 0$, $p(z, \beta)$ defined by (3.3) satisfies*

$$p(z, \beta) = \beta^{-1} \lim_{\Lambda \to \infty} N(\Lambda)^{-1} \log \mathscr{P}^{N(\Lambda)}(z)$$

$$= \max_{\rho} \left[\beta^{-1} \rho \log z + g(\rho) \right] \tag{3.10}$$

$$p(z, \beta) = \beta^{-1} \log z + p(z^{-1}, \beta) \tag{3.11}$$

These results are of the type discussed at length in Chapter 3 and we refrain from a formal proof.[11] We shall show that, for sufficiently small β^{-1}, a horizontal segment is present in the graph of g (see Fig. 14); using (3.10) we can see that this implies the existence of a first-order phase transition at $z = 1$.

5.3.3 PROPOSITION.[12] *For each cube $\Lambda \subset \mathbf{Z}^\nu$, we choose a set \mathscr{M} of subsets of Λ and define*

$$Z(\Lambda, n) = \sum_{S \in \mathscr{M}, N(S) = n} \exp\left[\frac{\beta}{2} \sum_{x \in S} \sum_{y \in \Lambda \setminus S} \Phi^2(\{x, y\}) \right] \tag{3.12}$$

If

$$\beta^{-1} \lim_{\Lambda \to \infty} N(\Lambda)^{-1} \log \sum_n Z(\Lambda, n) = p(1, \beta) \tag{3.13}$$

$$\liminf_{\Lambda \to \infty} \left[\sum_n Z(\Lambda, n) \right]^{-1} \sum_n [Z(\Lambda, n) N(\Lambda)^{-1} n] \le \rho_0 < \tfrac{1}{2} \tag{3.14}$$

then g reduces to a constant on the interval $[\rho_0, 1 - \rho_0]$.

[11] The spin system partition function \mathscr{Q} may be replaced by a lattice gas partition function, using (4.25) of Chapter 2. We may then apply Theorem 3.4.4 and 3.4.6. These theorems hold for spin systems, temperedness of the interaction being replaced by $\|\Phi^2\| < +\infty$.
[12] Instead of cubes we might take another sequence of regions for which the thermodynamic limit holds. We shall use this possibility in the proof of Theorem 5.4.6 below.

Given $\varepsilon > 0$ we choose a sequence (Λ_i) such that $\Lambda_i \to \infty$ and

$$\left[\sum_n Z(\Lambda_i, n)\right]^{-1} \sum_n [Z(\Lambda_i, n)N(\Lambda_i)^{-1}n] \le \rho_0 + \varepsilon/2$$

Then

$$\left[\sum_n Z(\Lambda_i, n)\right]^{-1} \sum_{n \ge N(\Lambda_i)(\rho_0 + \varepsilon)} Z(\Lambda_i, n)(\rho_0 + \varepsilon) \le \rho_0 + \varepsilon/2$$

therefore

$$1 \ge \left[\sum_n Z(\Lambda_i, n)\right]^{-1} \sum_{n < N(\Lambda_i)(\rho_0 + \varepsilon)} Z(\Lambda_i, n) \ge \frac{\varepsilon/2}{\rho_0 + \varepsilon}$$

so that

$$\lim_{i \to \infty} N(\Lambda_i)^{-1} \log \sum_{n < N(\Lambda_i)(\rho_0 + \varepsilon)} Z(\Lambda_i, n) = \beta p(1, \beta) \qquad (3.15)$$

If $n_i < N(\Lambda_i)(\rho_0 + \varepsilon)$ is chosen such that

$$Z(\Lambda_i, n_i) = \max_{n < N(\Lambda_i)(\rho_0 + \varepsilon)} Z(\Lambda_i, n)$$

then (3.15) yields

$$\lim_{i \to \infty} N(\Lambda_i)^{-1} \log Z(\Lambda_i, n_i) = \beta p(1, \beta) \qquad (3.16)$$

Going to a subsequence of (Λ_i) we assume that

$$N(\Lambda_i)^{-1} n_i \to \rho_1 \le \rho_0 + \varepsilon$$

then (3.8), (3.16), and (3.10) yield

$$p(1, \beta) \le g(\rho_1) \le g(\rho_0 + \varepsilon) \le g(\tfrac{1}{2}) = p(1, \beta)$$

which proves Proposition 5.3.3.

5.3.4 CHOICE OF THE SET \mathcal{M}. Let the cube Λ have side k:

$$\Lambda = \{x \in \mathbf{Z}^\nu : 0 \le x^i < k \quad \text{for} \quad i = 1, \ldots, \nu\} \qquad (3.17)$$

and define

$$\Lambda' = \{x \in \mathbf{Z}^\nu : 1 \le x^i < k - 1 \quad \text{for} \quad i = 1, \ldots, \nu\} \qquad (3.18)$$

We choose as set \mathcal{M} in Proposition 5.3.3 the set of all subsets of Λ'. With this choice, Proposition 5.3.3 can be interpreted as follows. We modify the grand canonical ensemble (at $z = 1$) by introducing a "boundary effect" (i.e., requiring that the sites on the boundary of Λ be empty). If this modification produces a change of the density ($\tfrac{1}{2}$ being replaced by a value less than or equal to ρ_0), then a first-order phase transition occurs.

We shall have thus proved Theorem 5.3.1 if we can verify (3.14) [that (3.13) holds is clear]. In order to do this we introduce some new terminology and notation.

We imagine \mathbf{Z}^ν to be imbedded in \mathbf{R}^ν in the usual manner. Given $S \in \mathcal{M}$, i.e., $S \subset \Lambda'$, we draw around each $x \in S$ the 2ν faces of the unit cube centered at x and suppress the faces which occur twice. The closed polyhedron obtained in this way is called $\Gamma(S)$. Each face of $\Gamma(S)$ separates a point $x \in S$ and a point $y \notin S$ in \mathbf{Z}^ν. Along a $(\nu - 2)$-dimensional edge of $\Gamma(S)$ either two or four faces meet. In the case of four faces we slightly deform the polyhedron, "chopping off" the edge from the cubes containing a point of S (see Fig. 15). When this is done $\Gamma(S)$ splits into connected components $\gamma_1, \ldots, \gamma_r$ which we call *cycles*.

Given a cycle γ, we denote by $N(\gamma)$ the number of points of \mathbf{Z}^ν inside of γ and by $|\gamma|_i$ the number of faces of γ orthogonal to the ith coordinate axis. We call the origin site of γ the point $x \in \mathbf{Z}^\nu$, contained inside of γ, and which is the first in lexicographic order with this property.

We prove now three easy lemmas which give estimates of the entropy, number of particles, and energy associated with cycles in terms of $|\gamma|_1, \ldots, |\gamma|_\nu$.

5.3.5 LEMMA. *The number of cycles γ with prescribed $|\gamma|_1, \ldots, |\gamma|_\nu$ is smaller than $N(\Lambda)\Pi_{i=1}^\nu 3^{|\gamma|_i - 1}$.*

First let the origin site of γ be fixed; ν of the faces of the unit cube around the origin site belong to γ. Building up γ face by face, with three possible orientations for each face, we find that there are at most $\Pi_{i=1}^\nu 3^{|\gamma|_i - 1}$ cycles with prescribed $|\gamma|_i$ and origin site, and the lemma follows.

5.3.6 LEMMA. *If $\Gamma(S)$ is composed of the cycles $\gamma_1, \ldots, \gamma_r$, then*

$$N(S) \le \sum_{j=1}^r N(\gamma_j) \tag{3.19}$$

and for any cycle γ

$$N(\gamma) \le \prod_{i=1}^\nu (\tfrac{1}{2}|\gamma|_i)^{1/(\nu-1)} \tag{3.20}$$

FIGURE 15 The polyhedron $\Gamma(S)$.

Since a γ_j may be contained in another one, we have in general

$$N(S) = \sum_{j=1}^{r} N(\gamma_j)\varepsilon_j$$

where $\varepsilon_j = \pm 1$, proving (3.19). If γ is a cycle, (3.20) expresses that its volume $N(\gamma)$ is less than or equal to the volume of a parallelepiped, the faces of which have areas $\frac{1}{2}|\gamma|_i$.[13]

5.3.7 LEMMA. If $S \subset \Lambda'$ we define

$$Z(S) = \exp\left[\frac{\beta}{2} \sum_{x \in S} \sum_{y \in \Lambda \setminus S} \Phi^2(\{x, y\})\right] \qquad (3.21)$$

Let $\Gamma(S)$ consist of the cycles $\gamma_1, \ldots, \gamma_r$ and $\Gamma(S')$ of $\gamma_1, \ldots, \gamma_{r-1}$ then

$$Z(S)/Z(S') \leq \exp\left[\frac{\beta}{2} \sum_{i=1}^{v} |\gamma_r|_i(\Phi_i + D_i)\right] \qquad (3.22)$$

The sets S, S' are such that $\Gamma(S) = \Gamma(S') \cup \gamma_r$. Notice now that two points $x, y \in \Lambda$ which are both inside or both outside of γ_r yield the same contribution to $Z(S)$ and $Z(S')$. Each face of γ_r separates two "nearest neighbor" points $x, y \in \Lambda$ such that $x \in S$, $y \notin S$; on the other hand either $x \in S'$ and $y \in S'$ or $x \notin S'$ and $y \notin S'$, so this yields the factor $\exp[(\beta/2)\Sigma_{i=1}^{v}|\gamma_r|_i\Phi_i]$ in (3.22). The number of ways in which $\xi = (\xi^1, \ldots, \xi^v)$ may occur as the difference $\xi = x - y$ or $\xi = y - x$ with x inside of γ_r and y outside of γ_r is less than or equal to $\Sigma_{i=1}^{v}|\gamma|_i|\xi^i|$ (draw between x and y a broken line constituted of v segments parallel to the coordinate axes; it must cross a face of γ and if this face is orthogonal to the ith coordinate axis, it can do so in only $|\xi^i|$ ways). Therefore, apart from the factor due to nearest neighbors, $Z(S)$ and $Z(S')$ differ by at most a factor

$$\exp\left[\frac{\beta}{2}\sum_{\xi}' |\Phi(\xi)| \sum_{i=1}^{v} |\gamma_r|_i|\xi^i|\right] = \exp\left[\frac{\beta}{2} \sum_{i=1}^{v} |\gamma_r|_i D_i\right]$$

which concludes the proof of the lemma.

We come now to the proof of (3.14). From (3.19) and (3.21) we get

$$\sum_{S \in \mathcal{M}} Z(S)N(\Lambda)^{-1}N(S) \leq \sum_{S \in \mathcal{M}} Z(S) \sum_{j=1}^{r} N(\Lambda)^{-1}N(\gamma_j)$$

$$= \sum_{\gamma} N(\Lambda)^{-1}N(\gamma)Z(\gamma) \qquad (3.23)$$

[13] Drop the condition that the length of the sides of γ are integers; (3.20) is preserved if we multiply the coordinates in \mathbf{R}^v by factors λ_i such that $\Pi_{i=1}^{v}\lambda_i = 1$. In particular we may achieve that $|\gamma|_1 = \cdots = |\gamma|_v$, and (3.20) results then from the fact that the volume of γ is less than that of a cube with the same total area.

where $Z(\gamma)$ is the sum of all $Z(S)$ such that the polyhedron $\Gamma(S)$ contains the cycle γ. We have

$$\left[\sum_{S' \in \mathcal{M}} Z(S')\right]^{-1} Z(\gamma) \leq \exp\left[\frac{\beta}{2} \sum_{i=1}^{v} |\gamma|_i(\Phi_i + D_i)\right] \tag{3.24}$$

as can be seen by retaining in the denominator of the left-hand side only the sum over those S' such that $\Gamma(S') = \Gamma(S)\backslash\gamma$ for some S, and then applying (3.22). From (3.23), (3.24), and (3.20) we get

$$\left[\sum_{S \in \mathcal{M}} Z(S)\right]^{-1} \sum_{S \in \mathcal{M}} Z(S) N(\Lambda)^{-1} N(S)$$

$$\leq \sum_{\gamma} N(\Lambda)^{-1} N(\gamma) \exp\left[\frac{\beta}{2} \sum_{i=1}^{v} |\gamma|_i(\Phi_i + D_i)\right]$$

$$\leq \sum_{\gamma} N(\Lambda)^{-1} \prod_{i=1}^{v} \left\{(\tfrac{1}{2}|\gamma|_i)^{1/(v-1)} \exp\left[\frac{\beta}{2}|\gamma|_i(\Phi_i + D_i)\right]\right\} \tag{3.25}$$

Replacing the sum over γ by a sum over $|\gamma|_1 = 2l_1, \ldots, |\gamma|_v = 2l_v$, and using Lemma 5.3.5, we obtain from (3.25)

$$\left[\sum_{n} Z(\Lambda, n)\right]^{-1} \sum_{n} [Z(\Lambda, n) N(\Lambda)^{-1} n]$$

$$= \left[\sum_{S \in \mathcal{M}} Z(S)\right]^{-1} \sum_{S \in \mathcal{M}} Z(S) N(\Lambda)^{-1} N(S)$$

$$\leq \sum_{l_1=1}^{\infty} \cdots \sum_{l_v=1}^{\infty} \prod_{i=1}^{v} \{l_i^{1/(v-1)} 3^{2l_i-1} \exp[\beta l_i(\Phi_i + D_i)]\}$$

$$= \prod_{i=1}^{v} \left\{\sum_{l=1}^{\infty} l^{1/(v-1)} 3^{2l-1} \exp[l\beta(\Phi_i + D_i)]\right\} \tag{3.26}$$

From this inequality it follows that, for sufficiently large β, (3.14) holds. This concludes the proof of Theorem 5.3.1.

5.4 THE GRIFFITHS INEQUALITIES

The essential point in the proof of Theorem 5.3.1 is the inequality (3.14). Formulated in the language of spin systems this inequality means the following: we require that the spins along the boundary of a cube Λ be oriented "down"; then the other spins tend also to be oriented "down." We can thus say that if condition (3.5) is satisfied, spins tend to be aligned in the same direction. The spins will in particular tend to be aligned in the case of negative Φ^2, and we expect that the more negative Φ^2 is, the more aligned the spins will be. These considerations lead to the very general Theorem 5.4.1 below,

which will be used in particular to extend the results of Section 5.3 on the existence of a first-order phase transition in lattice systems.

5.4.1 THEOREM.[14] *Consider a finite set* $\Lambda = \{x_1, \ldots, x_V\}$ *and let* σ_x *take the values* -1 *and* $+1$ *for each* $x \in \Lambda$. *For each* $X \subset \Lambda$, *let*

$$\sigma^X = \prod_{x \in X} \sigma_x \tag{4.1}$$

and let $J_X \geq 0$ *or* $J_X = +\infty$, $J_\varnothing = 0$. *We define*

$$U'(\sigma_{x_1}, \ldots, \sigma_{x_V}) = - \sum_{X \subset \Lambda} J_X \sigma^X \tag{4.2}$$

$$Z = \sum_\sigma \exp[-U'(\sigma_{x_1}, \ldots, \sigma_{x_V})] \tag{4.3}$$

$$\langle \sigma^X \rangle = Z^{-1} \sum_\sigma \sigma^X \exp[-U'(\sigma_{x_1}, \ldots, \sigma_{x_V})] \tag{4.4}$$

where the sums in (4.3) *and* (4.4) *extend over the* 2^V *choices of* $\sigma_{x_1}, \ldots, \sigma_{x_V}$. *With this notation we have*

$$\langle \sigma^X \sigma^Y \rangle - \langle \sigma^X \rangle \langle \sigma^Y \rangle = \frac{\partial \langle \sigma^X \rangle}{\partial J_Y} \tag{4.5}$$

for all $X, Y \subset \Lambda$ *and the following inequalities hold*

(I) $\langle \sigma^X \rangle \geq 0$ \hfill (4.6)

(II) $\langle \sigma^X \sigma^Y \rangle - \langle \sigma^X \rangle \langle \sigma^Y \rangle \geq 0$ \hfill (4.7)

The identity (4.5) follows immediately from the definitions.

In the proof of (I) and (II) we shall use the fact that the average (4.4) is not changed when a constant is added to U'. If we let $\sigma = (\sigma_{x_1}, \ldots, \sigma_{x_V})$ and

$$U^*(\sigma) = U'(\sigma) + \sum_{X \subset \Lambda} J_X = - \sum_{X \subset \Lambda} J_X(\sigma^X - 1) \tag{4.8}$$

$$Z^* = \sum_\sigma \exp[-U^*(\sigma)] \tag{4.9}$$

we can write

$$\langle \sigma^X \rangle = Z^{*-1} \sum_\sigma \sigma^X \exp[-U^*(\sigma)] \tag{4.10}$$

Clearly $\sigma^X - 1$ takes only the values 0 and -2 and (4.8) yields

$$U^*(\sigma) = 2 \sum_{X \subset \Lambda : \sigma^X = -1} J_X \tag{4.11}$$

[14] This theorem, due to Kelly and Sherman [1], embodies various results of Griffiths [4], I and II. We shall call (4.6) and (4.7) the "Griffiths inequalities". Besides the proof reproduced here, Kelly and Sherman give a very interesting "exponential-expansion" proof of the Griffiths inequalities.

If we define

$$A_X = \exp(-2J_X) \qquad Z(\sigma) = \exp[-U^*(\sigma)] \qquad (4.12)$$

we have $0 \le A_X \le 1$ and we may write

$$Z(\sigma) = \prod_{X \subset \Lambda : \sigma^X = -1} A_X \qquad (4.13)$$

$$Z^* = \sum_\sigma Z(\sigma) \qquad \langle \sigma^X \rangle = Z^{*-1} \sum_\sigma \sigma^X Z(\sigma) \qquad (4.14)$$

5.4.2 PROOF OF (I). Suppose first that $J_Y = 0$ for all $Y \subset \Lambda$; then (4.11) and (4.12) give $Z(\sigma) = 1$ for all σ, and by (4.14)

$$\langle \sigma^X \rangle = 2^{-V} \sum_\sigma \sigma^X = \begin{cases} 1 & \text{if} \quad X = \varnothing \\ 0 & \text{if} \quad X \ne \varnothing \end{cases} \qquad (4.15)$$

We have thus proved (I) in this simple case.

We shall prove (I) in the general case by induction on the number of subsets Y of Λ such that $J_Y \ne 0$.

Suppose that the expression

$$Z^* \langle \sigma^X \rangle = \sum_\sigma \sigma^X Z(\sigma) \qquad (4.16)$$

is positive if $J_Y \ne 0$ for only k subsets Y of Λ. Let $S \subset \Lambda$ be a $(k+1)$st subset, we want to show that (4.16) remains positive for all $J_S \ge 0$. It is convenient to consider the expression (4.16) as a function $f(A_S)$ of A_S because, by (4.13) and (4.14), f is linear inhomogeneous. Since the range of A_S is $[0, 1]$, it suffices to show that $f(0) \ge 0$, $f(1) \ge 0$.

The induction assumption yields $f(1) \ge 0$ because $A_S = 1$ corresponds to $J_S = 0$. On the other hand $A_S = 0$ corresponds to $J_S = +\infty$ and therefore $Z(\sigma) = 0$ if $\sigma^S = -1$:

$$f(0) = \sum_{\sigma : \sigma^S = +1} \sigma^X Z(\sigma) \qquad (4.17)$$

Notice that the $Z(\sigma)$ which occur in (4.17) and correspond to $\sigma^S = +1$ are independent of A_S and are therefore the same as for $J_S = 0$. We may thus write

$$f(0) = \sum_\sigma \sigma^X \cdot \frac{\sigma^S + 1}{2} \cdot Z(\sigma)$$

$$= \tfrac{1}{2}[\langle \sigma^X \sigma^S \rangle + \langle \sigma^X \rangle] Z^* \qquad (4.18)$$

where the expectation values are taken at $J_S = 0$ and are therefore positive by the induction assumption. We have thus $f(0) \ge 0$ and this concludes the proof of (I).

5.4.3 PRELIMINARIES TO THE PROOF OF (II). We introduce the notation

$$F = \frac{Z^{*2}}{4} [\langle \sigma^X \sigma^Y \rangle - \langle \sigma^X \rangle \langle \sigma^Y \rangle] \tag{4.19}$$

$$p = \sum_{\sigma: \sigma^X = 1, \sigma^Y = 1} Z(\sigma) \qquad q = \sum_{\sigma: \sigma^X = 1, \sigma^Y = -1} Z(\sigma) \tag{4.20}$$

$$r = \sum_{\sigma: \sigma^X = -1, \sigma^Y = 1} Z(\sigma) \qquad s = \sum_{\sigma: \sigma^X = -1, \sigma^Y = -1} Z(\sigma) \tag{4.21}$$

We have thus

$$F = \tfrac{1}{4}[(p + q + r + s)(p - q - r + s) - (p + q - r - s)(p - q + r - s)]$$
$$= \tfrac{1}{4}[(p + s)^2 - (q + r)^2 - (p - s)^2 + (q - r)^2]$$
$$= ps - qr$$
$$\tag{4.22}$$

From (4.13) and (4.22) we see that F is a polynomial in the A_S, $S \subset \Lambda$, which is of second degree (at most) in each variable. Furthermore A_X and A_Y appear homogeneously with exponent 1 in F (because r, s are proportional to A_X and q, s are proportional to A_Y).

For any polynomial $P(x_1, \ldots, x_M)$ we may in a unique manner write

$$P = P_0 + \sum_\alpha P_\alpha x^\alpha \tag{4.23}$$

where the summation runs over the $2^M - 1$ nonempty subsets $\alpha = \{i_1, \ldots, i_r\}$ of $\{1, \ldots, M\}$ and

$$x^\alpha = x_{i_1} \cdots x_{i_r} \tag{4.24}$$

while the polynomials P_0, P_α are even in every variable x_i, $i = 1, \ldots, M$. If we replace the x_i, $i = 1, \ldots, M$, by the A_S, $S \subset \Lambda$, and write A^α instead of x^α we can represent F as

$$F = \sum_\alpha F_\alpha A^\alpha \tag{4.25}$$

where $F_0 = 0$ because F is homogeneous of degree one in A_X and A_Y. The following lemma gives information about the structure of the A^α.

5.4.4 LEMMA. *Let $\sigma \neq \sigma'$ and put the monomial $Z(\sigma)Z(\sigma')$ into the form* (4.23):

$$Z(\sigma)Z(\sigma') = P_\alpha A^\alpha \tag{4.26}$$

then there exists $T_\alpha \subset \Lambda$, $T_\alpha \neq \varnothing$ such that

$$A_S \text{ appears in } A^\alpha \quad \Leftrightarrow \quad N(S \cap T_\alpha) \text{ is odd} \tag{4.27}$$

Let us write

$$D = \{x \in \Lambda : \sigma_x = -1\} \qquad D' = \{x \in \Lambda : \sigma'_x = -1\}$$

$$T_\alpha = (D \cup D') \backslash (D \cap D') = \{x \in \Lambda : \sigma_x \neq \sigma'_x\} \neq \varnothing \qquad (4.28)$$

By (4.13), the following statements are equivalent

$$A_S \text{ appears in } Z(\sigma) \qquad \Leftrightarrow \qquad \sigma^S = -1$$

$$\Leftrightarrow \qquad N(S \cap D) \text{ is odd}$$

Therefore

A_S appears in A^α

$$\Leftrightarrow \qquad A_S \text{ appears with exponent 1 in } Z(\sigma)Z(\sigma')$$

$$\Leftrightarrow \qquad N(S \cap D) + N(S \cap D') \text{ is odd} \qquad (4.29)$$

On the other hand

$$N(S \cap D) + N(S \cap D')$$

$$= N(S \cap (D \cup D')) + N(S \cap (D \cap D'))$$

$$= N(S \cap (D \cup D')) - N(S \cap (D \cap D')) \qquad (\text{mod } 2)$$

$$= N(S \cap [(D \cup D') \backslash (D \cap D')])$$

$$= N(S \cap T_\alpha) \qquad (4.30)$$

and (4.27) follows from (4.29) and (4.30), proving the lemma.

5.4.5 REMARKS

(a) Exactly half of the subsets S of Λ are such that $N(S \cap T_\alpha)$ is odd. Therefore, by (4.27), *half of the variables A_S appear in A^α.*
(b) Let $\sigma_1 \neq \sigma'_1$ and

$$Z(\sigma_1)Z(\sigma'_1) = P_{\alpha_1} A^{\alpha_1} \qquad (4.31)$$

be the decomposition corresponding to (4.26). *If $\alpha_1 \neq \alpha$, there is necessarily some variable A_S contained in A^{σ_1} and not in A^σ.*
(c) Because of Eqs. (4.20), (4.21), and (4.22), F is a sum of terms $\pm Z(\sigma)Z(\sigma')$ with $\sigma \neq \sigma'$. Therefore, *if $F_\alpha \neq 0$ in (4.25) there exists T_α such that (4.27) holds.* In particular the above remarks (a) and (b) hold.

5.4.6 PROOF OF (II). Since $0 \leq A_S \leq 1$, the monomials A^α in (4.25) are

positive. In order to prove (II) (i.e., $F \geq 0$) it suffices thus to show that $F_\alpha \geq 0$ for each α.

We have up to now considered F as a polynomial in all A_S, $S \subset \Lambda$. For fixed α, let \bar{F} be the polynomial in those A_S such that $N(S \cap T_\alpha)$ is even,

obtained by setting $A_{S'} = 1$ in F if $N(S' \cap T_\alpha)$ is odd. We can put \bar{F} in the form (4.23), obtaining

$$\bar{F} = \bar{F}_0 + \sum_\beta \bar{F}_\beta A^\beta \qquad (4.32)$$

We have set $A_{S'} = 1$ for precisely those variables $A_{S'}$ which occur in A^α so that, using 5.4.5,(b), we find

$$F_\alpha = \bar{F}_0 \qquad (4.33)$$

We shall denote by $\bar{Z}(\sigma), \bar{p}, \bar{q}, \bar{r}, \bar{s}$ the polynomials obtained by setting $A_{S'} = 1$ in $Z(\sigma), p, q, r, s$ if $N(S' \cap T)$ is odd.

A permutation $\sigma \to \sigma^*$ of the configurations σ is defined by

$$\sigma_x^* = \begin{cases} \sigma_x & \text{if } x \notin T_\alpha \\ -\sigma_x & \text{if } x \in T_\alpha \end{cases} \qquad (4.34)$$

Notice that, by (4.13) and (4.34),

$$\bar{Z}(\sigma^*) = \prod [A_S : (\sigma^*)^S = -1, N(S \cap T_\alpha) \text{ is even}]$$
$$= \prod [A_S : (\sigma)^S = -1, N(S \cap T_\alpha) \text{ is even}]$$
$$= \bar{Z}(\sigma) \qquad (4.35)$$

Since A_X appears in A^α, Lemma 5.4.4 shows that $N(X \cap T_\alpha)$ is odd; hence

$$\sigma^X = -(\sigma^*)^X \qquad (4.36)$$

and similarly

$$\sigma^Y = -(\sigma^*)^Y \qquad (4.37)$$

Introducing (4.35), (4.36), and (4.37) into the definitions (4.20) and (4.21) of p, q, r, s yields

$$\bar{p} = \bar{s} \qquad \bar{q} = \bar{r} \qquad (4.38)$$

Therefore

$$\bar{F} = \bar{p}^2 - \bar{q}^2 \qquad (4.39)$$

We insert (4.20) into (4.39) and separate the square terms and the cross terms:

$$\bar{F} = \bar{Q} + \bar{R} \qquad (4.40)$$

$$Q = \sum_{\sigma : \sigma^X = 1, \sigma^Y = 1} Z(\sigma)^2 - \sum_{\sigma : \sigma^X = 1, \sigma^Y = -1} Z(\sigma)^2$$

$$= \sum_{\sigma : \sigma^X = 1} \sigma^Y Z(\sigma)^2 \qquad (4.41)$$

The polynomial R is a sum of contributions $\pm Z(\sigma_1)Z(\sigma_1') = R_{\alpha_1}A^{\alpha_1}$ where $\sigma_1 \neq \sigma_1'$ and A_X does not appear as factor [see (4.13) and (4.20)]. In view of 5.4.5,(b) there is some variable A_{S_1} present in A^{α_1} and not in A^α; therefore \bar{R} does not contribute to the term \bar{F}_0 in (4.32), and we have

$$\bar{F}_0 = \bar{Q} \tag{4.42}$$

Notice now that by (4.41) \bar{Q} is of the form of the right-hand side of (4.17), where some of the A_T have been replaced by A_T^2 (we have again $0 \leq A_T^2 \leq 1$), and some by 1. These replacements preserve the conclusion that the right-hand side of (4.17) is positive. We have thus shown that $\bar{Q} \geq 0$; hence by (4.42) and (4.33), $F_\alpha \geq 0$, and this concludes the proof of (II).

As a first application of Theorem 5.4.1 we prove a variant of Theorem 5.3.1, improving a result by F. A. Berezin and Ya. G. Sinai, Existence of Phase Transitions for Lattice Gases with Attraction between Particles, *Trud. Moskovsk. Mat. Obshch.* **17**, 197–212 (1967).

5.4.7 THEOREM. *We use the same notation and assumptions as in Theorem 5.3.1 but replace (3.5) by $\Phi^2 \leq 0$ and $\Phi^2(\xi_1) < 0, \Phi^2(\xi_2) < 0$ for some linearly independent ξ_1, ξ_2. Then a first-order phase transition occurs at $z = 1$ for sufficiently small β^{-1}.*

After a change of coordinates, we may assume that $\xi_1 = (\xi_1{}^1, 0, 0, \dots)$, $\xi_2 = (\xi_2{}^1, \xi_2{}^2, 0, \dots)$. We define

$$M_k = \{n^1\xi_1 + n^2\xi_2 \in \mathbf{R}^2 : n^1, n^2 = 0, 1, \dots, k-1\} \tag{4.43}$$

We shall prove the existence of a first-order phase transition by using Proposition 5.3.3 but, instead of cubic regions Λ, we shall use regions Λ_k of the form

$$\Lambda_k = \{x : (x^1, x^2) \in \bar{M}_k, \quad 0 \leq x^i < k \quad \text{for} \quad i = 3, \dots, v\} \tag{4.44}$$

where \bar{M}_k is the set of points of \mathbf{R}^2 which can be written as convex linear combinations (with rational coefficients) of points of M_k (see Fig. 16). We let Λ_k' be the set of $x \in \Lambda_k$ such that (x^1, x^2) is not of the form $n^1\xi_1 + n^2\xi_2$ with n^1, n^2 equal to either 0 or $k-1$.

FIGURE 16 Example of sets M_k (circles) and \bar{M}_k (circles and crosses) for $k = 3$. A line is drawn around Λ_k'.

We choose as set \mathcal{M} in Proposition 5.3.3 the set of all subsets of Λ'_k. Clearly then (3.13) is satisfied and it remains to check (3.14). The left-hand side of (3.14) can be rewritten as

$$\liminf_{k \to \infty} \left[\sum_{S \in \mathcal{M}} Z(S) \right]^{-1} \sum_{S \in \mathcal{M}} Z(S) N(\Lambda_k)^{-1} N(S) \qquad (4.45)$$

where

$$Z(S) = \exp\left[\frac{\beta}{2} \sum_{x \in S} \sum_{y \in \Lambda \setminus S} \Phi^2(\{x, y\}) \right] \qquad (4.46)$$

We shall now express (4.45) and (4.46) in the language of Theorem 5.4.1. For each subset S of Λ_k and $x \in \Lambda_k$ we let $\sigma_x = +1$ if $x \notin S$, $\sigma_x = -1$ if $x \in S$ and we write

$$J_{\{x\}} = \begin{cases} +\infty & \text{if } x \in \Lambda_k \setminus \Lambda'_k \\ 0 & \text{otherwise} \end{cases} \qquad (4.47)$$

$$J_{\{x,y\}} = -\frac{\beta}{4} \Phi^2(\{x, y\}) \qquad (4.48)$$

$$J_X = 0 \quad \text{if } N(X) > 2 \qquad (4.49)$$

With this notation we have [see (4.11) and (4.12)]

$$Z(\sigma) = \begin{cases} Z(S) & \text{if } S \subset \Lambda'_k \\ 0 & \text{if } S \not\subset \Lambda'_k \end{cases} \qquad (4.50)$$

and the expression (4.45) becomes

$$\liminf_{k \to \infty} Z^{*-1} \sum_{\sigma} Z(\sigma) N(\Lambda_k)^{-1} \sum_x \frac{1 - \sigma_x}{2}$$

$$= \liminf_{k \to \infty} N(\Lambda_k)^{-1} \sum_x \left\langle \frac{1 - \sigma_x}{2} \right\rangle$$

$$= \tfrac{1}{2} - \limsup_{k \to \infty} \tfrac{1}{2} N(\Lambda_k)^{-1} \sum_x \langle \sigma_x \rangle \qquad (4.51)$$

We now use Theorem 5.4.1 to compare (4.51) with the analogous expression computed with a different choice of J_X. Let $J'_{\{x,y\}} = J_{\{x,y\}}$ if $(x^1, x^2) \in M_k$, $(y^1, y^2) \in M_k$, and $x^3 = y^3, \ldots, x^\nu = y^\nu$; otherwise let $J'_{\{x,y\}} = 0$, and also let $J'_X = J_X$ if $N(X) \neq 2$. Clearly $J'_X \leq J_X$; therefore, (4.5) and (4.7) yield

$$\langle \sigma_x \rangle \geq \langle \sigma_x \rangle'$$

where the J'_X have been used in computing $\langle \ \rangle'$. We have thus

$$N(\Lambda_k)^{-1} \sum_x \langle \sigma_x \rangle \geq N(\Lambda_k)^{-1} \sum_x \langle \sigma_x \rangle'$$

$$= N(\overline{M}_k)^{-1} \sum_{x \in M_k} \langle \sigma_x \rangle'$$

$$= N(\overline{M}_k)^{-1} N(M_k) \cdot N(M_k)^{-1} \sum_{x \in M_k} \langle \sigma_x \rangle' \qquad (4.52)$$

where we have used the fact that $\langle \sigma_x \rangle' = 0$ if $x \in \overline{M}_k \setminus M_k$. The interaction J'_X, restricted to M_k corresponds to a pair potential Φ'^2 such that $\Phi'_1 = \Phi^2(\xi_1) < 0$, $\Phi'_2 = \Phi^2(\xi_2) < 0$, and $D'_1 = D'_2 = 0$. Therefore (3.5) is satisfied and (3.14) holds for the interaction Φ'; since the left-hand side of (3.14) is given by (4.51) we have

$$\tfrac{1}{2} - \limsup_{k \to \infty} \tfrac{1}{2} N(M_k)^{-1} \sum_{x \in \overline{M}_k} \langle \sigma_x \rangle' < \tfrac{1}{2}$$

and because of (4.52) this implies

$$\tfrac{1}{2} - \limsup_{k \to \infty} \tfrac{1}{2} N(\Lambda_k)^{-1} \sum_{x \in \Lambda_k} \langle \sigma_x \rangle < \tfrac{1}{2} \qquad (4.53)$$

Therefore (3.14) holds for the interaction Φ proving Theorem 5.4.7.

5.4.8 PHASE TRANSITIONS IN LATTICE GASES WITH NEGATIVE INTER-ACTION. It is remarkable that for lattice gases with negative pair interactions we have a fairly detailed picture of the occurrence of phase transitions. If z is given by (1.17) we know from Section 5.1 that a phase transition may occur only at $z = 1$. We know from Section 5.2 that for large β^{-1} there is no phase transition. Theorem 5.4.7 shows that a first-order phase transition is present for small β^{-1} unless the interaction is essentially one-dimensional. In the latter case we can show[15] that, if the pair potential decreases sufficiently fast at infinity, no phase transition occurs for any value of β^{-1}.

5.4.9 SPIN SYSTEMS. Let the set Λ which occurs in Theorem 5.4.1 be a finite subset of \mathbf{Z}^ν. If $J_X = 0$ except for $N(X) = 2$, the "potential energy" (4.2) takes the form

$$U'(\sigma_{x_1}, \ldots, \sigma_{x_V}) = - \sum_{\{x,y\} \subset \Lambda} J_{\{x,y\}} \sigma^x \sigma^y \qquad (4.54)$$

which is that of a spin system without magnetic field. We assume translation invariance

$$J_{\{x+a, y+a\}} = J_{\{x,y\}}$$

If $J_{\{x,y\}} = 0$ except when x and y are nearest neighbors,[16] then (4.54) is the

[15] See Section 5.6.5.
[16] That is, x, y differ only by one unit in one coordinate.

potential energy of the *Ising model*. As is well known, the thermodynamic functions can be computed exactly and in closed form in the thermodynamic limit for the Ising model in one and two dimensions.[17] For $v = 2$ a phase transition occurs at a "Curie temperature" β_C^{-1} as a singularity of the thermodynamic functions. Below this temperature a *long range order* is present[18] in the sense that

$$L = \lim_{x-y \to \infty} \lim_{\Lambda \to \infty} \langle \sigma_x \sigma_y \rangle > 0 \qquad (4.55)$$

Theorem 5.4.1 implies that $\langle \sigma_x \sigma_y \rangle$ increases when $J_{\{x,y\}}$ increases; therefore L and the "Curie temperature" at which the long range order appears also increase with $J_{\{x,y\}}$. We prove in this way the existence of a long range order for a variety of systems.[19]

A magnetic field term can be added to (4.54); the potential energy then becomes

$$U'(\sigma_{x_1}, \ldots, \sigma_{x_\Lambda}) = -\sum_{x \in \Lambda} J_{\{x\}} \sigma_{\dot{x}} - \sum_{\{x,y\} \subset \Lambda} J_{\{x,y\}} \sigma_x \sigma_y \qquad (4.56)$$

If we assume translation invariance, then $J_{\{x\}}$ (the "magnetic field") is independent of x. We define a thermodynamic *magnetization per spin* by

$$\sigma = \lim_{\Lambda \to \infty} V(\Lambda)^{-1} \sum_{x \in \Lambda} \langle \sigma_x \rangle \qquad (4.57)$$

The study of a spin system in the presence of a magnetic field, with potential energy (4.56), is mathematically equivalent to the study of a lattice gas with potential energy (3.1).[20] In this equivalence the chemical potential of the lattice gas corresponds to a linear function of the magnetic field of the spin system, and the density corresponds to $(1 + \sigma)/2$. Theorem 5.4.7 establishes that for low values of β^{-1} the lattice gas undergoes a first-order phase transition, i.e., the density has a jump discontinuity as function of the activity. Translated into the language of spin systems this means that σ has a jump discontinuity as function of $J_{\{x\}}$, or if we put

$$\sigma_S = \lim_{J_{\{x\}} \to 0+} \sigma \qquad (4.58)$$

then $\sigma_S > 0$ for low values of β^{-1}. We call σ_S the *spontaneous magnetization*; it follows from Theorem 5.4.1 that it is an increasing function of the interaction $J_{\{x,y\}}$.[21] Theorem 5.2.1 shows that there is no spontaneous magnetization (i.e., $\sigma_S = 0$) at high β^{-1}; this result can also be obtained directly.[22]

[17] See Onsager [2], Hurst and Green [1].
[18] See, e.g., Schultz, Mattis, and Lieb [1].
[19] See Griffiths [4],I. The reader may convince himself that the occurrence of a long range order here is related to the existence of the first-order phase transition discussed in Theorem 5.4.7.
[20] See Section 2.4.3.
[21] The spontaneous magnetization of the two-dimensional Ising model can be computed exactly; see Yang [1].
[22] See Griffiths [4],III. See also Weng, Griffiths, and Fisher [1] and Fisher [4].

5.5 THE THEOREM OF MERMIN AND WAGNER

We have seen in Section 5.4 that classical spin systems with $v \geq 2$ and suitable interactions exhibit spontaneous magnetization at low temperature β^{-1}. For a corresponding class of quantum systems a negative result has been proved by Mermin and Wagner: for $v \leq 2$ there is no spontaneous magnetization. The difference in behavior between classical and quantum systems for $v = 2$ may be expressed by saying that quantum systems exhibit less than classical systems the tendency to have their spins oriented in the same direction coherently over the whole lattice \mathbf{Z}^2. That this is so is understandable because quantum spins may vary continuously between the "up" and "down" positions; it is thus possible to have neighboring spins approximately aligned but no coherent orientation over the whole lattice. This possibility does not exist for the classical spin systems which we have considered.[23]

We now define precisely the class of quantum spin systems which we shall consider ("isotropic Heisenberg models"). As discussed in Section 2.2 we have for each $x \in \mathbf{Z}^v$ a copy \mathscr{H}_x of a finite-dimensional Hilbert space \mathscr{H}. We assume that we have in \mathscr{H} an irreducible unitary representation of $SU(2)$[24] with the (self-adjoint) generators S_1, S_2, S_3 satisfying the commutation relations

$$[S_1, S_2] = iS_3 \qquad [S_2, S_3] = iS_1 \qquad [S_3, S_1] = iS_2 \qquad (5.1)$$

We define

$$S_\pm = 2^{-1/2}(S_1 \pm iS_2) \qquad (5.2)$$

we have then

$$\sum_{i=1}^{3} S_i^2 = S_+S_- + S_-S_+ + S_3^2 = s(s + 1)\mathbf{1} \qquad (5.3)$$

where $2s + 1$ is the dimensionality of \mathscr{H}. We note the commutation relations

$$[S_+, S_-] = S_3 \qquad [S_3, S_\pm] = \pm S_\pm \qquad (5.4)$$

Let $S_i(x)$ be the operator corresponding to S_i in \mathscr{H}_x. We shall use as Hamiltonian of our quantum spin system (for a finite $\Lambda \subset \mathbf{Z}^v$):

$$H(\Lambda) = - \sum_{x,x' \in \Lambda} J(x - x') \sum_{i=1}^{3} S_i(x)S_i(x') - h \sum_{x \in \Lambda} S_3(x) \qquad (5.5)$$

In this expression h is the "magnetic field" oriented in the 3-direction;

[23] It is possible to consider classical systems with continuously variable spin. These systems behave like quantum systems with respect to spontaneous magnetization. See Mermin [1].
[24] That is, a single- or double-valued representation of the three-dimensional rotation group.

we assume that $J(0) = 0$, $J(-x) = J(x)$, and

$$M = \sum_{x \in \mathbf{Z}^\nu} x^2 |J(x)| < +\infty \tag{5.6}$$

To prove the absence of spontaneous magnetization for the interaction (5.5) if $\nu \leq 2$ we shall need an inequality which we first establish.

5.5.1 LEMMA (*Bogoliubov inequality*). *Let the Hilbert space \mathfrak{H} have finite dimension, H be self-adjoint in \mathfrak{H}, $\beta > 0$, and write*

$$\langle X \rangle = (\mathrm{Tr}\, e^{-\beta H})^{-1} \,\mathrm{Tr}\, X\, e^{-\beta H} \tag{5.7}$$

if X is an operator in \mathfrak{H}. Then for all operators A, C in \mathfrak{H} we have

$$\tfrac{1}{2}\beta \langle AA^* + A^*A \rangle \langle [[C, H], C^*] \rangle \geq |\langle [C, A] \rangle|^2 \tag{5.8}$$

We define a positive semidefinite scalar product in the space of operators in \mathfrak{H} by

$$(A, B) = \sum_{\varphi\psi}{}' (\psi, A^*\varphi)(\varphi, B\psi) \frac{W_\varphi - W_\psi}{E_\psi - E_\varphi} \tag{5.9}$$

where the sum extends over the pairs (φ, ψ) of elements of an orthonormal system of eigenvectors of H, omitting pairs such that $E_\varphi = E_\psi$; we have written

$$E_\varphi = (\varphi, H\varphi) \qquad W_\varphi = (\mathrm{Tr}\, e^{-\beta H})^{-1} \exp[-\beta E_\varphi] \tag{5.10}$$

If $E_\varphi < E_\psi$ we have

$$\frac{\exp[-\beta E_\varphi] - \exp[-\beta E_\psi]}{\exp[-\beta E_\varphi] + \exp[-\beta E_\psi]} = \tanh \frac{\beta E_\psi - \beta E_\varphi}{2} < \tfrac{1}{2}\beta(E_\psi - E_\varphi)$$

hence

$$0 \leq \frac{W_\varphi - W_\psi}{E_\psi - E_\varphi} = (\mathrm{Tr}\, e^{-\beta H})^{-1} \frac{\exp[-\beta E_\varphi] - \exp[-\beta E_\psi]}{E_\psi - E_\varphi}$$

$$< (\mathrm{Tr}\, e^{-\beta H})^{-1} \tfrac{1}{2}\beta(\exp[-\beta E_\varphi] + \exp[-\beta E_\psi])$$

$$= \tfrac{1}{2}\beta(W_\varphi + W_\psi)$$

Therefore

$$(A, A) \leq \tfrac{1}{2}\beta \langle AA^* + A^*A \rangle \tag{5.11}$$

The Schwarz inequality for the scalar product (5.9) yields

$$(A, A)(B, B) \geq |(A, B)|^2 \tag{5.12}$$

Choosing $B = [C^*, H]$, we may write

$$(A, B) = \langle [C^*, A^*] \rangle \tag{5.13}$$

$$(B, B) = \langle [C^*, [H, C]] \rangle \tag{5.14}$$

The inequality (5.8) results from insertion of (5.11), (5.13), and (5.14) in (5.12) proving the lemma.

5.5.2 THEOREM. (*Mermin–Wagner*).[25] *If we define*

$$g(h) = \beta^{-1} \lim_{\Lambda \to \infty} N(\Lambda)^{-1} \log \mathrm{Tr}_{\mathscr{H}(\Lambda)} \exp[-\beta H(\Lambda)] \tag{5.15}$$

$$\bar{\sigma}(h) = \frac{d}{dh} g(h) \tag{5.16}$$

where $H(\Lambda)$ is given by (5.5), and (5.6) is satisfied, then

$$\lim_{h \to 0} \bar{\sigma}(h) = 0 \tag{5.17}$$

if $v \le 2$.

We shall first assume that J has finite range[26] and that the system is contained in a periodic box $\Lambda(a)$ as described in Section 2.3.4. Let Δ be the set of vectors $k = (k^1, \ldots, k^v) \in \mathbf{R}^v$ such that $k^i = 2\pi n^i / a^i$ where n^i is an integer and

$$-\frac{a^i}{2} < n^i \le \frac{a^i}{2} \tag{5.18}$$

We write $k \cdot x = \Sigma_1^v k^i x^i$ and

$$S_i(k) = \sum_{x \in \Lambda(a)} e^{-ik \cdot x} S_i(x) \qquad J(k) = \sum_{x \in \Lambda(a)} e^{-ik \cdot x} J(x) \tag{5.19}$$

so that,[27] if $x \in \Lambda(a)$,

$$S_i(x) = V(a)^{-1} \sum_{k \in \Delta} e^{ik \cdot x} S_i(k) \qquad J(x) = V(a)^{-1} \sum_{k \in \Delta} e^{ik \cdot x} J(k) \tag{5.20}$$

The quantity corresponding to $\bar{\sigma}(h)$ for our finite system is

$$\frac{d}{dh} \beta^{-1} V(a)^{-1} \log \mathrm{Tr} \exp[-\beta \hat{H}(\Lambda)] = \left\langle V(a)^{-1} \sum_{x \in \Lambda(a)} S_3(x) \right\rangle \tag{5.21}$$

where $\langle \rangle$ is defined by (5.7), $\hat{H}(\Lambda(a))$ replacing H. It is interesting, for later purposes, to use instead of (5.21) the more general expression

$$\sigma(h) = \left\langle V(a)^{-1} \sum_{x \in \Lambda(a)} S_3(x) e^{iK \cdot x} \right\rangle \tag{5.22}$$

[25] See Mermin and Wagner [1].
[26] That is, $J(x) = 0$ except for a finite number of points $x \in \mathbf{Z}^v$.
[27] We write $V(a) = N(\Lambda(a))$.

where K is chosen such that $e^{iK \cdot x} = \pm 1$ for every $x \in \Lambda(a)$ and $\hat{H}(\Lambda(a))$ is now defined by

$$\hat{H}(\Lambda(a)) = -\sum_{x,x' \in \Lambda(a)} \hat{J}(x - x') \sum_{i=1}^{3} S_i(x)S_i(x') - h \sum_{x \in \Lambda(a)} e^{-iK \cdot x} S_3(x) \quad (5.23)$$

We shall obtain an inequality for $\sigma(h)$ by letting $A = S_-(-k - K)$, $C = S_+(k)$, $H = \hat{H}(\Lambda(a))$ in the Bogoliubov inequality (5.8). We first evaluate the quantities which appear in (5.8). We have

$$\langle AA^* + A^*A \rangle = \langle S_+(k + K)S_-(-k - K) + S_-(-k - K)S_+(k + K) \rangle \quad (5.24)$$

$$\langle [C, A] \rangle = \sum_{x \in \Lambda(a)} \sum_{y \in \Lambda(a)} e^{-ik \cdot x} e^{i(k+K) \cdot y} \langle [S_+(x), S_-(y)] \rangle$$

$$= \sum_{x \in \Lambda(a)} e^{iK \cdot x} \langle S_3(x) \rangle = V(a)\sigma(h) \quad (5.25)$$

A straightforward but longer calculation yields successively

$$[C, H] = 2 \sum_{xx'} (e^{-ik \cdot x'} - e^{-ikx}) \hat{J}(x - x') S_3(x) S_+(x')$$

$$+ h \sum_{x} e^{-i(k+K) \cdot x} S_+(x)$$

$$\langle [[C, H], C^*] \rangle = 2 \sum_{xx'} (1 - e^{-ik \cdot (x - x')}) \hat{J}(x - x') \langle S_-(x')S_+(x) + S_3(x')S_3(x) \rangle$$

$$+ hV(a)\sigma(h) \quad (5.26)$$

The quantity (5.26) is positive by (5.8), so adding the same quantity with k replaced by $-k$ we find

$$\langle [[C, H], C^*] \rangle \leq 2 \sum_{xx'} (1 - \cos k \cdot (x - x')) \hat{J}(x - x')$$

$$\times \langle S_-(x')S_+(x) + S_+(x')S_-(x) + 2S_3(x')S_3(x) \rangle$$

$$+ 2hV(a)\sigma(h) \quad (5.27)$$

Notice now that by the Schwarz inequality, invariance under translations, and (5.3), we have

$$|\langle S_-(x')S_+(x) + S_+(x')S_-(x) + 2S_3(x')S_3(x) \rangle|$$

$$\leq \langle S_-(0)S_+(0) + S_+(0)S_-(0) + 2S_3(0)^2 \rangle \leq 2s(s + 1)$$

Inserting this in (5.27) and using $1 - \cos t \leq t^2/2$ and (5.6) we obtain

$$\langle [[C, H], C^*] \rangle \leq 2 \sum_{x} (1 - \cos k \cdot x) |J(x)| V(a) \cdot 2s(s + 1) + 2V(a)|h\sigma(h)|$$

$$\leq 2V(a) \left[\sum_{x} k^2 x^2 |J(x)| s(s + 1) + |h\sigma(h)| \right]$$

$$= 2V(a)[Ms(s + 1)k^2 + |h\sigma(h)|] \quad (5.28)$$

Insertion of (5.24), (5.25), and (5.28) in (5.8) yields

$$\langle S_+(k + K)S_-(-k - K) + S_-(-k - K)S_+(k + K)\rangle$$

$$\geq \frac{2\beta^{-1}V(a)^2\sigma(h)^2}{2V(a)[Ms(s + 1)k^2 + |h\sigma(h)|]} \quad (5.29)$$

If we sum (5.29) over $k \in \Delta$, using

$$\sum_{k\in\Delta} \langle S_+(k + K)S_-(-k - K) + S_-(-k - K)S_+(k + K)\rangle$$

$$= V(a) \sum_{x\in\Lambda(a)} \langle S_+(x)S_-(x) + S_-(x)S_+(x)\rangle \leq V(a)^2 s(s + 1)$$

we obtain

$$s(s + 1) \geq \frac{\beta^{-1}\sigma(h)^2}{V(a)} \sum_{k\in\Delta} [Ms(s + 1)k^2 + |h\sigma(h)|]^{-1} \quad (5.30)$$

When $a \to \infty$, the inequality (5.30) becomes

$$s(s + 1) \geq \beta^{-1}\bar\sigma(h)^2(2\pi)^{-\nu} \int_{-\pi}^{\pi} dk^1 \cdots \int_{-\pi}^{\pi} dk^\nu[Ms(s + 1)k^2 + |h\bar\sigma(h)|]^{-1}$$

$$\geq \beta^{-1}\bar\sigma(h)^2(2\pi)^{-\nu}\Omega_\nu \int_0^\pi \frac{k^{\nu-1} dk}{Ms(s + 1)k^2 + |h\bar\sigma(h)|} \quad (5.31)$$

In particular

$$s(s + 1) \geq \frac{\beta^{-1}\bar\sigma(h)^2}{\pi[|h\bar\sigma(h)|Ms(s + 1)]^{1/2}} \arctan\left(\frac{\pi^2 Ms(s + 1)}{|h\bar\sigma(h)|}\right)^{1/2} \quad (\nu = 1)$$

$$s(s + 1) \geq \frac{\beta^{-1}\bar\sigma(h)^2}{4\pi Ms(s + 1)} \log\left(1 + \frac{\pi^2 Ms(s + 1)}{|h\bar\sigma(h)|}\right) \quad (\nu = 2)$$

which yield, respectively,

$$|\bar\sigma(h)|^3 \leq \pi^2 M[s(s + 1)]^3\beta^2|h| \left[\arctan\left(\frac{\pi^2 Ms(s + 1)}{|h\bar\sigma(h)|}\right)^{1/2}\right]^{-2} \quad (\nu = 1)$$

$$(5.32)$$

$$\bar\sigma(h)^2 \leq 4\pi M[s(s + 1)]^2\beta \left[\log\left(1 + \frac{\pi^2 Ms(s + 1)}{|h\bar\sigma(h)|}\right)\right]^{-1} \quad (\nu = 2) \quad (5.33)$$

In view of Theorem 2.3.3,(a), every Φ satisfying (5.6) can be approximated by a finite range potential, and the inequalities (5.32), (5.33) retain their validity. These inequalities imply that $\bar\sigma(h) \to 0$ when $h \to 0$ ($\nu = 1, 2$), proving Theorem 5.5.2.

5.5.3 ABSENCE OF ANTIFERROMAGNETISM. Theorem 5.5.2 shows that the systems with interaction described by (5.5) and (5.6) have no spontaneous

magnetization or, in other words, do not exhibit *ferromagnetism*. A pheno-
menon related to ferromagnetism is *antiferromagnetism* where there are two
classes of sites (sublattices) and the spins of one class point in one direction,
those of the other class in the opposite direction. Consider now the sub-
lattices $\{x: e^{iK\cdot x} = +1\}$, $\{x: e^{iK\cdot x} = -1\}$; in view of the introduction of the
factor $e^{iK\cdot x}$ in (5.22) and (5.23) the inequalities (5.32) and (5.33) exclude
"spontaneous sublattice magnetization" and therefore antiferromagnetism.[28]

5.6 ONE–DIMENSIONAL SYSTEMS

The statistical mechanics of classical one-dimensional systems are rela-
tively tractable and at the same time relatively uninteresting; this is due to the
fact that no phase transition occurs in one-dimensional systems unless
"unnatural" long range forces are introduced. We shall successively consider
the case of lattice systems and continuous systems with hard core.

5.6.1 CLASSICAL LATTICE SYSTEMS. Let $(a, b]$ be a finite interval of \mathbf{Z}:

$$(a, b] = \{x \in \mathbf{Z}: a < x \le b\} \tag{6.1}$$

we write

$$Z_{b-a}(\Phi) = \sum_{n_{a+1}=0}^{N} \cdots \sum_{n_b=0}^{N} \exp[-U_{ab}(n_{a+1}, \ldots, n_b)] \tag{6.2}$$

$$P_{b-a}(\Phi) = (b - a)^{-1} \log Z_{b-a}(\Phi) \tag{6.3}$$

where

$$U_{ab}(n_{a+1}, \ldots, n_b) = \sum_{k \ge 1} \sum_{a < x_1 < \cdots < x_k \le b} \Phi_{\{x_1, \ldots, x_k\}}(n_{x_1}, \ldots, n_{x_k}) \tag{6.4}$$

and Φ is assumed to be translation invariant:

$$\Phi_{\{x_1+l, \ldots, x_k+l\}}(n_1, \ldots, n_k) = \Phi_{\{x_1, \ldots, x_k\}}(n_1, \ldots, n_k) \tag{6.5}$$

We shall consider interactions Φ with finite range; we suppose thus that
L exists such that, for all $k, y_1 < \cdots < y_k, n_1, \ldots, n_k$,

$$\Phi_{\{y_1, \ldots, y_k\}}(n_1, \ldots, n_k) = 0 \quad \text{if} \quad y_k - y_1 > L \tag{6.6}$$

We define vectors $\varphi_a \in \mathbf{R}^{(N+1)L}$ by

$$\varphi_a(n_1, \ldots, n_L) = \sum_{n-a+1=0}^{N} \cdots \sum_{n_0=0}^{N} \exp[-U_{-a,L}(n_{-a+1}, \ldots, n_L)] \tag{6.7}$$

[28] Examples of classical spin systems with antiferromagnetic behavior are given in Exercise 5.F.

then

$$Z_{a+L} = \sum_{n_1 = 0}^{N} \cdots \sum_{n_L = 0}^{N} \varphi_a(n_1, \ldots, n_L) \qquad (6.8)$$

$$\varphi_{a+1}(n_1, \ldots, n_L) = \sum_{n_0 = 0}^{N} \varphi_a(n_0, \ldots, n_{L-1}) F(n_1, \ldots, n_L) \qquad (6.9)$$

where

$$F(n_1, \ldots, n_L) = \exp\left[-\sum_{k \geq 1} \sum_{0 < x_1 < \cdots < x_{k-1} < L} \Phi_{\{x_1, \ldots, L\}}(n_{x_1}, \ldots, n_L) \right] \qquad (6.10)$$

We define a scalar product in $\mathbf{R}^{(N+1)L}$ by

$$(\psi, \varphi) = \sum_{n_1 = 0}^{N} \cdots \sum_{n_L = 0}^{N} \psi(n_1, \ldots, n_L) \varphi(n_1, \ldots, n_L) \qquad (6.11)$$

and let χ have all its components equal to one; then (6.8) may be written as

$$Z_{a+L} = (\chi, \varphi_a) \qquad (6.12)$$

We rewrite (6.9) as

$$\varphi_{a+1} = A\varphi_a \qquad (6.13)$$

where A is an $(N+1)^L \times (N+1)^L$ matrix. From (6.12) and (6.13) we obtain

$$Z_{a+L} = (\chi, A^a \varphi_0) \qquad (6.14)$$

We notice the following facts.

(a) *The matrix elements of A are greater than or equal to zero and the matrix elements of the Lth power A^L of A are strictly greater than zero.*

(b) *The components of χ and of φ_0 are strictly greater than zero.*

(c) *The matrix elements of A are real analytic functions of the variables* $\Phi_{\{0, x_2, \ldots, x_k\}}(n_1, \ldots, n_k)$.

Because of (a) we may apply Proposition 5.6.3 below; using also (b) we obtain

$$\lim_{a \to \infty} \lambda(\Phi)^{-a} Z_{a+l} = (\chi, \xi)(\eta, \varphi_0) > 0 \qquad (6.15)$$

and therefore

$$\lim_{b-a \to \infty} (b - a)^{-1} \log Z_{b-a}(\Phi) = \log \lambda(\Phi) \qquad (6.16)$$

Since the eigenvalue $\lambda(\Phi)$ remains simple, and because of (c), $\lambda(\Phi)$ is a real analytic function of the variables $\Phi_{\{0, x_2, \ldots, x_k\}}(n_1, \ldots, n_k)$. Thus we have proved the following theorem.

5.6.2 THEOREM. *Let \mathcal{B}_0 be the space of finite range interactions for a one-dimensional classical lattice system. If we let*

$$P(\Phi) = \lim_{b-a \to \infty} (b-a)^{-1} \log Z_{b-a}(\Phi) \tag{6.17}$$

then $P(\Phi)$ is a real analytical function on finite-dimensional subspaces of \mathcal{B}_0. Therefore no phase transition can occur for one-dimensional classical lattice systems with finite range interaction.

For completeness we establish now the property of matrices with positive elements which was used to obtain (6.15).

5.6.3 PROPOSITION.[29] *Let $A = (A_{ij})$ be an $n \times n$ real matrix and A^* be its adjoint, $(A_{ij}^*) = A_{ji}$. We assume that:*

(a) *$A_{ij} \geq 0$ for all i,j;*
(b) *there is an integer $L > 0$ such that $(A^L)_{ij} > 0$ for all i,j.*

Then there exists $\lambda > 0$ which is a simple eigenvalue for both A and A^ corresponding to eigenvectors ξ, η with strictly positive components and for all $x \in \mathbf{R}^n$*

$$\lim_{k \to \infty} \lambda^{-k} A^k x = (\eta, x)\xi \tag{6.18}$$

If $x \in \mathbf{R}^n$ and $x \neq 0$, a half-line $\gamma_x = \{\lambda x : \lambda > 0\}$ is defined; let $\Gamma = \{\gamma_x : x \neq 0$ and $x_1 \geq 0, \ldots, x_n \geq 0\}$ be the set of positive half-lines. By (a), (b) the mapping $A^* : \mathbf{R}^n \to \mathbf{R}^n$ induces a mapping of Γ into itself and, by Brouwer's fixed point theorem,[30] there exist $\lambda > 0$ and $\eta \neq 0$ such that $\eta_1 \geq 0, \ldots, \eta_n \geq 0$ and

$$A^*\eta = \lambda\eta \tag{6.19}$$

(b) implies that $\eta_1 > 0, \ldots, \eta_n > 0$. In particular the set

$$\Delta = \left\{ x : \sum_{i=1}^n \eta_i x_i = 1 \quad \text{and} \quad x_1 \geq 0, \ldots, x_n \geq 0 \right\} \tag{6.20}$$

is compact.

We have

$$(\eta, Ax) = \sum_{i=1}^n \sum_{j=1}^n \eta_j A_{ij} x_j = (A^*\eta, x) = \lambda(\eta, x) \tag{6.21}$$

Therefore $\lambda^{-1}A$ maps $H = \{x : (\eta, x) = 1\}$ into H and Δ into Δ. Again using Brouwer's fixed point theorem we see that there exists $\xi \in \Delta$ such that

$$A\xi = \lambda\xi \tag{6.22}$$

[29] Proposition 5.6.3 is a special case of the Perron–Frobenius theorem; see Gantmacher [1], Chapter XIII.
[30] Brouwer's fixed point theorem asserts that if T is a continuous mapping of the closed unit ball B_n of \mathbf{R}^n into itself, then T has a fixed point, i.e., there exists $x \in B$ such that $Tx = x$. The theorem applies here because Γ with the natural topology is homeomorphic to B_{n-1}.

By (b), $\xi_1 > 0, \ldots, \xi_n > 0$. Let \mathscr{A} be the restriction of $\lambda^{-1}A$ to H considered as a vector space with the origin ξ. The mapping \mathscr{A} is linear, $\mathscr{A}\Delta \subset \Delta$ and, by (b), $\mathscr{A}^L\Delta$ is contained in the interior of Δ. This implies that the (complex) eigenvalues of \mathscr{A} are strictly smaller than one in absolute value: \mathscr{A} is a contracting linear map. This gives immediately (6.18).

5.6.4 LATTICE GASES. We can specialize the lattice system described in 5.6.1 to a lattice gas by letting $N = 1$ and $\Phi_{\{x_1,\ldots,x_k\}}(n_1, \ldots, n_k) = 0$ unless $n_1 = \cdots = n_k = 1$. We then write

$$\Phi^k(x_1, \ldots, x_k) = \Phi_{\{x_1,\ldots,x_k\}}(1, \ldots, 1) \tag{6.23}$$

and (6.4) becomes

$$U(X) = \sum_{k \geq 1} \sum_{\{x_1,\ldots,x_k\} \subset X} \Phi^k(x_1, \ldots, x_k) \tag{6.24}$$

Let us write $K = \{0, 1\}$; then, for any set $S \subset \mathbf{Z}$, the product[31] K^S of one copy of $\{0, 1\}$ for every point of S may be considered as the space of configurations of occupied and empty sites in S. We define the Gibbs measure γ_{ab} to be the measure on $K^{(a,b]}$ which associates to every $(n_{a+1}, \ldots, n_b) \in K^{(a,b]}$ the mass $e^{-U(X)}$ where $X = \{x \in (a, b] : n_x = 1\}$. We let

$$Z_{b-a} = \sum_{n_{a+1}=0}^{1} \cdots \sum_{n_b=0}^{1} \gamma_{ab}(n_{a+1}, \ldots, n_b) \tag{6.25}$$

$$\bar{\gamma}_{ab} = Z_{b-a}^{-1}\gamma_{ab} \tag{6.26}$$

If $S \subset T \subset \mathbf{Z}$ we may write

$$K^T = K^S \times K^{T \setminus S} \tag{6.27}$$

If μ is a measure on K^T we obtain by partial summation with respect to the factor $K^{T \setminus S}$ a measure v on K^S; we write

$$v = \alpha_{ST}\mu$$

Theorem 5.6.2 shows that a one-dimensional lattice gas with finite range interaction exhibits no phase transition. We state now without proof a result valid without the assumption of finite range.

5.6.5 THEOREM.[32] *Let \mathscr{E} be the space of sequences $\Phi - (\Phi^k)_{k \geq 1}$ such that*

$$\sum_{l > 0} \sum_{0 < i_1 < \cdots < i_l} i_l|\Phi^{l+1}(0, i_1, \ldots, i_l)| < +\infty \tag{6.28}$$

[31] We give K the discrete topology and K^S the product topology (which makes it compact).
[32] See Ruelle [10].

If $\Phi \in \mathscr{E}$, then:

(a) *The following limit exists and is finite*

$$P(\Phi) = \lim_{b-a \to \infty} \frac{1}{b-a} \log Z_{ab} \qquad (6.29)$$

It is continuously differentiable on any finite-dimensional subspace of \mathscr{E}.

(b) *For every finite $S \subset \mathbf{Z}$ there exists a measure ρ_S on K^S such that*

$$\lim_{a \to -\infty, b \to \infty} \alpha_{S,(a,b]} \bar{\gamma}_{ab} = \rho_S \qquad (6.30)$$

There is a measure ρ on $K^{\mathbf{Z}}$ such that

$$\rho_S = \alpha_{SZ}\rho \qquad (6.31)$$

for all finite $S \subset \mathbf{Z}$, and ρ depends continuously on Φ on any finite-dimensional subspace of \mathscr{E} for the vague topology of measures.[33]

Notice that the measure ρ (on the space $K^{\mathbf{Z}}$ of all configurations of occupied and empty sites of the whole lattice \mathbf{Z}) describes an "equilibrium state" of the infinite system.[34] By part (b) of the theorem, ρ depends continuously on the temperature and chemical potential and this implies, in a certain sense, that no phase transition can occur.[35] For a pair potential ($\Phi^k = 0$ for $k > 2$), (6.28) becomes

$$\sum_{i > 0} i|\Phi^2(0, i)| < +\infty \qquad (6.32)$$

This condition ensures that the energy of interaction of all particles to the left of a point \mathbf{Z} with all the particles to the right is bounded.[36]

5.6.6 CLASSICAL CONTINUOUS SYSTEMS WITH HARD CORES. We conclude this section with a study of one-dimensional systems of particles with hard cores and finite range interaction. This study is analogous to that of lattice systems with finite range interaction. If Q denotes the configurational canonical partition function as in 3.4.3, we let[37]

$$\Delta(p, n, \beta) = \int_0^\infty d\lambda \, e^{-\beta p \lambda} Q((0, \lambda), n, \beta) \qquad (6.33)$$

[33] This last statement means that for every real continuous function φ on $K^{\mathbf{Z}}$, $\rho(\varphi)$ is a real continuous function of Φ on finite-dimensional subspaces of \mathscr{E}.
[34] The description given by the measure ρ is equivalent to the description provided by the correlation functions (see 7.1.1).
[35] See the discussion in Section 5.7. The continuity of ρ implies that the correlation functions depend continuously on z, β.
[36] If $\Phi^2 \le 0$ and (6.32) is violated, the existence of a phase transition has been conjectured by Fisher [3] and Kac (private communication).
[37] Δ is the "isothermal-isobaric" partition function.

and write

$$\lim_{n \to \infty} n^{-1} \log \Delta(p, n, \beta) = -\beta\gamma(p, \beta) \tag{6.34}$$

With the techniques of Chapter 3 we check that this limit exists and that

$$\gamma(p, \beta) = \inf_{\rho}(\rho^{-1}p + \rho^{-1}f(\rho, \beta)) \tag{6.35}$$

Comparison with (4.18) of Chapter 3 shows that if we put $\log z = \beta\gamma(p, \beta)$, then $p = p(z, \beta)$. Notice that $\gamma(p, \beta)$ is an increasing function of p.

5.6.7 THEOREM (*Van Hove*).[38] *We consider a one-dimensional continuous system with pair interaction. We assume that the pair potential Φ has a hard core [$\Phi(x) = +\infty$ for $|x| < R > 0$], finite range [$\Phi(x) = 0$ for $|x| \geq R_0$], and that $\Phi(x)$ is continuous for $|x| \geq R$. Under these conditions $p(z, \beta)$ is a real analytic function of $z > 0$. Therefore the system exhibits no phase transition.*

According to the above remarks it suffices to prove that $\gamma(p, \beta)$ is an analytic function of $p > 0$.

Let the integer k satisfy $k + 1 > R_0/R$; then in the expression

$$U(x_1, \ldots, x_n) = \sum_{i < j} \Phi(x_j - x_i)$$

the sum extends effectively only to the k nearest neighbors of each particle. We have thus

$$Q((0, \lambda), n, \beta) = \frac{1}{n!} \int_{(0,\lambda)^n} dx_1 \cdots dx_n \exp[-\beta U(x_1, \ldots, x_n)]$$

$$= \int_0^\lambda dx_n \int_0^{x_n} dx_{n-1} \cdots \int_0^{x_2} dx_1 \prod_{0 < j-i \leq k} \exp[-\beta\Phi(x_j - x_i)]$$

$$\tag{6.36}$$

Let $y_1 \leq y_2 \leq \cdots \leq y_k = 0$, $\lambda = y'_1 \leq y'_2 \leq \cdots \leq y'_k$, and $\eta_i = y_{i+1} - y_i$, $\eta'_i = y'_{i+1} - y'_i$ (see Fig. 17). For $n \geq k$ we write[39]

$$Q_n^\lambda(\eta, \eta') = Q_n^\lambda(\eta_1, \ldots, \eta_{k-1}; \eta'_1, \ldots, \eta'_{k-1})$$

$$= \frac{1}{n!} \int_{(0,\lambda)^n} dx_1 \cdots dx_n \left[\prod_{i=1}^k \prod_{j=1}^n \exp[-\beta\Phi(x_j - y_i)] \right]$$

$$\times \exp[-\beta U(x_1, \ldots, x_n)] \left[\prod_{i=1}^k \prod_{j=1}^n \exp[-\beta\Phi(y'_i - x_j)] \right]$$

[38] See Van Hove [2].
[39] The condition $n \geq k$ is really introduced only to simplify the notation.

FIGURE 17 Definition of the variables ξ_i, η_i, η_i'.

$$= \int_0^\lambda dx_n \int_0^{x_n} dx_{n-1} \cdots \int_0^{x_2} dx_1 \left[\prod_{i=1}^k \prod_{j=1}^i \right.$$

$$\times \exp[-\beta\Phi(x_j - y_i)] \left] \left[\prod_{0 < j-i \le k} \exp[-\beta\Phi(x_j - x_i)] \right] \right.$$

$$\times \left[\prod_{i=1}^k \prod_{j=n-k+i}^n \exp[-\beta\Phi(y_i' - x_j)] \right] \tag{6.37}$$

Letting $-M$ be a lower bound to Φ, we have the inequalities

$$Q((R_0, \lambda - R_0), n, \beta) \le Q_n^\lambda(\eta, \eta') \le e^{k(k+1)\beta M} Q((0, \lambda), n, \beta) \tag{6.38}$$

Let us define

$$\Delta_n^p(\eta, \eta') = \int_0^\infty d\lambda\, e^{-\beta p\lambda} Q_n^\lambda(\eta, \eta') \tag{6.39}$$

It follows from (6.37) and (6.39) that Δ_n^p is positive, continuous, and bounded; from (6.38) we obtain

$$\exp[-2\beta p R_0]\Delta(p, n, \beta) \le \Delta_n^p(\eta, \eta') \le e^{k(k+1)\beta M}\Delta(p, n, \beta) \tag{6.40}$$

which, together with (6.34), implies

$$\lim_{n \to \infty} n^{-1} \log \Delta_n^p(\eta, \eta') = -\beta\gamma(p, \beta) \tag{6.41}$$

Introducing the variables $\xi_0 = x_1 - y_k = x_1$, $\xi_1 = x_2 - x_1, \ldots, \xi_{n-1} = x_n - x_{n-1}, \xi_n = y_1' - x_n = \lambda - x_n$ (see Fig. 17) we may write

$$\int_0^\infty d\lambda\, e^{-\beta p\lambda} \int_0^\lambda dx_n \int_0^{x_n} dx_{n-1} \cdots \int_0^{x_2} dx_1 \cdots$$

$$= \int_0^\infty dx_n \int_0^{x_n} dx_{n-1} \cdots \int_0^{x_2} dx_1 \int_{x_n}^\infty d\lambda\, e^{-\beta p\lambda} \cdots$$

$$= \int_0^\infty d\xi_0 \int_0^\infty d\xi_1 \cdots \int_0^\infty d\xi_{n-1} \int_0^\infty d\xi_n \exp\left[-\beta p \sum_0^n \xi_i\right] \cdots$$

$$= \int_0^\infty d\xi_0 \exp[-\beta p\xi_0] \int_0^\infty d\xi_1 \exp[-\beta p\xi_1] \cdots$$

$$\cdots \int_0^\infty d\xi_{n-1} \exp[-\beta p\xi_{n-1}] \int_0^\infty d\xi_n \exp[-\beta p\xi_n] \cdots$$

Therefore

$$\Delta^p_{n_1+n_2+k}(\eta,\eta') = \int_0^\infty d\zeta_1 \exp[-\beta p \zeta_1] \cdots \int_0^\infty d\zeta_{k-1} \exp[-\beta p \zeta_{k-1}]$$

$$\times \exp\left[-\beta \sum_{1 \le i < j \le k} \Phi(\zeta_i + \cdots + \zeta_{j-1})\right] \Delta^p_{n_1}(\eta,\zeta)\Delta^p_{n_2}(\zeta,\eta')$$

$$(6.42)$$

In this formula the integrations can be restricted to the interval $(R, +\infty)$. Going over to the variables $\tilde{\eta}_i = \exp[-\beta p \eta_i]/\beta p$, the integration interval becomes finite:

$$\int_R^\infty d\eta_i \exp[-\beta p \eta_i] \cdots = \int_0^{\beta^{-1}p^{-1}e^{\beta pR}} d\tilde{\eta}_i \cdots$$

We may thus write

$$\Delta^p_{n_1+n_2+k}(\tilde{\eta},\tilde{\eta}') = \int d\tilde{\zeta}\Delta^p_{n_1}(\tilde{\eta},\tilde{\zeta})\rho(\tilde{\zeta})\Delta^p_{n_2}(\tilde{\zeta},\tilde{\eta}') \qquad (6.43)$$

where the functions $\Delta_n^p(\tilde{\eta},\tilde{\zeta})$ and $\rho(\tilde{\zeta})$ are continuous and strictly positive.

In view of (6.41) we can determine $\gamma(p,\beta)$ by studying the iterates of the kernel $\Delta_n^p(\tilde{\xi},\tilde{\eta})$, multiplication being defined by (6.43).

Consider now the Fredholm equation

$$f(\tilde{\xi}) = t \int d\tilde{\eta}\Delta_n^p(\tilde{\xi},\tilde{\eta})\rho(\tilde{\eta})f(\tilde{\xi}) \qquad (6.44)$$

The corresponding resolvent series

$$\sum_{q=1}^\infty t^q \Delta^p_{qn+(q-1)k}(\tilde{\xi},\tilde{\eta}) = [(1-t\Delta_n^p)^{-1} - 1](\tilde{\xi},\tilde{\eta})$$

converges in a circle with radius t_0 given, according to (6.41), by

$$t_0^{-1} = \lim_{q\to\infty} \sup [\Delta^p_{qn+(q-1)k}]^{1/q} = \exp[-(n+k)\beta\gamma(p,\beta)] \qquad (6.45)$$

On the other hand t_0 is given by the zero of the Fredholm determinant which is closest to the origin. Using the fact that the kernel is positive, we can show that the zero of the Fredholm determinant which is closest to the origin is simple and positive.[40] Furthermore, since the kernel is an analytic function of p, this zero is also. We see thus that t_0, and hence $\gamma(p,\beta)$, is an analytic function of $p > 0$, which concludes the proof of Theorem 5.6.7.

[40] This is an extension of the Perron–Frobenius theory [see (6.3)] to infinite-dimensional spaces; see Jentzsch [1].

5.7 THE NOTION OF PHASE TRANSITION

We have seen in this chapter and in the preceding one a certain number of situations where the thermodynamic functions are analytic in parts of the physical domain of the thermodynamic parameters. This supports the idea that the thermodynamic functions are piecewise analytic,[41] with singularities corresponding to phase transitions (see in particular 5.4.8). However, no general mechanism is known which would explain this piecewise analyticity.

Another view consists in defining a phase transition as a discontinuity in the correlation functions when the thermodynamic parameters are varied. This discontinuity may separate regions where the correlation functions have qualitatively different properties (e.g., ferromagnetic or antiferromagnetic behavior may appear in a spin system when β^{-1} is decreased below a certain Curie temperature).

It is unlikely that the two definitions of a phase transition that we have indicated are strictly equivalent. We may however hope for equivalence "in general" on the basis of an idea that we sketch now very briefly. It is known that the functional derivative of the thermodynamic function $P(\Phi)$ with respect to the k-body potential Φ^k is, formally, the k-body correlation function.[42] Suppose now that P is piecewise analytic in Φ in some sense; then passing from one piece of analytic function to another by modifying z, β will, in general, produce both a singularity of the thermodynamic function P and a discontinuity of the correlation functions, corresponding to a phase transition. Unfortunately we are far from being able to prove the analyticity needed for this argument to work.

One possible way of attacking the problem of phase transitions is to first try to define a phase. We shall use this approach in the next chapter.

BIBLIOGRAPHICAL NOTE

The theorem of Lee and Yang was published in 1952, the idea of Peierls developed in Section 5.3 in 1936, and the Theorem 5.6.7 of Van Hove in 1950. The rest of the material in this chapter is much more recent, many of the ideas being due to Griffiths (for Sections 5.3 and 5.4) and to Dobrushin (for Sections 5.2 and 5.3).

We have followed Lee and Yang [1] in Section 5.1, Gallavotti, Miracle, and Robinson [1] in Section 5.2, Ginibre, Grossmann, and Ruelle [1] in Section 5.3, Kelly and Sherman [1] (in part) in Section 5.4, Mermin and Wagner [1] in Section 5.5, and Van Hove [2] (in part) in Section 5.6.

[41] We have in this chapter looked only at the analyticity in z, disregarding the variable β. For the analyticity in β, see Lebowitz and Penrose [2].

[42] See 7.3.1.

EXERCISES

5.A. With the notation and assumptions of Proposition 5.1.1 show that

$$\mathscr{P}_n(-1,\ldots,-1,1,\ldots,1) \geq 0$$

Show that if $\langle \sigma^X \rangle$ is computed with respect to the interaction (4.54), $\langle \sigma^X \rangle$ can be put in the form

$$\frac{\mathscr{P}_n(-1,\ldots,-1,1,\ldots,1)}{\mathscr{P}_n(1,\ldots,1)} \geq 0$$

This proves (4.6) for the interaction (4.54). Extend this result to the case of the interaction (4.56).

5.B. The inequalities (4.7) contain as special cases

$$\langle \sigma_x \sigma_y \sigma_u \sigma_v \rangle - \langle \sigma_x \sigma_y \rangle \langle \sigma_u \sigma_v \rangle \geq 0$$

$$\langle \sigma_x \sigma_y \rangle - \langle \sigma_x \sigma_y \sigma_u \sigma_v \rangle \langle \sigma_u \sigma_v \rangle \geq 0$$

$$\langle \sigma_x \sigma_y \rangle - \langle \sigma_x \sigma_u \rangle \langle \sigma_u \sigma_y \rangle \geq 0$$

5.C. With the notation of Section 5.4 show that

$$\frac{\partial \langle \sigma^X \rangle}{\partial J_X} - \frac{\partial \langle \sigma^Y \rangle}{\partial J_X} = 1 - \langle \sigma^X \rangle^2 - \langle \sigma^X \sigma^Y \rangle + \langle \sigma^X \rangle \langle \sigma^Y \rangle$$

$$= Z^{*-2}[4(pr + 2qr + qs)] \geq 0$$

5.D.[43] For each finite $X \subset \mathbf{Z}^\nu$ let $0 \leq J_X \leq +\infty$, and let J_X be translation invariant. Given a finite $\Lambda \subset \mathbf{Z}^\nu$ we define the average $\langle \ \rangle_\Lambda$ by (4.14). Show that if $X \subset \Lambda \subset \Lambda'$, then $\langle \sigma^X \rangle_\Lambda \leq \langle \sigma^X \rangle_{\Lambda'}$. Therefore if $\Lambda \to \infty$ in the sense that every finite set is eventually contained in Λ, then the following limit exists

$$\lim_{\Lambda \to \infty} \langle \sigma^X \rangle_\Lambda = \langle \sigma^X \rangle_\infty \tag{*}$$

and is translation invariant. Show that the magnetization per spin defined by (4.57) is $\langle \sigma_x \rangle_\infty$.

From (*) we deduce the existence of the infinite volume correlation functions for the lattice gas corresponding to the above spin system:

$$\rho(x_1,\ldots,x_m) = \left\langle \frac{1 + \sigma_{x_1}}{2} \cdots \cdots \frac{1 + \sigma_{x_m}}{2} \right\rangle_\infty$$

5.E. For each $x \in \mathbf{Z}^\nu$ let $\varepsilon_x = \pm 1$ so that

$$\varepsilon_x \varepsilon_y = \varepsilon_{x+y}$$

If $(J_{\{x,y\}})$ is translation invariant interaction (for a classical spin system), so is

[43] See Griffiths [4],II.

$(\tilde{J}_{\{x,y\}})$ with $\tilde{J}_{\{x,y\}} = \varepsilon_{x-y}J_{\{x,y\}}$. If we also set $\tilde{\sigma}_x = \varepsilon_x\sigma_x$, then

$$\sum_{\{x,y\}\subset\Lambda} J_{\{x,y\}}\sigma_x\sigma_y + h\sum_{x\in\Lambda}\sigma_x = \sum_{\{x,y\}\subset\Lambda} \tilde{J}_{\{x,y\}}\tilde{\sigma}_x\tilde{\sigma}_y + h\sum_{x\in\Lambda}\varepsilon_x\tilde{\sigma}_x$$

Check that if the spin system with interaction J exhibits ferromagnetism, the spin system with interaction \tilde{J} exhibits antiferromagnetism in the sense of 5.5.3 (assuming that ε_x is not identically 1).

5.F. Consider a spin system with potential energy

$$U'(\sigma) = -\tfrac{1}{2}\sum_{x:\sigma_x=+1}\sum_{y:\sigma_y=-1}\Phi(\{x, y\})$$

Translate Theorems 5.3.1 and 5.4.7 into statements about ferromagnetism of this system. Using Exercise 5.E obtain corresponding results for anti-ferromagnetism. In particular show that if the interaction satisfies $\Phi_i > D_i$, $i = 1,\ldots,\nu \geq 2$ (with the notation of 5.3.1), then the system is antiferro-magnetic at low β^{-1}.

GROUP INVARIANCE OF PHYSICAL STATES

The states of a physical system may be described by positive linear functionals on a suitable B*-algebra. An important property of the equilibrium states of infinite systems in statistical mechanics is their invariance under some group (translations in \mathbf{Z}^ν or \mathbf{R}^ν, Euclidean transformations, etc.). In this chapter we analyze the general properties of states which are invariant under a group of automorphisms of a B*-algebra.

6.1 DESCRIPTION OF THE STATE OF AN INFINITE SYSTEM

The state of an infinite system in statistical mechanics is traditionally described by correlation functions in the classical case, by reduced density matrices in the quantum case (see Chapter 4). In the classical case we may also use a probability measure on the "space of infinite configurations" of the system (see Section 5.6.5). Another possibility (see Section 1.4) is to describe the restriction of the infinite system to each bounded space region Λ by a measure on the finite configurations in this region (classical case) or by a density matrix on the Fock space of this region (quantum case).

To unify these various descriptions we shall in this chapter adopt the point of view that a state is an expectation value functional (as indicated in Section 4.7). We admit thus that our physical theory contains a certain number of observables A_i to which the state associates an expectation value $\langle A_i \rangle$. To obtain a mathematically useful notion it is necessary to give some structure to the set $\{A_i\}$ of observables. In the cases to be studied we shall see that it is possible to consider the observables as self-adjoint elements of some B*-*algebra* \mathfrak{A} with identity 1. The expectation value will extend to a

linear functional on \mathfrak{A} which is positive (i.e., if $A \in \mathfrak{A}$ and $A \geq 0$, then $\langle A \rangle \geq 0$) and normalized (i.e., $\langle 1 \rangle = 1$); such a linear functional is in fact called a *state* on the B*-algebra \mathfrak{A}.

For the part which will interest us, the theory of B*-algebras may be viewed as a noncommutative extension of classical measure theory on a compact space. While this extension is not extremely difficult, it is more sophisticated than the mathematical tools used earlier in this monograph. The reader will in fact have to face in this chapter a few other "less classical" notions. For convenience we have collected the relevant results (without proofs) in the Appendix. It is desirable at this point that the reader acquaint himself with the contents of Sections A.2 and A.3 which concern B*-algebras.

The use of B*-algebras in physics has been promoted by Segal [2], Haag, and others.[1] Let us come back to the physical interpretation of the B*-algebra \mathfrak{A} as "the algebra generated by the observables." It should be said that what the "observables" are in a given physical problem is often not obvious and that there is generally some ambiguity in the choice of the B*-algebra \mathfrak{A}. Furthermore, usually, not all states on \mathfrak{A} are physically interesting. For a given physical problem we have thus not only to choose a B*-algebra \mathfrak{A} but also to impose restrictions of physical origin on the states. It follows that the mathematical structure that we associate with a physical problem is practically always richer than just that of a B*-algebra.

In many physical problems there is a naturally defined *invariance group*. We shall admit that this group acts as a group of automorphisms[2] on \mathfrak{A}. It is often natural to restrict our attention to *invariant states*, i.e., to states which are invariant under the group operations. Another type of condition consists in requiring that the restriction of a state to a subalgebra of \mathfrak{A} be of norm 1; a condition of this sort will be used to express that the number of particles in a bounded region $\Lambda \subset \mathbf{R}^\nu$ is essentially finite.

The ensemble averages of Chapter 1 define expectation value functionals, that is, states on suitable B*-algebras. Such a state ρ_Λ describes a finite system enclosed in a region Λ. It is reasonable to expect that when $\Lambda \to \infty$, ρ_Λ tends (in some sense) to a state ρ describing an infinite system and invariant under the translations of \mathbf{R}^ν or \mathbf{Z}^ν or the Euclidean transformations of \mathbf{R}^ν. It is naturally important to describe the B*-algebra \mathfrak{A} on which ρ is a state (see Section 7.1). It is also important to study the existence of the limit $\rho_\Lambda \to \rho$ (see Section 7.3). In this chapter we shall however concentrate on the implications of the group invariance of ρ, making very few further assumptions.

[1] There is an abundant literature on the use of B*-algebras and von Neumann algebras in elementary particle physics. Of relevance to statistical mechanics we mention the work of Gårding and Wightman [1], [2] on commutation relations, and the papers by Araki and Woods [1] and Araki and Wyss [1].

[2] See A.2.9; conjugate automorphisms α could also be considered: $\alpha(A_1 + \lambda A_2) = \alpha(A_1) + \bar{\lambda}\alpha(A_2)$, $\alpha(A_1 A_2) = \alpha(A_1)\alpha(A_2)$; they will play no role in what follows.

6.2 INVARIANT STATES

6.2.1 NOTATION AND DEFINITIONS. Let \mathfrak{A} be a B*-algebra and aut(\mathfrak{A}) the group of its automorphisms. We call *B*-algebra with a group of automorphisms* a triple (\mathfrak{A}, G, τ) where G is a group and $\tau : G \to$ aut(\mathfrak{A}) is a (group-) homomorphism.

If $g \in G$ we denote by τ_g the corresponding automorphism of \mathfrak{A}. We denote by τ_g' the linear operator on the dual \mathfrak{A}' of \mathfrak{A} such that if $A \in \mathfrak{A}$, $f \in \mathfrak{A}'$, then

$$\tau_g' f(A) = f(\tau_{g-1}A) \tag{2.1}$$

Let \mathscr{L}_G be the subspace of \mathfrak{A} generated by the elements of the form $A - \tau_g A$ with $A \in \mathfrak{A}$, $g \in G$, then

$$\mathscr{L}_G^\perp = \{f \in \mathfrak{A}' : A \in \mathscr{L}_G \Rightarrow f(A) = 0\} \tag{2.2}$$

is the space of G-invariant continuous linear forms on \mathfrak{A}.

If E is the set of states on \mathfrak{A}, $E \cap \mathscr{L}_G^\perp$ is the set of *G-invariant states*. \mathscr{L}_G^\perp is (weakly) closed in \mathfrak{A}'; hence, if \mathfrak{A} has an identity, $E \cap \mathscr{L}_G^\perp$ is (weakly) compact and convex.

If we perform the Gel'fand–Segal construction A.3.5 corresponding to an invariant state ρ, an interesting situation is obtained.

6.2.2 THEOREM (*Segal*).[3] *Let (\mathfrak{A}, G, τ) be a B*-algebra with a group of automorphisms. If $\rho \in E \cap \mathscr{L}_G^\perp$ and $(\mathfrak{H}_\rho, \pi_\rho, \Omega_\rho)$ is the corresponding cyclic representation of \mathfrak{A}, there exists a unique unitary representation U_ρ of G in \mathfrak{H}_ρ such that for all $g \in G$, $A \in \mathfrak{A}$,*

$$U_\rho(g)\pi_\rho(A)U_\rho(g)^{-1} = \pi_\rho(\tau_g A) \tag{2.3}$$

$$U_\rho(g)\Omega_\rho = \Omega_\rho \tag{2.4}$$

If G is a topological group and the functions $g \to \sigma(\tau_g A)$ on G are continuous (for all $A \in \mathfrak{A}$, $\sigma \in E$),[4] then U_ρ is a strongly continuous unitary representation of G in \mathfrak{H}_ρ.

If $g \in G$, the isometric operator $U_\rho(g)$ is defined on the dense set $\pi_\rho(\mathfrak{A})\Omega_\rho$ by

$$U_\rho(g)\pi_\rho(A)\Omega_\rho = \pi_\rho(\tau_g A)\Omega_\rho \tag{2.5}$$

and extends by continuity to \mathfrak{H}_ρ. This definition satisfies the theorem and is unique in view of (2.3) and (2.4). In the topological case the continuity of the representation results from

$$\lim_{g \to 1} \|(U_\rho(g) - U_\rho(1))\pi_\rho(A)\Omega_\rho\|^2$$

$$= 2\rho(A^*A) - \lim_{g \to 1} \rho(\tau_g A^* \cdot A) - \lim_{g \to 1} \rho(A^* \cdot \tau_g A) = 0 \tag{2.6}$$

[3] See Segal [1].

[4] It is in fact sufficient to assume the continuity of the functions $g \to \rho(A^* \cdot \tau_g A)$ for all $A \in \mathfrak{A}$.

6.2.3 THE ABELIAN SITUATION. Suppose that \mathfrak{A} is Abelian and has an identity; then the Gel'fand isomorphism A.3.9 identifies \mathfrak{A} with the algebra of complex continuous functions on a compact set K (the spectrum). An automorphism τ of \mathfrak{A} produces a permutation $\tilde{\tau}$ of its maximal ideals and it is readily seen that $\tilde{\tau}$ is a homeomorphism of K. Given a triple (\mathfrak{A}, G, τ) and $\rho \in E \cap \mathscr{L}_G^\perp$, let μ be the probability measure on K corresponding to ρ by the Gel'fand isomorphism. The G-invariance of ρ is equivalent to the invariance of μ under the homeomorphisms $\tilde{\tau}_g, g \in G$. It follows that the study of invariant states on an Abelian B*-algebra is the study of invariant measures, which is part of ergodic theory.

6.2.4 "QUASILOCAL" ALGEBRAS. The B*-algebras that we shall use to describe the infinite systems of statistical mechanics are typically obtained in the following manner. We associate B*-algebras \mathfrak{A}_Λ with bounded subsets Λ of \mathbf{R}^ν (or finite subsets Λ of \mathbf{Z}^ν) for $\nu > 0$, and we assume that:

(a) If $\Lambda \subset \Lambda'$, an isomorphism $\alpha_{\Lambda'\Lambda}$ of \mathfrak{A}_Λ into $\mathfrak{A}_{\Lambda'}$ is given such that if $\Lambda \subset \Lambda' \subset \Lambda''$, then $\alpha_{\Lambda''\Lambda} = \alpha_{\Lambda''\Lambda'}\,\alpha_{\Lambda'\Lambda}$.

(b) An isomorphism τ_x of \mathfrak{A}_Λ onto $\mathfrak{A}_{\Lambda+x}$ is given for each translation x of \mathbf{R}^ν or \mathbf{Z}^ν and each Λ, such that $\tau_{x+y} = \tau_x \cdot \tau_y, \tau_{-x} = (\tau_x)^{-1}$; and if $\Lambda \subset \Lambda'$, then $\tau_x \alpha_{\Lambda'\Lambda} = \alpha_{\Lambda'+x,\Lambda+x} \tau_x$.

(c) If $\Lambda \cap \Lambda' = \varnothing$ and $A \in \mathfrak{A}_\Lambda$, $A' \in \mathfrak{A}_{\Lambda'}$, then $\alpha_{\Lambda \cup \Lambda',\Lambda} A$ commutes with $\alpha_{\Lambda \cup \Lambda',\Lambda'} A'$.

Because of assumption (a) the \mathfrak{A}_Λ may be identified with subalgebras of an algebra $\tilde{\mathfrak{A}}$ such that $\tilde{\mathfrak{A}}$ is the union of the \mathfrak{A}_Λ. A norm is naturally defined on $\tilde{\mathfrak{A}}$, but $\tilde{\mathfrak{A}}$ is generally not complete; by completing $\tilde{\mathfrak{A}}$ we obtain a B*-algebra \mathfrak{A}. Furthermore, because of (b) there is an obviously defined (group-) homomorphism $\tau : G \to \mathrm{aut}(\mathfrak{A})$ where G is \mathbf{R}^ν or \mathbf{Z}^ν. The triple (\mathfrak{A}, G, τ) so obtained is sometimes called the "quasilocal" algebra constructed from the "local" algebras \mathfrak{A}_Λ.

We may also define a quasilocal algebra as a triple (\mathfrak{A}, G, τ), where G is \mathbf{R}^ν (or \mathbf{Z}^ν), together with a family of subalgebras \mathfrak{A}_Λ of \mathfrak{A}, labeled by bounded subsets of \mathbf{R}^ν (or \mathbf{Z}^ν). We assume that the following conditions are satisfied:

(a') If $\Lambda \subset \Lambda'$, then $\mathfrak{A}_\Lambda \subset \mathfrak{A}_{\Lambda'}$.

(b') $\tau_x \mathfrak{A}_\Lambda = \mathfrak{A}_{\Lambda+x}$.

(c') If $\Lambda \cap \Lambda' = \varnothing$, and $A \in \mathfrak{A}_\Lambda$, $A' \in \mathfrak{A}_{\Lambda'}$, then $[A, A'] = 0$.

(d') The union of the \mathfrak{A}_Λ is dense in \mathfrak{A}.

6.2.5 G-ABELIAN B*-ALGEBRAS. Consider a triple (\mathfrak{A}, G, τ) and let $\rho \in E \cap \mathscr{L}_G^\perp$. We shall denote by P_ρ the projection on the subspace of \mathfrak{H}_ρ formed by the vectors invariant under $U_\rho(G)$, where U_ρ is the representation of G introduced in Theorem 6.2.2. We shall investigate in Sections 6.3 and 6.4 the situation where the von Neumann algebra $[P_\rho \pi_\rho(\mathfrak{A}) P_\rho]''$ is Abelian,

i.e., where, for all $A_1, A_2 \in \mathfrak{A}$,

$$[P_\rho \pi_\rho(A_1)P_\rho, P_\rho \pi_\rho(A_2)P_\rho] = 0 \tag{2.7}$$

6.2.6 DEFINITION.[5] \mathfrak{A} is said to be G-Abelian if for all $\rho \in E \cap \mathscr{L}_G^\perp$, the von Neumann algebra generated by $P_\rho \pi_\rho(\mathfrak{A})P_\rho$ is Abelian. In other words \mathfrak{A} is G-Abelian if (2.7) holds for all $\rho \in E \cap \mathscr{L}_G^\perp$ and $A_1, A_2 \in \mathfrak{A}$.

We shall see that the Abelian and quasilocal algebras are examples of G-Abelian algebras. An important part of the theory of invariant measures extends to invariant states on G-Abelian algebras.

For the convenience of the reader we restate here the following classical result.

6.2.7 THEOREM (Alaoglu–Birkhoff).[6] Let \mathscr{U} be a semigroup of contractions of a Hilbert space \mathfrak{H}, i.e., a set of operators on \mathfrak{H} such that:

(a) $\|U\| \le 1$ for all $U \in \mathscr{U}$.
(b) If $U, U' \in \mathscr{U}$, then $U \cdot U' \in \mathscr{U}$.

Let P be the orthogonal projection on the space of vectors of \mathfrak{H} invariant under all the $U \in \mathscr{U}$. Then P is in the strong closure of the convex hull of \mathscr{U}.

We now give a usable characterization of G-Abelian algebras.

6.2.8 THEOREM.[7] The B*-algebra \mathfrak{A} is G-Abelian if and only if, for all self-adjoint $A_1, A_2 \in \mathfrak{A}$ and all $\rho \in E \cap \mathscr{L}_G^\perp$,

$$\inf_{A_1'} |\rho([A_1', A_2])| = 0 \tag{2.8}$$

where A_1' runs over the convex hull of $\{\tau_g A_1 : g \in G\}$.

It follows from the definition that \mathfrak{A} is G-Abelian if and only if for all $\rho \in E \cap \mathscr{L}_G^\perp$, for all $\Psi \in P_\rho \mathfrak{H}_\rho$ with $\|\Psi\| = 1$, and for all self-adjoint $A_1, A_2 \in \mathfrak{A}$ with $\|A_1\| \le 1$, $\|A_2\| \le 1$, we have

$$(\Psi, \pi_\rho(A_1)P_\rho \pi_\rho(A_2)\Psi) = (\Psi, \pi_\rho(A_2)P_\rho \pi_\rho(A_1)\Psi) \tag{2.9}$$

We first show that (2.8) implies (2.9). By Theorem 6.2.7, given $\varepsilon > 0$ there exist $\lambda_1 \ge 0, \ldots, \lambda_m \ge 0$ with $\sum \lambda_i = 1$ and $g_1, \ldots, g_m \in G$ such that

$$\|(\sum \lambda_i U_\rho(g_i) - P_\rho)\pi_\rho(A_1)\Psi\| < \varepsilon/2 \tag{2.10}$$

Therefore if we write

$$A_1' = \sum \lambda_i \tau_{g_i} A_1 \tag{2.11}$$

[5] See Lanford and Ruelle [1].
[6] See Riesz and Nagy [1].
[7] See Lanford and Ruelle [1].

we get

$$\|P_\rho \pi_\rho(A_1)\Psi - U_\rho(g)\pi_\rho(A_1')\Psi\| = \|U_\rho(g)[P_\rho \pi_\rho(A_1)\Psi - \pi_\rho(A_1')\Psi]\| \leq \varepsilon/2$$

$$(2.12)$$

for all $g \in G$. Given $\lambda_1' \geq 0, \ldots, \lambda_n' \geq 0$ such that $\sum \lambda_i' = 1$ and $g_1', \ldots, g_n' \in G$, we find, using (2.12),

$$|(\Psi, \pi_\rho(A_1)P_\rho \pi_\rho(A_2)\Psi) - (\Psi, \pi_\rho(A_2)P_\rho \pi_\rho(A_1)\Psi)|$$

$$\leq 2 \sum_i \lambda_i' \|\pi_\rho(A_2)\Psi\| \cdot \|P_\rho \pi_\rho(A_1)\Psi - U_\rho(g_i')\pi_\rho(A_1')\Psi\|$$

$$+ |(\Psi, [\sum \lambda_i' \pi_\rho(\tau_{g_i}A_1'), \pi_\rho(A_2)]\Psi)|$$

$$\leq \varepsilon + |(\Psi, \pi_\rho([\sum \lambda_i' \tau_{g_i}A_1', A_2])\Psi)| \qquad (2.13)$$

Since $(\Psi, \pi_\rho(\cdot)\Psi)$ is an invariant state, (2.8) implies that by appropriate choice of the λ_i', g_i, we have

$$|(\Psi, \pi_\rho([\sum \lambda_i' \tau_{g_i}A_1', A_2])\Psi)| < \varepsilon \qquad (2.14)$$

The left-hand side of (2.13) is thus smaller than 2ε, proving (2.8).

We show now that (2.9) implies (2.8). Using (2.9) we have, for λ_i, g_i satisfying (2.10),

$$|(\Psi, \pi_\rho([\sum \lambda_i \tau_{g_i}A_1, A_2])\Psi)| \leq 2\|\pi_\rho(A_2)\Psi\| \cdot \left\| \left[\sum_i \lambda_i U_\rho(g_i) - P_\rho \right] \pi_\rho(A_1)\Psi \right\|$$

$$+ |(\Psi, \pi_\rho(A_1)P_\rho \pi_\rho(A_2)\Psi)$$

$$- (\Psi, \pi_\rho(A_2)P_\rho \pi_\rho(A_1)\Psi)|$$

$$\leq \varepsilon \qquad (2.15)$$

Therefore, taking $\Psi = \Omega_\rho$, we obtain

$$|\rho([A_1', A_2])| \leq \varepsilon \qquad (2.16)$$

which proves (2.8).

6.2.9 COROLLARY. *Let H be a subgroup of G. Then, if \mathfrak{A} is H-Abelian, \mathfrak{A} is also G-Abelian.*

This follows from Theorem 6.2.8 if we observe that $\mathscr{L}_G^\perp \subset \mathscr{L}_H^\perp$ and the convex hull of $\{\tau_g A_1 : g \in G\}$ contains the convex hull of $\{\tau_g A_1 : g \in H\}$.

6.2.10 COROLLARY. *\mathfrak{A} is G-Abelian whenever one of the following conditions is satisfied.*

(a) *$E \cap \mathscr{L}_G^\perp$ is empty.*
(b) *\mathfrak{A} is Abelian.*

(c) *For all $\rho \in E \cap \mathscr{L}_G^{\perp}$ and all self-adjoint $A_1, A_2 \in \mathfrak{A}$,*

$$\inf_{g \in G} |\rho([\tau_g A_1, A_2])| = 0 \qquad (2.17)$$

(d) *G is a locally compact noncompact group and \mathfrak{A} is "weakly asymptotically Abelian" in the sense that for all $A_1, A_2 \in \mathfrak{A}$ and $\sigma \in E$*

$$\lim_{g \to \infty} |\sigma([\tau_g A_1, A_2])| = 0 \qquad (2.18)$$

(e) *\mathfrak{A} is "asymptotically Abelian," i.e., (2.18) is replaced in (d) by*

$$\lim_{g \to \infty} \|[\tau_g A_1, A_2]\| = 0$$

(f) *\mathfrak{A} is quasilocal in the sense of 6.2.4.*

Condition (a) is trivial and (c) follows from Theorem 6.2.5. Furthermore we have the implications

$$(b) \Rightarrow (c) \Leftarrow (d) \Leftarrow (e) \Leftarrow (f)$$

which concludes the proof.

6.2.11 COROLLARY.[8] *If G is the group of unitary[9] elements of the B*-algebra \mathfrak{A} (with an identity) and if G acts by inner automorphisms, $\tau_U A = UAU^{-1}$, then \mathfrak{A} is G-Abelian.*

This follows because the G-invariant states (called "traces") satisfy $\rho(A_1 \cdot A_2) = \rho(A_2 \cdot A_1)$ for all $A_1, A_2 \in \mathfrak{A}$.

6.2.12 THE CASE OF A GROUP WITH AN INVARIANT MEAN. The case where, in the triple (\mathfrak{A}, G, τ), the group G has an invariant mean (see Section A.6) and τ is continuous in some sense is of special interest.[10] In particular, if \mathfrak{A} has an identity, the set $E \cap \mathscr{L}_G^{\perp}$ of G-invariant states is not empty[11] (see Proposition 6.2.13). Furthermore, the projection P_ρ can be obtained more explicitly in terms of U_ρ (see Proposition 6.2.14) than through the use of the Alaoglu–Birkhoff theorem.

6.2.13 PROPOSITION. *Let G have an invariant mean and the functions $g \to \sigma(\tau_g A)$ on G be continuous (for all $A \in \mathfrak{A}, \sigma \in E$).*

(a) *If \mathfrak{A}_1 is a sub-B*-algebra of \mathfrak{A}, invariant under τ_G, and ρ_1 is a G-invariant state on \mathfrak{A}_1, then there exists a G-invariant state ρ on \mathfrak{A} which extends ρ_1.*

[8] See Størmer [1].
[9] U is unitary if $UU^* = U^*U = 1$.
[10] For a study of this situation see in particular Doplicher, Kadison, Kastler, and Robinson [1].
[11] There exist triples (\mathfrak{A}, G, τ) without invariant state. For instance, take K: the real projective line, $\mathfrak{A} = \mathscr{C}(K)$, G: the projective transformations of K; there is no G-invariant measure on K.

If ρ_1 is an extremal point of the set of invariant states on \mathfrak{A}_1, ρ may be chosen extremal in the set of invariant states on \mathfrak{A}.

(b) *If \mathfrak{A} has an identity, $E \cap \mathscr{L}_G^\perp \neq \emptyset$.*

(a) Let $\tilde{\rho} \in E$ be an extension of ρ_1 to \mathfrak{A} (see A.3.2). By assumption the functions $\varphi_A : g \to \tilde{\rho}(\tau_g A)$ are bounded and continuous on G. If \mathfrak{M} is a right-invariant mean of G, put $\rho(A) = \mathfrak{M}\varphi_A$; then ρ is a positive G-invariant linear functional on \mathfrak{A} and $\|\rho\| \leq 1$. The restriction of ρ to \mathfrak{A}_1 is ρ_1 and therefore $\|\rho\| = 1$.

Let $K \subset E \cap \mathscr{L}_G^\perp$ be the set of G-invariant extensions of ρ_1 to \mathfrak{A}; K is a nonempty convex weakly compact set. If ρ_1 is an extremal point of the set of invariant states on \mathfrak{A}_1 we check that every extremal point ρ of K[12] is an extremal point of $E \cap \mathscr{L}_G^\perp$ (if not, we could write $\rho = \frac{1}{2}\rho' + \frac{1}{2}\rho''$; $\rho', \rho'' \in E \cap \mathscr{L}_G^\perp$; $\rho' \notin K$, then ρ' would not be an extension of ρ_1, in contradiction with the extremality of ρ_1). This concludes the proof of (a).

The proof of (b) is obtained by taking for \mathfrak{A}_1 the multiples of 1 in (a).

6.2.14 PROPOSITION. *Let G have an invariant mean and the functions $g \to \sigma(\tau_g A)$ on G be continuous (for all $A \in \mathfrak{A}, \sigma \in E$). If $\rho \in E \cap \mathscr{L}_G^\perp$ and (χ_α) is an M-net*[13] *on G, then*

$$P_\rho = \underset{\alpha \to \infty}{\text{st lim}} \int dg\chi_\alpha(g)U_\rho(g) \tag{2.19}$$

By Theorem 6.2.2 the representation U_ρ is strongly continuous; (2.19) is thus a consequence of the following *mean ergodic theorem.*

6.2.15 PROPOSITION. *Let (χ_α) be an M-net on G, U a strongly continuous unitary representation of G in a complex Hilbert space \mathfrak{H}, and P the projection on the subspace of \mathfrak{H} formed by the vectors invariant under U, then*

$$\underset{\alpha \to \infty}{\text{st lim}} \int dg\chi_\alpha(g)U(g) = P \tag{2.20}$$

We have to show that for all $\Psi \in \mathfrak{H}$

$$\underset{\alpha \to \infty}{\lim} \left\| \left[\int dg\chi_\alpha(g)U(g) - P \right]\Psi \right\| = 0 \tag{2.21}$$

This is trivial if $P\Psi = \Psi$ [because $U(g)P\Psi = P\Psi$] and we may thus assume that $P\Psi = 0$. The kernel of P is the orthogonal complement of

$$P\mathfrak{H} = \bigcap_g \{\Phi \in \mathfrak{H} : U(g)\Phi = \Phi\}$$

$$= \bigcap_g \{U(g)\Phi' - \Phi' : \Phi' \in \mathfrak{H}\}^\perp$$

[12] Such an extremal point exists by the Krein–Milman theorem.
[13] See A.6.1.

Therefore the kernel of P is generated by the spaces $\{U(g)\Phi' - \Phi' : \Phi' \in \mathfrak{H}\}$. If $P\Psi = 0$, then for any $\varepsilon > 0$ we may choose $n \geq 0, g_1, \ldots, g_n \in G$ and $\Phi_1, \ldots, \Phi_n \in \mathfrak{H}$ such that

$$\left\| \Psi - \sum_{i=1}^{n} [U(g_i)\Phi_i - \Phi_i] \right\| < \varepsilon$$

This shows that it is sufficient to prove

$$\lim_{\alpha \to \infty} \left\| \left[\int dg \chi_\alpha(g) U(g) - P \right] \left[U(g_1)\Phi - \Phi \right] \right\| = 0 \qquad (2.22)$$

for all $g_1 \in G, \Phi \in \mathfrak{H}$ and this follows from

$$\left\| \left[\int dg \chi_\alpha(g) U(g) - P \right] \left[U(g_1)\Phi - \Phi \right] \right\|$$

$$= \left\| \int dg \chi_\alpha(g) U(g) [U(g_1)\Phi - \Phi] \right\|$$

$$= \left\| \int dg \chi_\alpha(g) [U(gg_1) - U(g)]\Phi \right\|$$

$$= \left\| \int dg [\chi_\alpha(gg_1^{-1}) - \chi_\alpha(g)] U(g)\Phi \right\|$$

$$\leq \int dg |\chi_\alpha(gg_1^{-1}) - \chi_\alpha(g)| \cdot \|\Phi\| \to 0$$

6.2.16 PROPOSITION. *Let G have an invariant mean and the functions $g \to \sigma(\tau_g A)$ on G be continuous (for all $A \in \mathfrak{A}, \sigma \in E$); also, let (χ_α) be an M-net on G. The B*-algebra \mathfrak{A} is G-Abelian if and only if, for all self-adjoint $A_1, A_2 \in \mathfrak{A}$ and all $\rho \in E \cap \mathscr{L}_G^\perp$,*

$$\lim_{\alpha \to \infty} \int dg \chi_\alpha(g) \rho([\tau_g A_1, A_2]) = 0 \qquad (2.23)$$

Let $\Psi \in P_\rho \mathfrak{H}_\rho, \|\Psi\| = 1$. Using Proposition 6.2.14 and replacing $\rho(\cdot)$ by $(\Psi, \pi_\rho(\cdot)\Psi)$ we see that (2.23) is equivalent to

$$(\Psi, \pi_\rho(A_1) P_\rho \pi_\rho(A_2)\Psi) = (\Psi, \pi_\rho(A_2) P_\rho \pi_\rho(A_1)\Psi) \qquad (2.24)$$

which is equivalent to (2.7). This proves that \mathfrak{A} is G-Abelian if and only if (2.23) holds.

6.3 ERGODIC STATES

In this section we begin a detailed analysis of invariant states on G-Abelian algebras by studying the ergodic states.

6.3.1 DEFINITION.[14] *Given a triple* (\mathfrak{A}, G, τ) *let* $\mathscr{E}(E \cap \mathscr{L}_G{}^{\perp})$ *be the set of extremal points of the set* $E \cap \mathscr{L}_G{}^{\perp}$ *of* G-*invariant states. The points of* $\mathscr{E}(E \cap \mathscr{L}_G{}^{\perp})$ *are called* G-*ergodic states.*

In the Abelian situation (see 6.2.3) ergodic states reduce to ergodic measures. Before characterizing the ergodic states in Theorem 6.3.3 below, we prove some technical results using von Neumann algebras (see Section A.4). In this proposition we do not assume that \mathfrak{A} is G-Abelian.

6.3.2 PROPOSITION. *Let* $\rho \in E \cap \mathscr{L}_G{}^{\perp}$; *then*

(a) $$[\pi_\rho(\mathfrak{A}) \cup \{P_\rho\}]' = [\pi_\rho(\mathfrak{A}) \cup U_\rho(G)]' \tag{3.1}$$

(b) *the mapping*

$$\alpha : [\pi_\rho(\mathfrak{A}) \cup U_\rho(G)]' \to P_\rho[\pi_\rho(\mathfrak{A}) \cup U_\rho(G)]' \tag{3.2}$$

is a *-isomorphism;*

(c) $$P_\rho[\pi_\rho(\mathfrak{A}) \cup U_\rho(G)]' = P_\rho[P_\rho\pi_\rho(\mathfrak{A})P_\rho]' \tag{3.3}$$

(d) *if the von Neumann algebra* $[P_\rho\pi_\rho(\mathfrak{A})P_\rho]''$ *is Abelian, then*

$$P_\rho[P_\rho\pi_\rho(\mathfrak{A})P_\rho]' = P_\rho[P_\rho\pi_\rho(\mathfrak{A})P_\rho]'' \tag{3.4}$$

(e) $P_\rho[P_\rho\pi_\rho(\mathfrak{A})P_\rho]''$ *is the strong closure of* $P_\rho\pi_\rho(\mathfrak{A})P_\rho$.

(a) We have $[\pi_\rho(\mathfrak{A}) \cup \{P_\rho\}]' \supset [\pi_\rho(\mathfrak{A}) \cup U_\rho(G)]'$ because $P_\rho \in [U_\rho(G)]''$. Conversely, if $C \in [\pi_\rho(\mathfrak{A}) \cup \{P_\rho\}]'$, then, for all $A \in \mathfrak{A}$ and $g \in G$

$$CU_\rho(g)\pi_\rho(A)\Omega_\rho = C\pi_\rho(\tau_g A)\Omega_\rho = \pi_\rho(\tau_g A)C\Omega_\rho$$
$$= U_\rho(g)\pi_\rho(A)P_\rho C\Omega_\rho = U_\rho(g)C\pi_\rho(A)\Omega_\rho$$

(b) If $C \in [\pi_\rho(\mathfrak{A}) \cup U_\rho(G)]'$, then $CP_\rho = P_\rho CP_\rho = P_\rho C$; hence, α is a *-homomorphism. If $CP_\rho = 0$, then $C\Omega_\rho = 0$ and $C\pi_\rho(A)\Omega_\rho = \pi_\rho(A)C\Omega_\rho = 0$; hence, $C = 0$ so that α is an isomorphism.

(c) This follows from the formula $(\mathscr{B}')_P + (\mathscr{B}_P)'$ (see A.4.3) with $\mathscr{B} = [\pi_\rho(\mathfrak{A}) \cup U_\rho(G)]''$, $P = P_\rho$.

(d) The vector Ω_ρ is cyclic for the restriction to $P_\rho\mathfrak{H}_\rho$ of $P_\rho[P_\rho\pi_\rho(\mathfrak{A})P_\rho]''$; hence if this von Neumann algebra is commutative, it is equal to its commutant (see A.4.5), namely, to $P_\rho[P_\rho\pi_\rho(\mathfrak{A})P_\rho]'$ restricted to $P_\rho\mathfrak{H}_\rho$.

(e) $P_\rho[P_\rho\pi_\rho(\mathfrak{A})P_\rho]''$ is the strong closure of the algebra generated by $P_\rho\pi_\rho(\mathfrak{A})P_\rho$, and $P_\rho\pi_\rho(\mathfrak{A})P_\rho$ is strongly dense in this algebra because $P_\rho \in [U_\rho(G)]''$.

[14] see Segal [1].

6.3.3 THEOREM. *Let* $\rho \in E \cap \mathscr{L}_G^{\perp}$. *Consider the conditions*

(a) $\rho \in \mathscr{E}(E \cap \mathscr{L}_G^{\perp})$;
(b) *the set* $\pi_\rho(\mathfrak{A}) \cup U_\rho(G)$ *is irreducible in* \mathfrak{H}_ρ;
(c) P_ρ *is one-dimensional.*[15]

We have (a) \Leftrightarrow (b) \Leftarrow (c). *If* \mathfrak{A} *is G-Abelian, then* (a), (b), *and* (c) *are equivalent.*

(a) \Leftrightarrow (b). The existence of a self-adjoint operator $C \in [\pi_\rho(\mathfrak{A}) \cup U_\rho(G)]'$, such that $0 \le C \le 1$ and C is not a multiple of 1, is equivalent (see A.3.5) to non(a) and non(b).

(b) \Leftarrow (c) follows from Proposition 6.3.2,(b).

If \mathfrak{A} is G-Abelian, then $P_\rho[\pi_\rho(\mathfrak{A}) \cup U_\rho(G)]' = P_\rho[P_\rho\pi_\rho(\mathfrak{A})P_\rho]''$ by Proposition 6.3.2,(c) and (d). Thus, (b) implies that $P_\rho[P_\rho\pi_\rho(\mathfrak{A})P_\rho]''$ reduces to the multiples of P_ρ, and, since Ω_ρ is cyclic in $P_\rho\mathfrak{H}_\rho$ with respect to $P_\rho[P_\rho\pi_\rho(\mathfrak{A})P_\rho]''$, (c) follows.

6.3.4 REMARK. In view of Proposition 6.3.2, condition (b) in Theorem 6.3.3 is also equivalent to the following conditions.

(b') *The set* $\pi_\rho(\mathfrak{A}) \cup \{P_\rho\}$ *is irreducible in* \mathfrak{H}_ρ.
(b") *The set* $P_\rho\pi_\rho(\mathfrak{A})P_\rho$ *is irreducible in* $P_\rho\mathfrak{H}_\rho$.

6.3.5 PROPOSITION. *Let* G *have an invariant mean and the functions* $g \to \sigma(\tau_g A)$ *on* G *be continuous (for all* $A \in \mathfrak{A}$, $\sigma \in E$); *also let* (χ_α) *be an M-net on* G. *If* $\rho \in E \cap \mathscr{L}_G^{\perp}$, *the following conditions are equivalent to the condition* (c) *of Theorem 6.3.3.*

(d) *For all* $A_1, A_2 \in \mathfrak{A}$,

$$\lim_{\alpha \to \infty} \int dg \chi_\alpha(g) \rho(\tau_g A_1 \cdot A_2) = \rho(A_1)\rho(A_2) \tag{3.5}$$

(e) *For all self-adjoint* $A \in \mathfrak{A}$,

$$\lim_{\alpha \to \infty} \int dg_1 \int dg_2 \chi_\alpha(g_1)\chi_\alpha(g_2) \rho(\tau_{g_1} A \cdot \tau_{g_2} A) = \rho(A)^2 \tag{3.6}$$

In view of Proposition 6.2.14, Eqs. (3.5) and (3.6) may be rewritten as

$$(\Omega_\rho, \pi_\rho(A_1)P_\rho\pi_\rho(A_2)\Omega_\rho) = (\Omega_\rho, \pi_\rho(A_1)\Omega_\rho)(\Omega_\rho, \pi_\rho(A_2)\Omega_\rho)$$

$$(\Omega_\rho, \pi_\rho(A)P_\rho\pi_\rho(A)\Omega_\rho) = (\Omega_\rho, \pi_\rho(A)\Omega_\rho)^2$$

respectively; both mean that P_ρ is the orthogonal projection on Ω_ρ.

[15] This property is known as "uniqueness of the vacuum"; it expresses that the only G-invariant vectors are multiples of Ω_ρ.

6.3.6 REMARK. Equation (3.5) holds in particular if G is noncompact and

$$\lim_{g \to \infty} \rho(\tau_g A_1 \cdot A_2) = \rho(A_1) \cdot \rho(A_2) \tag{3.7}$$

which is of the form (4.11) of Chapter 3, i.e., of the form of a cluster property. We may also consider (3.5) as a *weak cluster property*.

6.4 DECOMPOSITION OF INVARIANT STATES INTO ERGODIC STATES

We shall see in this section that an invariant state on a G-Abelian B*-algebra can be represented in a unique way as an integral (in some sense) over ergodic states. To discuss this situation we shall use the language and techniques of Choquet's theory of integral representations on convex compact sets (see Section A.5).

6.4.1 THEOREM.[16] *Let \mathfrak{A} have an identity. If $A \in \mathfrak{A}$, we denote by \hat{A} the function $\sigma \to \hat{A}(\sigma) = \sigma(A)$ on $E \cap \mathscr{L}_G{}^\perp$. We let $\rho \in E \cap \mathscr{L}_G{}^\perp$ and assume that $[P_\rho \pi_\rho(\mathfrak{A}) P_\rho]''$ is Abelian. Then there is a unique maximal measure μ_ρ on $E \cap \mathscr{L}_G{}^\perp$ with resultant ρ. This measure is determined by*

$$\mu_\rho\left(\prod_{i=1}^{l} \hat{A}_i\right) = \left(\Omega_\rho, \left[\prod_{i=1}^{l} (P_\rho \pi_\rho(A_i) P_\rho)\right]\Omega\right) \tag{4.1}$$

for all l and self-adjoint A_1, \ldots, A_l.

By assumption there is a projection-valued measure F on \mathbf{R}^l such that

$$P_\rho \pi_\rho(A_i) P_\rho = \int t_i \, dF(t_1, \ldots, t_l) \tag{4.2}$$

If \mathscr{P} is a complex polynomial of l variables, we have

$$|(\Omega_\rho, \mathscr{P}(P_\rho \pi_\rho(A_1) P_\rho, \ldots, P_\rho \pi_\rho(A_l) P_\rho) \Omega_\rho)|$$

$$= \left|\left(\Omega_\rho, \int \mathscr{P}(t_1, \ldots, t_l) \, dF(t_1, \ldots, t_l) \Omega_\rho\right)\right|$$

$$\leq \sup_{\|\Phi\| = 1, \Phi \in P_\rho \mathfrak{H}_\rho} |\mathscr{P}((\Phi, \pi_\rho(A_1)\Phi), \ldots, (\Phi, \pi_\rho(A_l)\Phi))|$$

$$\leq \sup_{\sigma \in E \cap \mathscr{L}_G{}^\perp} |\mathscr{P}(\sigma(A_1), \ldots, \sigma(A_l))|$$

$$= \sup_{\sigma \in E \cap \mathscr{L}_G{}^\perp} |\mathscr{P}(\hat{A}_1(\sigma), \ldots, \hat{A}_l(\sigma))|$$

Therefore (4.1) defines a linear functional on the polynomials in the \hat{A}, which is continuous for the topology of uniform convergence on $E \cap \mathscr{L}_G{}^\perp$. This

[16] See Lanford and Ruelle [1].

functional extends uniquely (use the Stone–Weierstrass theorem) to a measure $\mu_\rho \geq 0$ of norm 1 on $E \cap \mathscr{L}_G^\perp$.

Let $\rho = \Sigma_1^m \lambda_i \rho_i$ where $\rho_1, \ldots, \rho_m \in E \cap \mathscr{L}_G^\perp$, $\lambda_1, \ldots, \lambda_m > 0$, and $\Sigma \lambda_i = 1$. There exist (see A.3.5) uniquely defined self-adjoint operators $T_i \in [\pi_\rho(\mathfrak{A})]'$ such that $0 \leq T_i \leq 1$ and

$$\lambda_i \rho_i(\cdot) = (T_i \Omega_\rho, \pi_\rho(\cdot) T_i \Omega_\rho) \tag{4.3}$$

We have $\Sigma_i T_i^2 = 1$. From the G-invariance of ρ_i and the uniqueness of T_i we find also $U_\rho(g) T_i^2 U_\rho(g^{-1}) = T_i^2$, so that

$$T_i \in [U_\rho(G)]' \qquad [T_i, P_\rho] = 0 \tag{4.4}$$

We use now the uniqueness property of the Gel'fand–Segal construction (see A.3.5) to identify \mathfrak{H}_{ρ_i} with the closure of $\pi_\rho(\mathfrak{A}) T_i \Omega_\rho$, π_{ρ_i} with the restriction of π_ρ to \mathfrak{H}_{ρ_i} and Ω_{ρ_i} with $\lambda_i^{-1/2} T_i \Omega_\rho$. In this way U_{ρ_i}, P_{ρ_i} are also identified with the restrictions of U_ρ, P_ρ to \mathfrak{H}_{ρ_i}. In particular, $[P_{\rho_i} \pi_{\rho_i}(\mathfrak{A}) P_{\rho_i}]''$ is Abelian and μ_{ρ_i} is therefore defined. We may write

$$\mu_{\rho_i} \left(\prod_{i=1}^l \hat{A}_i \right) = \left(\Omega_{\rho_i}, \left[\prod_{i=1}^l (P_{\rho_i} \pi_{\rho_i}(A_i) P_{\rho_i}) \right] \Omega_{\rho_i} \right)$$

$$= \lambda_i^{-1} \left(T_i \Omega_\rho, \left[\prod_{i=1}^l (P_\rho \pi_\rho(A_i) P_\rho) \right] T_i \Omega_\rho \right)$$

$$= \lambda_i^{-1} \left(\Omega_\rho, \left[\prod_{i=1}^l (P_\rho \pi_\rho(A_i) P_\rho) \right] T_i^2 \Omega_\rho \right)$$

so that

$$\sum_i \lambda_i \mu_{\rho_i} \left(\prod_{i=1}^l \hat{A}_i \right) = \mu_\rho \left(\prod_{i=1}^l \hat{A}_i \right)$$

and therefore

$$\mu_\rho = \sum_i \lambda_i \mu_{\rho_i} \tag{4.5}$$

Let now μ be a measure on $E \cap \mathscr{L}_G^\perp$ with resultant ρ (i.e., $\mu \succ \delta_\rho$: see Section A.5), and let φ be a continuous function on $E \cap \mathscr{L}_G^\perp$. Given $\varepsilon > 0$ there exists a measure $\mu' = \Sigma_i \lambda_i \delta_{\rho_i}$ with finite support such that $\lambda_i > 0$, $\rho_i \in E \cap \mathscr{L}_G^\perp$, $\Sigma_i \lambda_i \rho_i = \rho$, and $|\mu(\varphi) - \mu'(\varphi)| < \varepsilon$ (see A.5.1).

If φ is convex we have

$$\mu(\varphi) - \varepsilon \leq \mu'(\varphi) = \sum \lambda_i \delta_{\rho_i}(\varphi) \leq \sum \lambda_i \mu_{\rho_i}(\varphi) = \mu_\rho(\varphi) \tag{4.6}$$

Therefore $\mu \succ \delta_\rho \Rightarrow \mu \prec \mu_\rho$, and μ_ρ is the unique maximal measure on $E \cap \mathscr{L}_G^\perp$ such that $\mu_\rho \succ \delta_\rho$.

6.4.2 COROLLARY. *If \mathfrak{A} has an identity and is G-Abelian, then $E \cap \mathscr{L}_G^{\perp}$ is a simplex.*

This follows from Theorem 6.4.1, the definition 6.2.6 of a G-Abelian algebra, and the definition of a simplex (see A.5.3). If \mathfrak{A} is Abelian we find that *the invariant probability measures on a compact set with a group of homeomorphisms form a simplex.*[17]

In the "good cases" the maximal measure μ_ρ is carried by the extremal points of $E \cap \mathscr{L}_G^{\perp}$, which are the G-ergodic states. This is so in particular if \mathfrak{A} is separable because E and $E \cap \mathscr{L}_G^{\perp}$ are then metrizable (see A.5.4).

6.4.3 PROPOSITION. *If \mathfrak{A} is a G-Abelian, separable B*-algebra with an identity, then every G-invariant state ρ is the resultant of a unique measure carried by the G-ergodic states.*

In the rest of this section we consider states which have a restriction of norm 1 to subalgebras of \mathfrak{A}. This situation is frequently met in physical applications.

6.4.4 PROPOSITION. *Let \mathfrak{A} have an identity and \mathfrak{B} be a self-adjoint subalgebra of \mathfrak{A}. Define*

$$\mathscr{F} = \{\sigma \in E : \text{the restriction of } \sigma \text{ to } \mathfrak{B} \text{ has norm } 1\} \tag{4.7}$$

Then:

(a) *\mathscr{F} is a countable intersection of open subsets of E.*
(b) *If μ is a probability measure on E with resultant ρ, then*

$$\rho \in \mathscr{F} \Leftrightarrow \mu \text{ is carried by } \mathscr{F} \tag{4.8}$$

We may assume that \mathfrak{B} is a sub-B*-algebra of \mathfrak{A} (replace \mathfrak{B} by its closure). If we write

$$B_h = \{A \in \mathfrak{B} : A = A^*, \|A\| \leq 1\} \tag{4.9}$$

then (a) results from

$$\mathscr{F} = \{\sigma \in E : \sup_{A \in B_h} \sigma(A) = 1\} = \bigcap_{m > 0} V_m \tag{4.10}$$

$$V_m = \bigcup_{A \in B_h} \left\{ \sigma \in E : \sigma(A) > 1 - \frac{1}{m} \right\} \tag{4.11}$$

We now prove (b). First assume $\rho \in \mathscr{F}$ and let $\mu = \mu' + \mu''$ where μ', μ'' are carried, respectively, by V_m, $E \setminus V_m$.

For all $A \in B_h$,

$$\mu''(\hat{A}) \leq \|\mu''\| \cdot \left(1 - \frac{1}{m} \right) \tag{4.12}$$

[17] For this result see Jacobs [1] and Phelps [1].

where \hat{A} is the function defined on E by $\hat{A}(\sigma) = \sigma(A)$. From (4.12) we obtain

$$\rho(A) = \mu(\hat{A}) = \mu'(\hat{A}) + \mu''(\hat{A}) \le \|\mu'\| + \|\mu''\|\left(1 - \frac{1}{m}\right) = 1 - \frac{1}{m}\|\mu''\|$$

and therefore $\|\mu''\| = 0$. Being carried by V_m for all m, the measure μ is carried by \mathcal{F}.

Assume now that μ is carried by \mathcal{F}; hence, by V_{2m}. There exists a compact set $K \subset V_{2m}$ such that $\mu(K) > 1 - 1/2m$. If (U_λ) is an increasing approximate identity of \mathfrak{B} we have (see A.3.3)

$$\sigma \in V_{2m} \quad \Rightarrow \quad \lim \sigma(U_\lambda) > 1 - \frac{1}{2m} \tag{4.13}$$

There exists thus a finite family $(U_{\lambda_i})_{1 \le i \le n}$ such that

$$\bigcup_{i=1}^{n}\left\{\sigma \in E : \sigma(U_{\lambda_i}) > 1 - \frac{1}{2m}\right\} \supset K$$

hence (see A.3.2) there exists λ_0 such that

$$\left\{\sigma \in E : \sigma(U_{\lambda_0}) > 1 - \frac{1}{2m}\right\} \supset K \tag{4.14}$$

Therefore

$$\rho(U_{\lambda_0}) = \mu(\hat{U}_{\lambda_0}) > \left(1 - \frac{1}{2m}\right)\mu(K) > 1 - \frac{1}{m} \tag{4.15}$$

hence $\rho \in V_m$ and, since this is true for all m, $\rho \in \mathcal{F}$.

6.4.5 PROPOSITION. *Let (\mathfrak{A}_α) be a countable family of sub-B*-algebras of \mathfrak{A} such that $\bigcup_\alpha \mathfrak{A}_\alpha$ is dense in \mathfrak{A}. Let \mathfrak{I}_α be a separable closed two-sided ideal of \mathfrak{A}_α for each α and define*
$$\mathcal{F}_\alpha = \{\sigma \in E : \text{the restriction of } \sigma \text{ to } \mathfrak{I}_\alpha \text{ has norm } 1\}, \quad \mathcal{F} = \bigcap_\alpha \mathcal{F}_\alpha$$

(a) *If $\rho \in \mathcal{F}$, then \mathfrak{H}_ρ is separable.*

(b) *There exists a sequence (A_i) of self-adjoint elements of \mathfrak{A} such that if $\rho \in \mathcal{F}$ and $\sigma \in E$, then $\rho(A_i) \ne \sigma(A_i)$ for some i.*

(c) *Let \mathfrak{A} have an identity and be G-Abelian; if μ is a probability measure on $E \cap \mathcal{L}_G^\perp$ with resultant $\rho \in \mathcal{F}$, then*

$$(\mu \text{ maximal on } E \cap \mathcal{L}_G^\perp) \Leftrightarrow (\mu \text{ carried by } \mathcal{E}(E \cap \mathcal{L}_G^\perp)) \tag{4.16}$$

Every $\rho \in \mathcal{F} \cap E \cap \mathcal{L}_G^\perp$ is thus the resultant of a unique probability measure μ_ρ carried by $\mathcal{F} \cap \mathcal{E}(E \cap \mathcal{L}_G^\perp)$.

(a) Let \mathfrak{H}_α be the closed subspace of \mathfrak{H}_ρ spanned by $\pi_\rho(\mathfrak{A}_\alpha)\Omega_\rho$. By the uniqueness of the Gel'fand–Segal construction (see A.3.5), the restriction of π_ρ to \mathfrak{A}_α is unitarily equivalent to the representation π_α constructed from

the restriction ρ_α of ρ to \mathfrak{A}_α. Since the restriction of ρ_α to \mathfrak{I}_α has norm 1, $\pi_\rho(\mathfrak{I}_\alpha)\Omega_\rho$ is dense in $\mathfrak{H}_\alpha{}^{18}$; hence, \mathfrak{H}_α is separable. Since \mathfrak{H}_ρ is spanned by the countable family \mathfrak{H}_α, \mathfrak{H}_ρ is separable.

(b) For each α let $(A_{\alpha\beta})$ be a countable dense subset of the unit ball of the subspace of self-adjoint elements of \mathfrak{I}_α, and let (A_i) be an ordering of $(A_{\alpha\beta})$ into a sequence. Let $\sigma \in E$; then

$$\sigma \in \mathscr{F} \quad \Leftrightarrow \quad \sup_\beta |\sigma(A_{\alpha\beta})| = 1 \qquad \text{for all} \quad \alpha \qquad (4.17)$$

Therefore if $\rho \in \mathscr{F}, \sigma \notin \mathscr{F}$ we have $\rho(A_i) \neq \sigma(A_i)$ for some i. On the other hand, if $\rho, \sigma \in \mathscr{F}$ and $\rho(A_i) = \sigma(A_i)$ for all i, then the restrictions of ρ, σ to \mathfrak{I}_α coincide for all α; hence, (see A.3.3) the restrictions of ρ, σ to \mathfrak{A}_α coincide for all α; hence, $\rho = \sigma$.

(c) We know (see A.5.4) that if μ is carried by $\mathscr{E}(E \cap \mathscr{L}_G{}^\perp)$, then μ is maximal. We prove now the converse.

We assume that μ is maximal; therefore $\mu = \mu_\rho$ (see Theorem 6.4.1) and μ_ρ is carried by \mathscr{F} [see Proposition 6.4.4,(b)]. Using the sequence (A_i) of (b), let p be the mapping which associates to $\sigma \in K = E \cap \mathscr{L}_G{}^\perp$ the sequence $(\sigma(A_i))$; p is a continuous mapping of K into the product of a sequence of copies of \mathbf{R}. A measure ν_ρ on the convex compact set $p(K)$ is defined by

$$\nu_\rho(\psi) = \mu_\rho(\psi \circ p) = \mu(\psi \circ p) \qquad \psi \in \mathscr{C}(p(K)) \qquad (4.18)$$

We prove that ν_ρ is maximal on $p(K)$. We have $\nu_\rho \succ \delta_{p\rho}$. Let $\nu \succ \delta_{p\rho}$ on $p(K)$; then, if $\varepsilon > 0$ and $\psi \in \mathscr{C}(p(K))$ we can find (see A.5.1) a measure ν' with finite support on $p(K)$ such that

$$\delta_{p\rho} \prec \nu' = \sum_{i=1}^{n} \alpha_i \delta_{\sigma_i} \qquad |\nu(\psi) - \nu'(\psi)| < \varepsilon$$

Let $\sigma_i = p\rho_i, \rho_i \in K$, then $p(\Sigma \alpha_i \rho_i) = \Sigma \alpha_i \sigma_i = p\rho$; hence, by (b), $\Sigma \alpha_i \rho_i = \rho$. If ψ is convex we have thus

$$\nu(\psi) - \varepsilon < \nu'(\psi) = \sum \alpha_i \psi(p\rho_i) \leq \sum \alpha_i \mu_{\rho_i}(\psi \circ p)$$
$$= \mu_\rho(\psi \circ p) = \nu_\rho(\psi) \qquad (4.19)$$

because the mapping $\sigma \to \mu_\sigma$ is affine. Therefore $\nu \prec \nu_\rho$ so ν_ρ is maximal.

Since $p(K)$ is metrizable, ν_ρ is carried by $\mathscr{E}(p(K))$ (see A.5.4); therefore $\mu = \mu_\rho$ is carried by $\mathscr{F} \cap p^{-1}(\mathscr{E}(p(K)))$. It is easily checked from (b) that $\mathscr{F} \cap p^{-1}(\mathscr{E}(p(K))) \subset \mathscr{E}(K) = \mathscr{E}(E \cap \mathscr{L}_G{}^\perp)$ which concludes the proof.

[18] If (U_λ) is an approximate identity of \mathfrak{I}_α, $\lim \rho_\alpha(U_\lambda) = 1$ implies $\|\pi_\rho(U_\lambda)\Omega_\rho - \Omega_\rho\| \to 0$. This shows that $\pi_\rho(\mathfrak{I}_\alpha)\Omega$ is dense in \mathfrak{H}_α.

6.5 PURE THERMODYNAMIC PHASES AS ERGODIC STATES

Let us assume that \mathfrak{A} is a quasilocal algebra (see 6.2.4) so that G is \mathbf{R}^ν or \mathbf{Z}^ν and \mathfrak{A} is G-Abelian. Also, let ρ be a G-invariant state on \mathfrak{A} with the physical interpretation of an equilibrium state of an infinite system. We shall show that ρ can be given the interpretation of a pure thermodynamic phase if and only if it is a G-ergodic state.[19]

The groups $\mathbf{R}^\nu, \mathbf{Z}^\nu$ are Abelian and have therefore an invariant mean. It is easy to construct M-nets (χ_α) for $\mathbf{R}^\nu, \mathbf{Z}^\nu$: we let χ_α be the characteristic function, divided by α^ν, of a cube with side α, and we let α run over the integers greater than zero.

6.5.1 LEMMA. *ρ is ergodic if and only if, for all self-adjoint $A \in \mathfrak{A}$,*

$$\lim_{\alpha \to \infty} \rho([\langle A \rangle_\alpha]^2) = [\rho(A)]^2 \tag{5.1}$$

where

$$\langle A \rangle_\alpha = \int dg \chi_\alpha(g) \tau_g A \tag{5.2}$$

assuming that the mapping $\tau_. A : G \to \mathfrak{A}$ is continuous.[20]

This is just a rephrasing of Proposition 3.5,(e). We may also rewrite (5.1) as

$$\lim_{\alpha \to \infty} \rho([\langle A \rangle_\alpha - \rho(A)]^2) = 0 \tag{5.3}$$

Let us now come back to the interpretation of self-adjoint elements of \mathfrak{A} as physical observables. An observable A will in general fluctuate, i.e., the measure determined by ρ on the spectrum of \mathfrak{A} is not carried by a point. Equation (5.3) means however that $\langle A \rangle_\alpha$ fluctuates very little if α is large, i.e., if ρ is ergodic the space average of an observable over a large region is almost constant. We can interpret the fact that the space average of an observable tends to a constant to mean that the state we consider is a pure thermodynamic phase. A phase is indeed characterized by the fact that coarse-grained (i.e., space-averaged) quantities do not fluctuate. On the contrary, for a mixture (say of water and ice) some coarse-grained quantities (say the density) fluctuate. Notice that the nth moment of the distribution of the coarse-grained average of the observable A is

$$\lim_{\alpha \to \infty} \rho([\langle A \rangle_\alpha]^n) \tag{5.4}$$

[19] This fact was noted and used in Ruelle [6]. It is unclear to what extent it may have been realized earlier.

[20] We introduce this strong continuity requirement to simplify the discussion.

Given a mixture ρ it should, on physical grounds, have a unique decomposition into pure phases, i.e., a unique integral representation on the set $\mathcal{E}(E \cap \mathcal{L}_G{}^{\perp})$ of ergodic states. The existence of this integral representation has been discussed in Section 6.4; it is given by a measure μ_ρ on $E \cap \mathcal{L}_G{}^{\perp}$ determined (see Theorem 6.4.1) by

$$\mu_\rho(\hat{A}_1 \cdots \hat{A}_l) = (\Omega_\rho, \pi_\rho(A_1)P_\rho\pi(A_2)P_\rho \cdots P_\rho\pi_\rho(A_l)\Omega_\rho)$$
$$= \lim_{\alpha \to \infty} \rho(\langle A_1 \rangle_\alpha \cdots \langle A_l \rangle_\alpha) \tag{5.5}$$

where the second equality results from Proposition 6.2.14 and relates the decomposition into pure phases to the moments (5.4).

In this section we have until now assumed that the symmetry group G of the theory was \mathbf{R}^ν (continuous systems) or \mathbf{Z}^ν (lattice systems). In physical applications G is usually larger, being, for instance, the Euclidean group. It is then natural to say that the state ρ represents a pure thermodynamic phase when ρ is G-ergodic.

6.5.2 SPONTANEOUS BREAKDOWN OF SYMMETRY. Let $\rho \in \mathcal{E}(E \cap \mathcal{L}_G{}^{\perp})$ represent a pure thermodynamic phase corresponding to the invariance group G. Let also H be a subgroup of G such that \mathfrak{A} is H-Abelian. If we have $\rho \notin \mathcal{E}(E \cap \mathcal{L}_H{}^{\perp})$, we may decompose ρ into H-ergodic states[21] and these have a stronger cluster property[22] than the original state ρ. This situation is common in physics and known as "spontaneous breakdown of symmetry": a state ρ with large symmetry group G and "long range order" (i.e., poor cluster property) decomposes into states with less symmetry but better cluster properties. We say that the symmetry corresponding to the group G is broken.

Here are a few examples

Group G	Group H	Elements of $\mathcal{E}(E \cap \mathcal{L}_G{}^{\perp})$	Elements of $\mathcal{E}(E \cap \mathcal{L}_H{}^{\perp})$
Euclidean group in \mathbf{R}^ν	Translation group in \mathbf{R}^ν	Isotropic fluid	Isotropic fluid (no decomposition)
Euclidean group in \mathbf{R}^ν	Translation group in \mathbf{R}^ν	Unoriented crystal	Oriented crystal
Translation group in \mathbf{R}^ν	Lattice group of translations	Oriented crystal	Oriented crystal with fixed mean positions of particles

[21] For the case where H is a normal subgroup of G and G/H is compact see Robinson and Ruelle [1].
[22] See Remark 6.3.6.

In quantum systems, the gauge group corresponding to the conservation of the particle number may also be broken (in Helium II).[23]

It is unfortunate that no equilibrium state of a realistic system has been proved up to now to be a crystal (in the sense of the above table). This reduces for the time being the interest of discussing in detail the spontaneous breakdown of symmetry in statistical mechanics.

From the mathematical point of view we may ask if there is a natural choice of nontrivial subgroup H of G. This question has been examined in particular for $G = \mathbf{R}^\nu$.[24] Let us write in that case

$$U_\rho(a) = \int_{\mathbf{R}^\nu} e^{ip \cdot a} \, dF(p) \tag{5.6}$$

and let $\mathscr{X} = \{p \in \mathbf{R}^\nu : F(\{p\}) \neq 0\}$. It can be shown that \mathscr{X} is a subgroup of \mathbf{R}^{ν}[25] and three cases may be distinguished.

E1. $\mathscr{X} = \{0\}$: the state has no natural decomposition, this situation corresponds to the weak mixing of ergodic theory.

E2. \mathscr{X} is a closed subgroup of \mathbf{R}^ν: there is then a natural choice of $H \neq \{0\}$ and a natural decomposition of ρ into H-ergodic states.

E3. \mathscr{X} is not closed: the state decomposes with respect to various subgroups without a best possible choice of H.[26]

BIBLIOGRAPHICAL NOTE

The study of invariant states on a B*-algebra \mathfrak{A} with a group G of automorphisms is a natural extension of the study of invariant measures on a compact set with a group of homeomorphisms, and the notion of ergodic state was introduced as early as 1951 by Segal [1]. However, to be interesting, the extension of ergodic theory to invariant states on a B*-algebra requires some restriction on \mathfrak{A} and G. We have developed such an extension in Sections 5.2–5.4 under the assumption that \mathfrak{A} is G-Abelian. G-Abelian or asymptotically Abelian algebras in one form or another have been studied by a number of people: first by Doplicher, Kastler, Robinson[27] and independently Ruelle [7], later also by Kadison, Lanford, and Størmer.[28] We have here mostly followed Lanford and Ruelle [1].

[23] For an analysis of this situation see in particular Ginibre [3].
[24] See Kastler and Robinson [1].
[25] See Exercise 6.F.
[26] The case E3 appears to arise in certain physical situations, e.g., in liquid crystals constituted of layers of elongated molecules where the direction of the molecules is rotated from one layer to the next by an angle $\alpha\pi$, α irrational.
[27] See Doplicher, Kastler, and Robinson [1], Kastler and Robinson [1].
[28] See Robinson and Ruelle [1], Lanford and Ruelle [1], Doplicher, Kadison, Kastler, and Robinson [1], Størmer [1], [2], [3].

EXERCISES[29]

6.A. Let \mathfrak{A} be G-Abelian and have an identity. A new demonstration that $E \cap \mathscr{L}_G^{\perp}$ is a simplex (see Corollary 6.4.2) may be obtained by showing directly that the cone of G-invariant positive linear forms on \mathfrak{A} is a lattice (see A.5.3). It is actually sufficient to prove this lattice property on the sets C_ρ of invariant positive linear forms bounded by an invariant state ρ. According to A.3.5, if $f \in C_\rho$, there is a unique self-adjoint positive $S_f \in [\pi_\rho(\mathfrak{A})]'$ such that

$$f(\cdot) = (\Omega_\rho, \pi_\rho(\cdot)S_f\Omega_\rho)$$

Check that the mapping $f \to S_f$ is order preserving, linear, and that $S_f \in [\pi_\rho(\mathfrak{A}) \cup U_\rho(G)]'$, which is Abelian by Proposition 6.3.2,(c), (d). To finish the proof, use the lattice property of the self-adjoint elements of an Abelian B*-algebra (see A.3.9).

6.B. Let \mathfrak{A} be G-Abelian and $\rho \in E \cap \mathscr{L}_G^{\perp}$.

(a) Show that, if $A \in \mathfrak{A}$, an operator \bar{A} on \mathfrak{H}_ρ is defined by

$$\bar{A}\pi_\rho(A_1)\Omega_\rho = \pi_\rho(A_1)P_\rho\pi_\rho(A)\Omega_\rho$$

[use $\|\bar{A}\pi_\rho(A_1)\Omega_\rho\|^2 = (C^{1/2}\Omega_\rho, \pi_\rho(A^*)P_\rho\pi_\rho(A)C^{1/2}\Omega_\rho)$ with $C = P_\rho\pi_\rho(A_1{}^*A_1)P_\rho$].

(b) Check that the linear mapping $\pi_\rho(A) \to \bar{A}$ is norm reducing and that its image is contained in $[\pi_\rho(\mathfrak{A}) \cup U_\rho(G)]'$.

(c) Check that $P_\rho\pi_\rho(A)P_\rho = \bar{A}P_\rho$

6.C. For $A \in \mathfrak{A}$, let conv $\tau_G A$ be the set of all convex combinations $\Sigma\lambda_i\tau_{g_i}A$. Show that the following implications hold between the conditions below:

(a) For all $\rho \in E \cap \mathscr{L}_G^{\perp}$ and self-adjoint $A \in \mathfrak{A}$ the weak closure of the set π_ρ (conv $\tau_G A$) contains an element in $[\pi_\rho(\mathfrak{A})]'$ [30]

(b) For all $\rho \in E \cap \mathscr{L}_G^{\perp}$, self-adjoint $A \in \mathfrak{A}$, and finite subset S of \mathfrak{A},

$$\inf_{A' \in \text{conv}\,\tau_G A} \sup_{A_1, A_2 \in S} |\rho(A_1[A', A_2])| = 0$$

(c) For all $\rho \in E \cap \mathscr{L}_G^{\perp}$ and self-adjoint $A \in \mathfrak{A}$, \bar{A} is in the weak closure of π_ρ (conv $\tau_G A$) where \bar{A} is defined by 6.B.

[29] The Exercises 6.A–6.D result from discussion with Lanford and are directly inspired by Størmer [1], [3].

[30] If this condition holds, G is called by Størmer a large group of automorphisms. Størmer [1] has proved that the inner automorphisms of a B*-algebra \mathfrak{A} with an identity form a large group.

[To prove (b) \Rightarrow (c) use Theorem 6.2.7 to approximate $P_\rho \pi_\rho(A)\Omega_\rho$ by $\pi_\rho(A'')\Omega_\rho$ with $A'' \in \operatorname{conv} \tau_G A$; then use (b) with $A' \in \operatorname{conv} \tau_G A''$.]

(d) For all $\rho \in E \cap \mathscr{L}_G^\perp$ and $A \in \mathfrak{A}$,

$$\bar{A} \in [\pi_\rho(\mathfrak{A})]''$$

(e) For all $\rho \in E \cap \mathscr{L}_G^\perp$,

$$[\pi_\rho(\mathfrak{A}) \cup U_\rho(G)]' \subset [\pi_\rho(\mathfrak{A})]''$$

$$(\text{or } [\pi_\rho(\mathfrak{A})]' \cap [U_\rho(G)]' = (\text{center of } [\pi_\rho(\mathfrak{A})]'') \cap [U_\rho(G)]')$$

[To prove (d) \Rightarrow (e) remark that by 6.B, (b), (c),

$$P_\rho[P_\rho \pi_\rho(\mathfrak{A})P_\rho]'' \subset P_\rho\{[\pi_\rho(\mathfrak{A}) \cap U_\rho(G)]' \cap [\pi_\rho(\mathfrak{A})]''\},$$

then use Proposition 6.3.2.]

6.D. If the condition (e) of 6.C is satisfied and if ρ_1, ρ_2 are quasiequivalent G-ergodic states, then $\rho_1 = \rho_2$ [31] [quasiequivalent means that there exists a $*$-isomorphism β of the weak closure of $\pi_{\rho_1}(\mathfrak{A})$ to the weak closure of $\pi_{\rho_2}(\mathfrak{A})$ such that $\pi_{\rho_2}(\cdot) = \beta \pi_{\rho_1}(\cdot)$].

[Suppose $\rho_1 \neq \rho_2$ and let $\rho = \frac{1}{2}\rho_1 + \frac{1}{2}\rho_2$. In view of the argument in 6.A there are two orthogonal projections $T_1, T_2 \in [\pi_\rho(\mathfrak{A}) \cup U_\rho(G)]'$ such that $T_1 \cdot T_2 = 0$, $\frac{1}{2}\rho_i(\cdot) = (\Omega_\rho, \pi_\rho(\cdot)T_i\Omega_\rho)$, and π_{ρ_i} is identified with the restriction of π_ρ to the range of T_i. Since $T_i \in [\pi_\rho(\mathfrak{A})]''$ let $\pi_\rho(A) \to T_1$; then $\pi_\rho(A)$ tends to 1 on the range of T and the existence of the isomorphism β shows that $\pi_\rho(A)$ tends to 1 on the range of T_2, a contradiction.]

6.E. If ρ is a state on \mathfrak{A} and the weak closure of $\pi_\rho(\mathfrak{A})$ in \mathfrak{H}_ρ is a factor (see A.4.5), then ρ is called a *factor state*.

(a) In the situation of 6.C,(e), a G-invariant factor state is G-ergodic.

(b) Let ρ be G-invariant and let the center of $[\pi_\rho(\mathfrak{A})]''$ be contained in $[U_\rho(G)]'$; then ρ is a factor state if it is ergodic. In particular in the situation of Corollary 6.2.11 the ergodic states are factor states (called factor traces).

(c)[32] Let \mathfrak{A} be weakly asymptotically Abelian [see Corollary 6.2.10,(d)] and ρ be a factor state on \mathfrak{A}; then

$$\lim_{g \to \infty} |\rho(\tau_g A \cdot B) - \rho(\tau_g A)\rho(B)| = 0$$

In particular, if ρ is invariant, $\rho(\tau_g A \cdot B) \to \rho(A)\rho(B)$. [The C*-algebra \mathfrak{B} generated by $\pi_\rho(\mathfrak{A}) \cup \pi_\rho(\mathfrak{A})'$ is irreducible; therefore, according to A.2.8 we can write $B - \rho(B) = B_1 + B_2$ where $B_1, B_2 \in \mathfrak{B}$, $B_1\Omega_\rho = 0$, $B_2^*\Omega_\rho = 0$. Insert this in $\rho(\tau_g A \cdot B)$ and use the fact that B_2 can be approximated in norm by a finite sum $\Sigma \pi_\rho(A_i) \cdot A_i'$ with $A_i \in \mathfrak{A}$, $A_i' \in [\pi_\rho(\mathfrak{A})]'$.]

[31] This applies in particular if ρ_1, ρ_2 are unitarily equivalent, i.e., if there exists an isometry V of \mathfrak{H}_{ρ_1} onto \mathfrak{H}_{ρ_2} such that $V\pi_{\rho_2}(\cdot)V^{-1} = \pi_{\rho_1}(\cdot)$.

[32] Due to Kastler and Robinson (private communication).

6.F.[33] Let G be Abelian locally compact noncompact and the functions $g \to \sigma(\tau_g A)$ on G be continuous (for all $A \in \mathfrak{A}$, $\sigma \in E$). Let $\rho \in E \cap \mathscr{L}_G^\perp$ and $E(\cdot)$ be the spectral measure on the character group \hat{G} such that

$$U_\rho(g) = \int_G X(g)\, dE(X)$$

Notice that $E(\{1\}) = P_\rho$.

(a) If (χ_α) is an M-net on G, then

$$E(\{X\}) = \operatorname*{st\,lim}_{\alpha \to \infty} \int dg \chi_\alpha(g) X(g)^{-1} U_\rho(g)$$

[Apply 6.2.14 to the representation $X(g)^{-1} U_\rho(g)$.]

(b) Suppose that \mathfrak{A} is weakly asymptotically Abelian [see Corollary 6.2.10,(d)]. Using (a) show that for $X_1, X_2, X_3 \in \hat{G}$ we have

$$E(\{X_1\})\pi_\rho(A_1)E(\{X_1 X_2\})\pi_\rho(A_2)E(\{X_3\})$$
$$= E(\{X_1\})\pi_\rho(A_2)E(\{X_2^{-1} X_3\})\pi_\rho(A_1)E(\{X_3\})$$

in particular,

$$[E(\{X_1\})\pi_\rho(A_1)E(\{X_1\}), E(\{X_1\})\pi_\rho(A_2)E(\{X_1\})] = 0$$

(c) Let $\mathscr{X} = \{X \in \hat{G} : E(\{X\}) \neq 0\}$; under the conditions of (b) show that

$$P_\rho \pi_\rho(A^*)E(\{X\})\pi_\rho(A)P_\rho = P_\rho \pi_\rho(A)E(\{X^{-1}\})\pi_\rho(A^*)P_\rho$$

hence $\mathscr{X} = \mathscr{X}^{-1}$.

(d) Let ρ be ergodic; then, under the conditions of (b), $\mathscr{X} \cdot \mathscr{X} \subset \mathscr{X}$ (therefore \mathscr{X} is a group) and $E(\{X\})$ is one-dimensional for $X \in \mathscr{X}$.

[To prove that $\mathscr{X} \cdot \mathscr{X} \subset \mathscr{X}$ use the fact that P_ρ is one-dimensional and

$$E(\{X_1\})\pi_\rho(A_1)E(\{X_1 X_2\})\pi_\rho(A_2)E(\{X_2\}) = E(\{X_1\})\pi_\rho(A_2)P_\rho \pi_\rho(A_1)E(\{X_2\})$$

From the irreducibility of $\pi_\rho(\mathfrak{A}) \cup U_\rho(G)$ prove that $E(\{X\})\pi_\rho(\mathfrak{A})E(\{X\})$ is irreducible on the range of $E(\{X\})$; since it is Abelian by (b), it is one-dimensional.]

6.G. We say that the complex function f on the locally compact group G is uniformly continuous if, for all $\varepsilon > 0$, we can find a neighborhood \mathscr{N} of the identity in G such that, for all $x \in G$,

$$|f(y_1 x y_2) - f(x)| < \varepsilon \qquad \text{if} \quad y_1, y_2 \in \mathscr{N}$$

[33] Adapted from Kastler and Robinson [1], see also Robinson and Ruelle [1].

Let $\mathscr{C}_U(G)$ be the B^*-algebra of bounded complex uniformly continuous functions on G; it can be shown[34] that if there exists a state on $\mathscr{C}_U(G)$ invariant under right translations, then G has an invariant mean. From this and Proposition 6.2.13 prove that the following property characterizes the groups with an invariant mean among locally compact groups:

For all \mathfrak{A} with an identity and τ such that the functions $g \to \sigma(\tau_g A)$ on G continuous (for all $A \in \mathfrak{A}$, $\sigma \in E$), there exists a G-invariant state on \mathfrak{A}.

[34] See Greenleaf [1].

CHAPTER 7

THE STATES OF
STATISTICAL MECHANICS

In this chapter we study the thermodynamic equilibrium states of infinite systems. First we describe the triples (\mathfrak{A}, G, τ) associated with these systems (Section 7.1). We then investigate the existence and properties of the thermodynamic limit for equilibrium states. In particular we prove a weak version of the Gibbs phase rule (Section 7.5), and we establish a remarkable analyticity property (KMS boundary condition) which holds for quantum systems (Section 7.6).

7.1 B*–ALGEBRAS FOR CLASSICAL AND QUANTUM STATISTICAL MECHANICS

In this section we describe the states which occur in the statistical mechanics of classical and quantum systems, and we exhibit B*-algebras on which these states are positive linear forms. In particular we check that the decomposition theory of Section 6.4 yields a unique integral representation in terms of ergodic states. We shall take as invariance group G in each case the translation group \mathbf{R}^v or \mathbf{Z}^v; it is always permissible to take a larger group (see 6.2.9).

7.1.1 CLASSICAL LATTICE SYSTEMS. For definiteness we take the case of a lattice gas. We assume thus that every point of \mathbf{Z}^v may be occupied by either zero or one particle. A *configuration* of the system is described by giving the set $X \subset \mathbf{Z}^v$ of occupied lattice sites. Alternatively, we may give the occupation number 0 or 1 of each point of \mathbf{Z}^v; the space of all configurations of the system is then identified with the product

$$K = \{0, 1\}^{\mathbf{Z}^v} \tag{1.1}$$

of one copy of the set $\{0, 1\}$ for each lattice site. Notice that K is a compact space with the product of the discrete topologies of the sets $\{0, 1\}$.[1]

A *state* of the system is naturally defined as a probability measure on the space K of all configurations, i.e., as a state on the B*-algebra $\mathscr{C}(K)$ of all complex continuous functions on K. If $A \in \mathscr{C}(K)$, we may regard A as a function of subsets of \mathbf{Z}^ν (a configuration $X \in K$ is identified with the set $X \subset \mathbf{Z}^\nu$ of occupied lattice sites). We define

$$\tau_x A(X) = A(X - x) \tag{1.2}$$

for $x \in \mathbf{Z}^\nu$. The triple (\mathfrak{A}, G, τ) of the present theory is $(\mathscr{C}(K), \mathbf{Z}^\nu, \tau)$. Since \mathfrak{A} is Abelian, it is G-Abelian.

Let Λ be a finite subset of \mathbf{Z}^ν and let \mathscr{C}_Λ be the space of complex functions of subsets of Λ.[2] We define \mathfrak{A}_Λ as the subalgebra of \mathfrak{A} formed by the elements A such that

$$A(X) = f(X \cap \Lambda) \tag{1.3}$$

for some $f \in \mathscr{C}_\Lambda$. By the Stone–Weierstrass theorem the union of the \mathfrak{A}_Λ is dense in \mathfrak{A}. This shows that \mathfrak{A}_Λ has the structure of a quasilocal algebra (see 6.2.4). Furthermore it follows that \mathfrak{A} is separable and therefore every invariant state ρ is the resultant of a unique measure carried by the ergodic states (see Proposition 6.4.3).

Let ρ be a state on \mathfrak{A}; using the notation of (1.3), we may write

$$\rho(A) = \sum_{X \subset \Lambda} \mu_\Lambda(X) f(X) \tag{1.4}$$

The number $\mu_\Lambda(X) \geq 0$ is the probability that the lattice sites in X are occupied by a particle and the lattice sites in $\Lambda \backslash X$ are empty. For every finite $\Lambda \subset \mathbf{Z}^\nu$ let $\mu_\Lambda(\cdot)$ be a function greater than or equal to zero of the subsets of Λ. We say that the $\mu_\Lambda(\cdot)$ form a *system of density distributions* if the following conditions are satisfied.

(a) *Normalization*

$$\sum_{X \subset \Lambda} \mu_\Lambda(X) = 1 \tag{1.5}$$

(b) *Compatibility*: if $\Lambda \subset \Lambda'$, then

$$\mu_\Lambda(X) = \sum_{Y \subset \Lambda' \backslash \Lambda} \mu_{\Lambda'}(X \cup Y) \tag{1.6}$$

There is clearly a one-to-one correspondence between the states ρ on \mathfrak{A} and the systems of density distributions $(\mu_\Lambda(\cdot))$. The \mathbf{Z}^ν-invariance of ρ is expressed in terms of $(\mu_\Lambda(\cdot))$ by

[1] It can be shown that K is homeomorphic to a Cantor set.
[2] \mathscr{C}_Λ is 2^V-dimensional if Λ contains V elements.

(c) *Invariance*

$$\mu_\Lambda(X) = \mu_{a+\Lambda}(a + X) \tag{1.7}$$

We may associate to every state ρ on \mathfrak{A} a *correlation function* $\rho(\cdot)$ defined on finite subsets of \mathbf{Z}^v and such that $\rho(X)$ is the probability that all lattice sites in X are simultaneously occupied. We have $\rho(X) = \mu_X(X)$ and therefore

$$\rho(X) = \sum_{Y:X \subset Y \subset \Lambda} \mu_\Lambda(Y) \tag{1.8}$$

$$\mu_\Lambda(X) = \sum_{Y:X \subset Y \subset \Lambda} (-1)^{N(Y)-N(X)} \rho(Y) \tag{1.9}$$

7.1.2 CLASSICAL CONTINUOUS SYSTEMS.[3] For continuous systems the *configurations* are countable subsets X of \mathbf{R}^v such that $X \cap \Lambda$ is finite for every bounded $\Lambda \subset \mathbf{R}^v$. A state should be a probability measure on the set of all such X. We shall, however, first approach the definition of states in a different manner by giving for each bounded open set $\Lambda \subset \mathbf{R}^v$ the probability of finding exactly n particles in Λ and the probability distribution of their positions.

For every bounded open set $\Lambda \subset \mathbf{R}^v$ and integer $n \geq 0$, let $\mu_\Lambda^n \geq 0$ be a measure on $\Lambda^n \subset \mathbf{R}^{nv}$. We assume that μ_Λ^n is invariant under permutation of the n factors of Λ^n. We shall say that the μ_Λ^n form a *system of density distributions* if they satisfy the following conditions.

(a) $$\mu_\varnothing^0(\mathbf{R}^0) = 1 \tag{1.10}$$

(b) If $\Lambda \subset \Lambda'$, then

$$\mu_\Lambda^n(x_1, \ldots, x_n) = \sum_{p=0}^\infty \frac{(n+p)!}{n!\,p!} \int_{\Lambda' \setminus \Lambda} dx_{n+1} \cdots \int_{\Lambda' \setminus \Lambda} dx_{n+p} \mu_{\Lambda'}^{n+p}(x_1, \ldots, x_{n+p}) \tag{1.11}$$

where we have used the functional notation $\mu_\Lambda^n(x_1, \ldots, x_n)\,dx_1 \cdots dx_n$ for $d\mu_\Lambda^n(x_1, \ldots, x_n)$. Notice that from (a) and (b) we obtain the normalization condition

$$\sum_n \int dx_1 \cdots dx_n \mu_\Lambda^n(x_1, \ldots, x_n) = 1 \tag{1.12}$$

Condition (b) is a compatibility condition.

The correlation functions corresponding to the density distribution μ_Λ^n are defined (if they exist; see Section 4.7) by

$$\rho(x_1, \ldots, x_n) = \sum_{p=0}^\infty \frac{(n+p)!}{p!} \int dx_{n+1} \cdots dx_{n+p} \mu_\Lambda^{n+p}(x_1, \ldots, x_{n+p}) \tag{1.13}$$

[3] We restrict ourselves to the description of configurations, ignoring momenta. This subsection is descriptive in nature; for proofs of the statements, see Ruelle [8].

where we assume $x_1, \ldots, x_n \in \Lambda$. We may invert (1.13) formally, obtaining

$$\mu_\Lambda{}^n(x_1, \ldots, x_n) = \frac{1}{n!} \sum_{p=0}^{\infty} \frac{(-1)^p}{p!} \int_{\Lambda^p} dx_{n+1} \cdots dx_{n+p} \rho(x_1, \ldots, x_{n+p}) \quad (1.14)$$

The series in the right-hand side of (1.14) may however be divergent; we know indeed that the correlation functions may not determine the state of a system.[4]

We shall associate with every system of density distributions $(\mu_\Lambda{}^n)$ a state on a B*-algebra \mathfrak{A} which we now construct.

For every bounded open set $\Lambda \subset \mathbf{R}^\nu$ let $\mathscr{K}_\Lambda{}^n$ be the space of real continuous function on $(\mathbf{R}^\nu)^n$ with support in Λ^n. We call \mathscr{K}_Λ the space of sequences $(f^n)_{n \geq 0}$ where $f^n \in \mathscr{K}_\Lambda{}^n$ and $f^n = 0$ for n large enough, and we call \mathscr{K} the union of the \mathscr{K}_Λ.

We denote by \mathscr{T} the (topological) sum $\Sigma_{n \geq 0}(\mathbf{R}^\nu)^n$ of disjoint copies of the spaces $(\mathbf{R}^\nu)^n$. If $f = (f^n)_{n \geq 0} \in \mathscr{K}$, a function Sf on \mathscr{T} is defined so that its restriction to $(\mathbf{R}^\nu)^n$ is

$$Sf(x_1, \ldots, x_n) = \sum_{p \geq 0} \sum_{i_1 = 1}^{n} \cdots \sum_{i_p = 1}^{n} f^p(x_{i_1}, \ldots, x_{i_p}) \quad (1.15)$$

We consider the functions on \mathscr{T} of the form $\varphi(Sf_1, \ldots, Sf_q)$ where $f_1, \ldots, f_q \in \mathscr{K}$ and φ is a bounded continuous complex function on \mathbf{R}^q. With respect to the usual operations on functions and the *-operation given by complex conjugation, the $\varphi(Sf_1, \ldots, Sf_q)$ form a commutative algebra $\tilde{\mathfrak{A}}$ with an involution. The closure \mathfrak{A} of $\tilde{\mathfrak{A}}$ with respect to the uniform norm is an Abelian B*-algebra.

Given a family of density distributions $(\mu_\Lambda{}^n)$, a state ρ on \mathfrak{A} is defined as follows. For each $\varphi(Sf_1, \ldots, Sf_q)$, let Λ be such that $f_1, \ldots, f_q \in \mathscr{K}_\Lambda$ and write

$$\rho(\varphi(Sf_1, \ldots, Sf_q)) = \sum_{n \geq 0} \int dx_1 \cdots dx_n \mu_\Lambda{}^n(x_1, \ldots, x_n)$$

$$\times \; \varphi(Sf_1(x_1, \ldots, x_n), \ldots, Sf_q(x_1, \ldots, x_n)) \quad (1.16)$$

This definition is independent of the choice of Λ and extends by continuity to a state on \mathfrak{A}. We call \mathscr{F} the set of states which are obtained as above from a family of density distributions; the mapping $(\mu_\Lambda{}^n) \to \rho$ is then one-to-one onto \mathscr{F}.

The way in which translations of \mathbf{R}^ν act as automorphisms of \mathfrak{A} is obvious. A family of density distributions defines an invariant state if it satisfies the condition

(c) $\quad \mu_\Lambda{}^n(x_1, \ldots, x_n) = \mu_{\Lambda + a}^n(x_1 + a, \ldots, x_n + a)$ \hfill (1.17)

for all $a \in \mathbf{R}^\nu$ and Λ.

[4] See Section 4.7. Notice that the series in (1.14) is convergent if (7.20) of Chapter 4 holds.

Since the algebra \mathfrak{A} is not separable, a decomposition of invariant states into ergodic states cannot be obtained from Proposition 6.4.3. However it can be shown that Proposition 6.4.5 applies, with suitably chosen \mathfrak{A}_α, \mathcal{T}_α,[5] yielding a decomposition of the invariant states in \mathcal{F} into ergodic states in \mathcal{F}. Moreover, every state $\rho \in \mathcal{F}$ has a unique integral representation in terms of pure states[6] in \mathcal{F}. These pure states may be identified to infinite configurations X defined as countable subsets X of \mathbf{R}^ν where the points of X are attributed a finite multiplicity and $X \cap \Lambda$ is finite for every bounded Λ.

We mention finally the relation between the Hilbert space construction in Section 4.7 and the Gel'fand–Segal construction corresponding to a state $\rho \in \mathcal{F}$. It can be shown that there exists for each $f \in \mathcal{K}$ a self-adjoint operator $A(f)$ (unbounded) such that for all $\sigma(Sf_1, \ldots, Sf_q)$

$$\pi_\rho(\sigma(Sf_1, \ldots, Sf_q)) = \varphi(A(f_1), \ldots, A(f_q)) \tag{1.18}$$

A "field" $A(x)$ may be introduced such that (7.16) of Chapter 4 holds and the existence of correlation functions [as given by (7.17) of Chapter 4] depends on whether Ω_ρ is contained in the domain of all operators $A(f)$.

7.1.3 QUANTUM LATTICE SYSTEMS. We have described in 1.3.3 how, for a quantum lattice system, a finite-dimensional Hilbert space $\mathcal{H}(\Lambda)$ is associated with every finite $\Lambda \subset \mathbf{Z}^\nu$. In Section 2.2 we have noticed that if $\Lambda \subset \Lambda'$, the natural isomorphism

$$\mathcal{H}(\Lambda') \cong \mathcal{H}(\Lambda) \otimes \mathcal{H}(\Lambda' \backslash \Lambda) \tag{1.19}$$

determines an isomorphism

$$\alpha_{\Lambda'\Lambda} : A \to A \otimes \mathbf{1} \tag{1.20}$$

of the C*-algebra \mathfrak{A}_Λ of all bounded operators in $\mathcal{H}(\Lambda)$ into the C*-algebra $\mathfrak{A}_{\Lambda'}$ of all bounded operators in $\mathcal{H}(\Lambda')$. Furthermore an isomorphism $\tau_x : \mathfrak{A}_\Lambda \to \mathfrak{A}_{\Lambda+x}$ is given for each translation x of \mathbf{Z}^ν by

$$\tau_a A = V_\Lambda(a) A V_{a+\Lambda}(-a) \tag{1.21}$$

where $V_\Lambda(a)$ is defined by Eq. (2.1) of Chapter 2. It is easily verified that the \mathfrak{A}_Λ, $\alpha_{\Lambda'\Lambda}$, τ_x satisfy the conditions (a'), (b'), (c') of 6.2.4 and therefore define a quasilocal B*-algebra \mathfrak{A}, which we take as B*-algebra of the quantum lattice system.

The algebra \mathfrak{A} is separable[7] and therefore Proposition 6.4.3 shows that there is a unique decomposition of the invariant states into ergodic states.

[5] See Ruelle [8].

[6] The pure states are the G-ergodic states for G reduced to the identity.

[7] We remark that \mathfrak{A} is also simple: since the \mathfrak{A}_Λ are simple, every nontrivial morphism $\mathfrak{A} \to \mathfrak{B}$ is an isomorphism (hence an isometry) on the \mathfrak{A}_Λ; therefore it is an isometry (hence an isomorphism) on \mathfrak{A}.

Any state ρ on \mathfrak{A} is determined by its restrictions to the \mathfrak{A}_Λ. Now, given a state ρ_Λ on \mathfrak{A}_Λ, there is a unique density matrix μ_Λ on $\mathscr{H}(\Lambda)$ such that, for all $A \in \mathfrak{A}_\Lambda$,

$$\rho_\Lambda(A) = \text{Tr}_{\mathscr{H}(\Lambda)}\mu_\Lambda A \tag{1.22}$$

Therefore it is equivalent to give a state on \mathfrak{A} or to give a *system of density matrices* (μ_Λ) consisting of a density matrix μ_Λ for each finite $\Lambda \subset \mathbf{Z}^\nu$ such that

(a) if $\Lambda \subset \Lambda'$ then [with the identification (1.19)]

$$\mu_\Lambda = \text{Tr}_{\mathscr{H}(\Lambda' \setminus \Lambda)}\mu_{\Lambda'} \tag{1.23}$$

The invariance of ρ is expressed by

(b) $$\mu_{a+\Lambda} = V_\Lambda(a)\mu_\Lambda V_{a+\Lambda}(a) \tag{1.24}$$

for all $a \in \mathbf{Z}^\nu$ and Λ.

7.1.4 THE FOCK REPRESENTATION. Before describing the states and B*-algebras of quantum continuous systems, we review briefly some facts concerning the Fock representation of the canonical commutation and anticommutation relations.

We have defined in 1.3.1,(c) the Fock space $\mathscr{H}_\varepsilon(\Lambda)$ associated with a Lebesgue measurable subset Λ of \mathbf{R}^ν:

$$\mathscr{H}_\varepsilon(\Lambda) = \bigoplus_{n=0}^{\infty} L_\varepsilon^2(\Lambda^n) \tag{1.25}$$

where $L_\varepsilon^2(\Lambda^n)$ consists of the square integrable functions on Λ^n which are either completely symmetric ($\varepsilon = +$) or completely antisymmetric ($\varepsilon = -$) in their n arguments.

Let f be a real square integrable function in Λ; we define an annihilation operator $a(f)$ and a creation operator $a^*(f)$ as follows: if $\Psi = (\Psi^{(n)})_{n \geq 0} \in \mathscr{H}_\varepsilon(\Lambda)$, then

$$(a(f)\Psi)^{(n)}(x_1, \ldots, x_n) = (n+1)^{1/2} \int_\Lambda dx f(x)\Psi^{(n+1)}(x, x_1, \ldots, x_n) \tag{1.26}$$

$$(a^*(f)\Psi)^{(n)}(x_1, \ldots, x_n) = n^{-1/2} \sum_{i=1}^{n} \varepsilon^{i-1} f(x_i)\Psi^{(n-1)}(x_1, \ldots, \hat{x}_i, \ldots, x_n) \tag{1.27}$$

where \hat{x}_i means that the variable x_i has to be omitted. The operators $a(f), a^*(f)$ are unambiguously defined on the domain D formed by the vectors Ψ which have only a finite number of nonvanishing components.

They satisfy

$$(\Psi_1, a(f)\Psi_2) = (a^*(f)\Psi_1, \Psi_2) \qquad \text{for} \quad \Psi_1, \Psi_2 \in D \qquad (1.28)$$

$$[a(f), a(g)]_{\bar{\varepsilon}} = [a^*(f), a^*(g)]_{\bar{\varepsilon}} = 0 \qquad (1.29)$$

$$[a(f), a^*(g)]_{\bar{\varepsilon}} = \mathbf{1} \int_\Lambda dx f(x) g(x) \qquad (1.30)$$

where we have written $[A, B]_{\bar{\varepsilon}} = AB - \varepsilon BA$. The identities (1.29) and (1.30) constitute the so-called canonical commutation relations (CCR) for $\varepsilon = 1$ or anticommutation relations (CAR) for $\varepsilon = -1$.

In the case of the CCR it is convenient to introduce the symmetric operators

$$\varphi(f) = 2^{-1/2}(a(f) + a^*(f)) \qquad \pi(f) = \frac{1}{i} 2^{-1/2}(a(f) - a^*(f)) \quad (1.31)$$

It can be shown that $\varphi(f)$ and $\pi(f)$ extend uniquely from the domain D to (unbounded) self-adjoint operators.

In the case of the CAR, we have for all $\Psi \in \mathscr{H}_-(\Lambda)$

$$\|a(f)\Psi\|^2 + \|a^*(f)\Psi\|^2 = \|\Psi\|^2 \int_\Lambda dx f(x)^2 \qquad (1.32)$$

so that the operators $a(f), a^*(f)$ are bounded.

The mapping $f \to a(f), a^*(f)$ being linear we write

$$a(f) = \int dx f(x) a(x) \qquad a^*(f) = \int dx f(x) a^*(x) \qquad (1.33)$$

introducing the "field" notation $a(x), a^*(x)$. The CCR and CAR may then be rewritten formally as

$$[a(x), a(y)]_{\bar{\varepsilon}} = [a^*(x), a^*(y)]_{\bar{\varepsilon}} = 0 \qquad [a(x), a^*(y)]_{\bar{\varepsilon}} = \delta(x - y) \quad (1.34)$$

The ensemble averages for a system contained in the bounded region Λ are defined by density matrices ρ_Λ on $\mathscr{H}_\varepsilon(\Lambda)$.

$$\langle A \rangle = \text{Tr} \, \rho_\Lambda A$$

For the explicit form of ρ_Λ see 1.3.1. The following quantities, if they exist, are the finite volume reduced density matrices (see Section 4.6).

$$\rho_\Lambda^{p,q}(x_1, \ldots, x_p; y_1, \ldots, y_q) = \text{Tr}(\rho_\Lambda a^*(x_1) \cdots a^*(x_p) a(y_1) \cdots a(y_q)) \quad (1.35)$$

Under suitable conditions, these reduced density matrices will have, when $\Lambda \to \infty$, limits which describe the state of an infinite system. Since for CCR the operators $a(f), a^*(f)$ are unbounded we encounter however the same difficulties with reduced density matrices as we found in the

classical case with correlation functions. To avoid these difficulties it is natural to use instead of (1.35) the expectation values of bounded functions of the $\varphi(f), \pi(f)$. For the CAR we may use bounded functions of the (bounded) $a(f), a^*(f)$. In both cases we have thus a state on a C*-algebra \mathfrak{A}_Λ of operators in $\mathscr{H}_\varepsilon(\Lambda)$.

The restriction of a state of an infinite system to the bounded region Λ yields a state μ_Λ on \mathfrak{A}_Λ. We shall always assume that μ_Λ is defined by a density matrix in $\mathscr{H}_\varepsilon(\Lambda)$. This condition means physically that the probability of finding simultaneously an infinite number of particles in Λ is zero. On the other hand, for an infinite system the probability of finding an infinite number of particles in \mathbf{R}^ν is in typical situations one; such states of the system cannot therefore be described by density matrices in $\mathscr{H}_\varepsilon(\mathbf{R}^\nu)$.[8]

It is necessary at this point to describe how an operator in \mathfrak{A}_Λ may be identified with an operator in $\mathfrak{A}_{\Lambda'}$ for $\Lambda \subset \Lambda'$. Let $\Lambda_1 \cap \Lambda_2 = \varnothing$, we introduce an isomorphism

$$\mathscr{H}_\varepsilon(\Lambda_1 \cup \Lambda_2) \cong \mathscr{H}_\varepsilon(\Lambda_1) \otimes \mathscr{H}_\varepsilon(\Lambda_2) \tag{1.36}$$

as follows. If $\Psi_1 = (\Psi_1^{(m)}) \in \mathscr{H}_\varepsilon(\Lambda_1)$, $\Psi_2 = (\Psi_2^{(n)}) \in \mathscr{H}_\varepsilon(\Lambda_2)$, and Ψ corresponds to $\Psi_1 \otimes \Psi_2$ by (1.36), we let

$$\Psi^{(m+n)}(x_1, \ldots, x_m, y_1, \ldots, y_n)$$

$$= \left(\frac{m! n!}{(m+n)!} \right)^{1/2} \Psi_1^{(m)}(x_1, \ldots, x_m) \Psi_2^{(n)}(y_1, \ldots, y_n) \tag{1.37}$$

for $x_1, \ldots, x_m \in \Lambda_1$, $y_1, \ldots, y_n \in \Lambda_2$, and then extend $\Psi^{(m+n)}$ to $(\Lambda_1 \cup \Lambda_2)^{m+n}$ by symmetry.

In particular if $\Lambda \subset \Lambda'$ we have

$$\mathscr{H}_\varepsilon(\Lambda') \cong \mathscr{H}_\varepsilon(\Lambda) \otimes \mathscr{H}_\varepsilon(\Lambda' \backslash \Lambda) \tag{1.38}$$

and therefore an isomorphism

$$\alpha_{\Lambda'\Lambda} : A \to A \otimes \mathbf{1} \tag{1.39}$$

of the C*-algebra of all bounded operators in $\mathscr{H}(\Lambda)$ into the C*-algebra of all bounded operators in $\mathscr{H}(\Lambda')$. The reader may check that (1.39) sends the creation and annihilation operators $a(f), a^*(f)$ on $\mathscr{H}_\varepsilon(\Lambda)$ to the creation and annihilation operators $a(f), a^*(f)$ on $\mathscr{H}_\varepsilon(\Lambda')$.

If $a \in \mathbf{R}^\nu$ we define an isometry $V_\Lambda(a)$ of $\mathscr{H}_\varepsilon(\Lambda)$ onto $\mathscr{H}_\varepsilon(a + \Lambda)$ by

$$(V_\Lambda(a)\Psi)^{(n)}(x_1, \ldots, x_n) = \Psi^{(n)}(x_1 - a, \ldots, x_n - a) \tag{1.40}$$

[8] For characterizations of the states which are represented by density matrices see Dell'Antonio, Doplicher, and Ruelle [1]. The state describing an infinite system yields, by the Gel'fand–Segal construction, a representation of the CCR or CAR; this representation is (in general) not unitarily equivalent to a multiple of the Fock representation. Exercise 6.D. shows that the states of infinite systems yield many inequivalent representations of the CCR and CAR.

An isomorphism τ_a of the C*-algebra of all bounded operators in $\mathcal{H}_\varepsilon(\Lambda)$ onto the C*-algebra of all bounded operators in $\mathcal{H}_\varepsilon(\Lambda + a)$ is then given by

$$\tau_a A = V_\Lambda(a) A V_{a+\Lambda}(-a) \tag{1.41}$$

7.1.5 QUANTUM CONTINUOUS SYSTEMS—BOSONS. Let \mathfrak{A}_Λ be the C*-algebra of all bounded operators in $\mathcal{H}_+(\Lambda)$. It is easily verified that the $\mathfrak{A}_\Lambda, \alpha_{\Lambda'\Lambda}$ [defined by (1.39)], τ_x [defined by (1.41)] satisfy the conditions (a'), (b'), (c') of 6.2.4 and therefore define a quasilocal B*-algebra \mathfrak{A}, which we take as the B*-algebra of a boson system.

We are interested in the set \mathscr{F} of states on \mathfrak{A} which are described by a density matrix μ_Λ on each \mathfrak{A}_Λ. It is equivalent to give $\rho \in \mathscr{F}$ or a *family of density matrices* (μ_Λ) consisting of a density matrix μ_Λ for each bounded measurable $\Lambda \subset \mathbf{R}^\nu$, such that

(a) if $\Lambda \subset \Lambda'$ then [with the identification (1.38)]

$$\mu_\Lambda = \mathrm{Tr}_{\mathscr{H}_+(\Lambda' \smallsetminus \Lambda)} \mu_{\Lambda'} \tag{1.42}$$

The invariance of ρ is expressed by

(b) $$\mu_{a+\Lambda} = V_\Lambda(a) \mu_\Lambda V_{a+\Lambda}(a) \tag{1.43}$$

Notice now that a state of \mathfrak{A}_Λ is described by a density matrix if and only if its restriction to the ideal \mathfrak{I}_Λ of compact (=completely continuous) operators has norm 1.[9] We may thus characterize the states in \mathscr{F} as those which have a restriction of norm 1 to every \mathfrak{I}_Λ, and it is in fact sufficient to take the countable family $(\mathfrak{I}_{\Lambda_n})$ where Λ_n is a sphere with integer radius n centered at the origin. Since \mathfrak{I}_Λ is separable,[10] Proposition 6.4.5 applies and shows that every invariant state in \mathscr{F} has a unique decomposition into ergodic states.

7.1.6 QUANTUM CONTINUOUS SYSTEMS—FERMIONS. Let \mathfrak{A}_Λ be the C*-algebra of all bounded operators in $\mathcal{H}_-(\Lambda)$. It is easily verified that the $\mathfrak{A}_\Lambda, \alpha_{\Lambda'\Lambda}$ [defined by (1.39)], τ_x [defined by (1.41)] satisfy the conditions (a'), (b') but not (c') of 6.2.4. An algebra \mathfrak{A} can be constructed as indicated in 6.2.4 but it is no longer quasilocal; we shall show that it is nevertheless \mathbf{R}^ν-Abelian.

We write $\mathcal{H}_-(\Lambda) = \mathcal{H}^e(\Lambda) \oplus \mathcal{H}^o(\Lambda)$ where

$$\mathcal{H}^e(\Lambda) = \bigoplus_{n\,\mathrm{even}} L_-^2(\Lambda^n) \qquad \mathcal{H}^o(\Lambda) = \bigoplus_{n\,\mathrm{odd}} L_-^2(\Lambda^n) \tag{1.44}$$

[9] The dual of the Banach space of compact operators consists of the operators of trace class; furthermore, a state on a two-sided ideal of a B*-algebra has a unique extension to a state of the whole algebra (see A.3.3).

[10] The operators of finite rank are dense among the compact operators in a Hilbert space.

and, for all $A \in \mathfrak{A}_\Lambda$ we let $A = A^e + A^o$

$$A^e \mathcal{H}^e(\Lambda) \subset \mathcal{H}^e(\Lambda) \qquad A^e \mathcal{H}^o(\Lambda) \subset \mathcal{H}^o(\Lambda)$$

$$A^o \mathcal{H}^e(\Lambda) \subset \mathcal{H}^o(\Lambda) \qquad A^o \mathcal{H}^o(\Lambda) \subset \mathcal{H}^e(\Lambda)$$

A^e and A^o are the even and odd parts of A. It follows from (1.38), (1.39) that if $\Lambda \subset \Lambda'$, then

$$(\alpha_{\Lambda'\Lambda}A)^e = \alpha_{\Lambda'\Lambda}A^e \qquad (\alpha_{\Lambda'\Lambda}A)^o = \alpha_{\Lambda'\Lambda}A^o$$

Notice also that

$$(\tau_x A)^e = \tau_x A^e \qquad (\tau_x A)^o = \tau_x A^o$$

Let Λ_1, Λ_2 be disjoint and $\Psi_i^e \in \mathcal{H}^e(\Lambda_i), \Psi_i^o \in \mathcal{H}^o(\Lambda_i), A_i \in \mathfrak{A}_{\Lambda_i}$; we check that

$$\Psi_1^e \otimes \Psi_2^e = \Psi_2^e \otimes \Psi_1^e \qquad \Psi_1^e \otimes \Psi_2^o = \Psi_2^o \otimes \Psi_1^e$$

$$\Psi_1^o \otimes \Psi_2^e = \Psi_2^e \otimes \Psi_1^o \qquad \Psi_1^o \otimes \Psi_2^o = -\Psi_2^o \otimes \Psi_1^o$$

and therefore

$$[A_1^e, A_2^e]_- = [A_1^e, A_2^o]_- = [A_1^o, A_2^e]_- = 0 \tag{1.45}$$

$$[A_1^o, A_2^o]_+ = 0 \tag{1.46}$$

where a number of canonical isomorphisms have not been made explicit.

Let ρ be an \mathbf{R}^ν-invariant state on \mathfrak{A} and $A_1 \in \mathfrak{A}_{\Lambda_1}, A_2 \in \mathfrak{A}_{\Lambda_2}$. Also let $x_1, \ldots, x_n \in \mathbf{R}^\nu$ be chosen such that the sets $(x_i + \Lambda_1)$ are pairwise disjoint and disjoint from Λ_2. Then,

$$\left\| \pi_\rho \left(\frac{1}{n} \sum_{i=1}^n \tau_{x_i} A_1^o \right) \Omega_\rho \right\|^2 \leq \frac{1}{n^2} \rho \left(\left[\sum_{i=1}^n (\tau_{x_i} A_1^o)^*, \sum_{i=1}^n \tau_{x_i} A_1^o \right]_+ \right)$$

$$\leq \frac{1}{n^2} \sum_{i=1}^n \rho([(\tau_{x_i} A_1^o)^*, \tau_{x_i} A_1^o]_+)$$

$$= \frac{1}{n} \rho([A_1^{o*}, A_1^o]_+) \tag{1.47}$$

and therefore

$$\left| \rho \left(\left[\frac{1}{n} \sum_{i=1}^n \tau_{x_i} A_1, A_2 \right] \right) \right| = \left| \rho \left(\left[\frac{1}{n} \sum_{i=1}^n \tau_{x_i} A_1^o, A_2 \right] \right) \right|$$

$$\leq 2 \| A_2 \| \cdot \left[\frac{1}{n} \rho([A_1^{o*}, A_1^o]_+) \right]^{1/2} \tag{1.48}$$

Letting $n \to \infty$ in (1.48) we see (Theorem 6.2.8) that \mathfrak{A} is \mathbf{R}^ν-Abelian.[11] We obtain also from (1.47) that $\rho(A^o) = 0$ for all odd A^o, i.e., an \mathbf{R}^ν-invariant state is necessarily an *even state*. There is a one-to-one correspondence between the even states on \mathfrak{A} and the states on the algebra \mathfrak{A}^e generated by the even elements of all the \mathfrak{A}_Λ. We may thus take the quasilocal algebra \mathfrak{A}^e instead of \mathfrak{A} as B*-algebra of a fermion system.

We are interested in the set \mathcal{F} of states on \mathfrak{A} which are described by a density matrix μ_Λ on each \mathfrak{A}_Λ. Each $\rho \in \mathcal{F}$ corresponds to a family of density matrices satisfying, as in the case of bosons, the compatibility condition (1.42). The invariance is expressed by (1.43) and implies that

$$\mathrm{Tr}_{\mathcal{H}_{-(\Lambda)}} \mu_\Lambda A^o = 0 \tag{1.49}$$

for all odd $A^o \in \mathfrak{A}_\Lambda$. Therefore μ_Λ commutes with the orthogonal projections on $\mathcal{H}^e(\Lambda)$ and $\mathcal{H}^o(\Lambda)$.

Just as in the case of bosons, we may characterize the states in \mathcal{F} as those which have a restriction of norm 1 to a family of algebras \mathfrak{I}_Λ. Every invariant state in \mathcal{F} has thus a unique decomposition into ergodic states.

7.2 ENTROPY [12]

We shall in this section attribute a mean entropy to the invariant states discussed in Section 7.1. The importance of this notion comes from the fact that the mean entropy of an equilibrium state of an infinite system will turn out to be the same as its physical entropy per unit volume. In fact, the physical entropy may be interpreted as a measure of the amount of randomness present in the description of a system, and it is natural that this can be computed from the knowledge of the state describing the system.[13]

The notion of mean entropy of a state occurs in other contexts than statistical mechanics, notably in information theory where it measures the quantity of information per symbol in a message.[14] The Kolmogorov–Sinaï invariant[15] of ergodic theory is also a mean entropy.

7.2.1 CLASSICAL LATTICE SYSTEMS. We treat the case of lattice gases. Let $(\mu_\Lambda(\cdot))$ be the family of density distributions associated with a state ρ. The entropy corresponding to a finite $\Lambda \subset \mathbf{Z}^\nu$ is defined by[16]

$$S(\Lambda) = - \sum_{X \subset \Lambda} \mu_\Lambda(X) \log \mu_\Lambda(X) \tag{2.1}$$

[11] This was first remarked in Lanford and Robinson [1]. See also Powers [1].

[12] The results in this section are mostly extracted from Robinson and Ruelle [2] for classical systems and from Lanford and Robinson [1] for quantum systems.

[13] For an early result in this direction see Nettleton and Green [1].

[14] See for instance Khinchin [1].

[15] See for instance Jacobs [1] and Arnold and Avez [1].

[16] Let m be a positive measure on Ω and $0 \le f \in L^1(\Omega, m)$, $\int f \, dm = 1$; we define an entropy $S(f) = -\int f \log f \, dm$. If m gives a weight 1 to every point of Ω then $S \ge 0$ [case (2.1)]; if $\int dm = 1$, then $S \le 0$ [case (2.7)].

7.2.2 PROPOSITION. *The following inequalities hold*

(a) $$0 \le S(\Lambda) \le N(\Lambda) \log 2 \tag{2.2}$$

(b) $\Lambda \subset \Lambda' \quad \Rightarrow \quad 0 \le S(\Lambda') - S(\Lambda) \le (N(\Lambda') - N(\Lambda)) \log 2 \tag{2.3}$

(c) *Strong subadditivity*

$$S(\Lambda_1 \cup \Lambda_2) - S(\Lambda_1) - S(\Lambda_2) + S(\Lambda_1 \cap \Lambda_2) \le 0 \tag{2.4}$$

Notice that $S(\emptyset) = 0$; therefore, (a) follows from (b). If $\Lambda \subset \Lambda'$, we have

$$S(\Lambda') - S(\Lambda) = - \sum_{X \subset \Lambda} \sum_{Y \subset \Lambda' \setminus \Lambda} \mu_{\Lambda'}(X \cup Y)[\log \mu_{\Lambda'}(X \cup Y) - \log \mu_{\Lambda}(X)] \tag{2.5}$$

Since $\mu_{\Lambda'}(X \cup Y) \le \mu_{\Lambda}(X)$, (2.5) gives

$$S(\Lambda') - S(\Lambda) \ge 0 \tag{2.6}$$

Writing $\tilde{\mu}_{\Lambda}(X) = 2^{N(\Lambda)} \mu_{\Lambda}(X)$, we obtain

$$\tilde{S}(\Lambda) = S(\Lambda) - \log 2^{N(\Lambda)} = - 2^{-N(\Lambda)} \sum_{X \subset \Lambda} \tilde{\mu}_{\Lambda}(X) \log \tilde{\mu}_{\Lambda}(X) \tag{2.7}$$

and therefore

$$\tilde{S}(\Lambda') - \tilde{S}(\Lambda) = - 2^{-N(\Lambda')} \sum_{X \subset \Lambda} \sum_{Y \subset \Lambda' \setminus \Lambda} \tilde{\mu}_{\Lambda'}(X \cup Y) \log \frac{\tilde{\mu}_{\Lambda'}(X \cup Y)}{\tilde{\mu}_{\Lambda}(X)} \tag{2.8}$$

The convexity of $t \to t \log t \ (t > 0)$ implies

$$t \log t \ge t - 1 \quad \text{or} \quad -\log t \le \frac{1}{t} - 1 \tag{2.9}$$

Inserting this with $t = \tilde{\mu}_{\Lambda'}(X \cup Y)/\tilde{\mu}_{\Lambda}(X)$ in (2.8) yields

$$S(\Lambda') - S(\Lambda) - [N(\Lambda') - N(\Lambda)] \log 2$$
$$= \tilde{S}(\Lambda') - \tilde{S}(\Lambda)$$
$$\le 2^{-N(\Lambda')} \sum_{X \subset \Lambda} \sum_{Y \subset \Lambda' \setminus \Lambda} \tilde{\mu}_{\Lambda'}(X \cup Y) \left[\frac{\tilde{\mu}_{\Lambda}(X)}{\tilde{\mu}_{\Lambda'}(X \cup Y)} - 1 \right] = 0 \tag{2.10}$$

(2.6) and (2.10) prove (b).

Again making use of (2.9), we obtain

$$S(\Lambda_1 \cup \Lambda_2) - S(\Lambda_1) - S(\Lambda_2) + S(\Lambda_1 \cap \Lambda_2)$$
$$= - \sum_{X \subset \Lambda_1 \cap \Lambda_2} \sum_{Y \subset \Lambda_1 \setminus \Lambda_2} \sum_{Z \subset \Lambda_2 \setminus \Lambda_1} \mu_{\Lambda_1 \cup \Lambda_2}(X \cup Y \cup Z)$$
$$\times \log \frac{\mu_{\Lambda_1 \cup \Lambda_2}(X \cup Y \cup Z) \cdot \mu_{\Lambda_1 \cap \Lambda_2}(X)}{\mu_{\Lambda_1}(X \cup Y) \cdot \mu_{\Lambda_2}(X \cup Z)}$$
$$\le \sum \mu_{\Lambda_1 \cup \Lambda_2} \left[\frac{\mu_{\Lambda_1} \cdot \mu_{\Lambda_2}}{\mu_{\Lambda_1 \cup \Lambda_2} \cdot \mu_{\Lambda_1 \cap \Lambda_2}} - 1 \right]$$

$$= \sum_{X,Y} \frac{\mu_{\Lambda_1}(X \cup Y)}{\mu_{\Lambda_1 \cap \Lambda_2}(X)} \sum_{Z} \mu_{\Lambda_2}(X \cup Z) - \sum_{X,Y,Z} \mu_{\Lambda_1 \cup \Lambda_2}(X \cup Y \cup Z)$$

$$= \sum_{X,Y} \mu_{\Lambda_1}(X \cup Y) - \sum_{X,Y,Z} \mu_{\Lambda_1 \cup \Lambda_2}(X \cup Y \cup Z) = 0$$

which proves (c).

7.2.3 PROPOSITION.

(a) *Let ρ be \mathbf{Z}^ν-invariant; then, if $\Lambda \to \infty$ in the sense of Van Hove, the following limit exists*

$$s(\rho) = \lim_{\Lambda \to \infty} N(\Lambda)^{-1} S(\Lambda) \tag{2.11}$$

and $0 \le s(\rho) \le \log 2$.

(b) *The functional $s(\cdot)$ is affine upper semicontinuous on $E \cap \mathscr{L}^1_{\mathbf{Z}^\nu}$.*

Let $\Lambda_1 \cap \Lambda_2 = \varnothing$ in (2.4), we obtain the *subadditivity property*

$$S(\Lambda_1 \cup \Lambda_2) \le S(\Lambda_1) + S(\Lambda_2) \qquad \text{if} \quad \Lambda_1 \cap \Lambda_2 = \varnothing \tag{2.12}$$

Using this we prove the special case of (a) where $\Lambda = \Lambda(a)$.[17] If we consider $S(\Lambda(a))$ as a function of a^1, \ldots, a^ν, it follows from (2.12) and \mathbf{Z}^ν-invariance that it is separately subadditive in each variable; hence, by Proposition 7.2.4 below,

$$\lim_{a \to \infty} \frac{S(\Lambda(a))}{V(a)} = \inf_a \frac{S(\Lambda(a))}{V(a)} \tag{2.13}$$

Furthermore $0 \le s(\rho) \le \log 2$ by Proposition 2.2,(a).

Let $0 < \alpha < 1$ and denote by $(\mu_{1\Lambda}(\cdot)), (\mu_{2\Lambda}(\cdot))$ the families of density distributions associated with the \mathbf{Z}^ν-invariant states ρ_1, ρ_2. Using the convexity of $t \log t$ and the increase of $\log t$, we have

$$- \sum_{X \subset \Lambda} [\alpha \mu_{1\Lambda}(X) \log \mu_{1\Lambda}(X) + (1-\alpha)\mu_{2\Lambda}(X) \log \mu_{2\Lambda}(X)]$$

$$\le - \sum_{X \subset \Lambda} [\alpha \mu_{1\Lambda}(X) + (1-\alpha)\mu_{2\Lambda}(X)] \log[\alpha \mu_{1\Lambda}(X) + (1-\alpha)\mu_{2\Lambda}(X)]$$

$$\le - \sum_{X \subset \Lambda} [\alpha \mu_{1\Lambda}(X) \log(\alpha \mu_{1\Lambda}(X)) + (1-\alpha)\mu_{2\Lambda}(X) \log((1-\alpha)\mu_{2\Lambda}(X))]$$

$$= - \sum_{X \subset \Lambda} [\alpha \mu_{1\Lambda}(X) \log \mu_{1\Lambda}(X) + (1-\alpha)\mu_{2\Lambda}(X) \log \mu_{2\Lambda}(X)]$$

$$\quad - \alpha \log \alpha - (1-\alpha) \log(1-\alpha)$$

$$\le - \sum_{X \subset \Lambda} [\alpha \mu_{1\Lambda}(X) \log \mu_{1\Lambda}(X) + (1-\alpha)\mu_{2\Lambda}(X) \log \mu_{2\Lambda}(X)] + \log 2 \tag{2.14}$$

[17] This will be sufficient for later applications; the proof of the general case uses the strong subadditivity property (see Robinson and Ruelle [2]). The notation $\Lambda(a)$, $V(a)$, $a \to \infty$ has the same meaning here as in Section 2.1.

Dividing by $V(a)$ and letting $a \to \infty$ yields

$$s(\alpha\rho_1 + (1 - \alpha)\rho_2) = \alpha s(\rho_1) + (1 - \alpha)s(\rho_2) \qquad (2.15)$$

This shows that $s(\cdot)$ is affine.

For fixed Λ, X, the number $\mu_\Lambda(X)$ depends continuously on ρ for the weak topology of $E \cap \mathscr{L}^{\perp 18}$ and therefore $S(\Lambda)$ is also a continuous function of ρ. By (2.13), $s(\cdot)$ is the lower bound of a family of continuous functions and therefore upper semicontinuous,[19] concluding the proof.

For completeness we establish now the property of subadditive functions which has been used above.

7.2.4 PROPOSITION. *Let* $F(a^1, \ldots, a^\nu)$ *be defined and real for* $a^1, \ldots, a^\nu \in \mathbf{Z}$, $a^1 > 0, \ldots, a^\nu > 0$. *If* F *is subadditive separately in* a^1, \ldots, a^ν, *i.e.*

$$F(a^1, \ldots, b^i + c^i, \ldots, a^\nu) \le F(a^1, \ldots, b^i, \ldots, a^\nu) + F(a^1, \ldots, c^i, \ldots, a^\nu) \quad (2.16)$$

then

$$\lim_{a \to \infty} \frac{F(a^1, \ldots, a^\nu)}{a^1 \cdot \cdots \cdot a^\nu} = \inf_a \frac{F(a^1, \ldots, a^\nu)}{a^1 \cdot \cdots \cdot a^\nu} \qquad (2.17)$$

Let us write

$$C = \inf_a \frac{F(a^1, \ldots, a^\nu)}{a^1 \cdot \cdots \cdot a^\nu}$$

and suppose that C is not $-\infty$. Clearly

$$\liminf_{a \to \infty} \frac{F(a^1, \ldots, a^\nu)}{a^1 \cdot \cdots \cdot a^\nu} \ge C \qquad (2.18)$$

Given $\varepsilon > 0$ choose $b = (b^1, \ldots, b^\nu)$ such that $(b^1 \cdot \cdots \cdot b^\nu)^{-1} \times F(b^1, \ldots, b^\nu) < C + \varepsilon$ and let $a^i = n^i b^i + c^i$ for $i = 1, \ldots, \nu$ where n^i is an integer greater than or equal to zero and $0 \le c^i < b^i$. Subadditivity yields

$$F(a^1, \ldots, a^\nu) \le \left(\prod_{i=1}^\nu n^i\right) F(b^1, \ldots, b^\nu)$$

$$+ \left[V(a) - \left(\prod_{i=1}^\nu n^i\right) V(b)\right] F(1, \ldots, 1)$$

Therefore

$$\limsup_{a \to \infty} \frac{F(a^1, \ldots, a^\nu)}{a^1 \cdot \cdots \cdot a^\nu} \le C + \varepsilon \qquad (2.19)$$

[18] For each Λ, X we can find $A_{\Lambda,X} \in \mathscr{C}(K)$ such that $\mu_\Lambda(X) = \rho(A_{\Lambda,X})$; see 7.1.1.
[19] See A.1.3.

The proposition results from (2.18) and (2.19) if C is finite; the case $C = -\infty$ is treated analogously.

7.2.5 CLASSICAL CONTINUOUS SYSTEMS. Let $(\mu_\Lambda{}^n)$ be a family of density distributions and suppose that $\mu_\Lambda{}^n$ is absolutely continuous with respect to Lebesgue measure. We then write

$$d\mu_\Lambda{}^n(x_1, \ldots, x_n) = f_\Lambda{}^n(x_1, \ldots, x_n)\frac{e^{-V(\Lambda)}}{n!}\, dx_1 \cdots dx_n \qquad (2.20)$$

where $V(\Lambda)$ is the Lebesgue measure (volume) of Λ.[20] The entropy corresponding to the bounded open region $\Lambda \subset \mathbf{R}^\nu$ is defined by

$$S(\Lambda) = -\sum_{n=0}^{\infty} \frac{e^{-V(\Lambda)}}{n!} \int dx_1 \cdots dx_n f_\Lambda{}^n(x_1, \ldots, x_n) \log f_\Lambda{}^n(x_1, \ldots, x_n) \qquad (2.21)$$

7.2.6 PROPOSITION. *The following inequalities hold*

(a) $$S(\Lambda) \leq 0 \qquad (2.22)$$

(b) $$\Lambda \subset \Lambda' \quad \Rightarrow \quad S(\Lambda') - S(\Lambda) \leq 0 \qquad (2.23)$$

(c) *Strong subadditivity*

$$S(\Lambda_1 \cup \Lambda_2) - S(\Lambda_1) - S(\Lambda_2) + S(\Lambda_1 \cap \Lambda_2) \leq 0 \qquad (2.24)$$

The proof of these inequalities is closely analogous to that of the corresponding inequalities for classical lattice systems (Proposition 7.2.2).

The following propositions, 7.2.7 and 7.2.8 give the definition and properties of a mean entropy for invariant states. The proofs will be omitted; they are analogous to those for the lattice case (Proposition 7.2.3) but technically more difficult.[21]

7.2.7. PROPOSITION. *If the family* $(\mu_\Lambda{}^n)$ *satisfies the invariance condition* (1.17), *then the following limit exists when* $\Lambda \to \infty$ *in the sense of Van Hove*

$$s = \lim_{\Lambda \to \infty} \frac{S(\Lambda)}{V(\Lambda)} \qquad (2.25)$$

and $-\infty \leq s \leq 0$.

A mean entropy $s(\rho)$ is defined for every $\rho \in E \cap \mathscr{L}^\perp$ by:

(a) $s(\rho) = -\infty$ if ρ is not given by a system of density distributions (i.e., $\rho \notin \mathscr{F}$).

(b) $s(\rho) = -\infty$ if ρ is given by a system of density distributions $(\mu_\Lambda{}^n)$ but the measures $\mu_\Lambda{}^n$ are not all absolutely continuous with respect to the Lebesgue measure.

(c) $s(\rho) = s$ as defined in Proposition 7.2.7 otherwise.

[20] The normalization in (2.20) and (2.21) is chosen such that the inequalities in 7.2.6 hold, but in applications other normalizations may be preferred.

[21] See Robinson and Ruelle [2].

7.2.8 PROPOSITION. *The functional* $s(\cdot)$ *is affine upper semicontinuous on* $E \cap \mathscr{L}^{\perp}_{\mathbf{R}^{\nu}}$.

7.2.9 QUANTUM LATTICE SYSTEMS. Let (μ_{Λ}) be the family of density matrices associated with a state ρ. The entropy corresponding to a finite $\Lambda \subset \mathbf{Z}^{\nu}$ is defined by

$$S(\Lambda) = - \operatorname{Tr}_{\mathscr{H}(\Lambda)} \mu_{\Lambda} \log \mu_{\Lambda} \tag{2.26}$$

7.2.10 PROPOSITION. *The following inequalities hold*[22]

(a) $$0 \le S(\Lambda) \le N(\Lambda) \log(N + 1) \tag{2.27}$$

(b) $\Lambda \subset \Lambda'$ \Rightarrow $S(\Lambda') - S(\Lambda) \le [N(\Lambda') - N(\Lambda)] \log(N + 1)$ (2.28)

(c) *Subadditivity*

$$S(\Lambda_1 \cup \Lambda_2) \le S(\Lambda_1) + S(\Lambda_2) \qquad \text{if} \quad \Lambda_1 \cap \Lambda_2 = \varnothing \tag{2.29}$$

The easy proof of (a) is left to the reader. If $\Lambda_1 \cap \Lambda_2 = \varnothing$, we may write

$$\mathscr{H}(\Lambda_1 \cup \Lambda_2) = \mathscr{H}(\Lambda_1) \otimes \mathscr{H}(\Lambda_2)$$

and $\mu_{\Lambda_1}, \mu_{\Lambda_2}$ are the partial traces of $\mu_{\Lambda_1 \cup \Lambda_2}$ with respect to $\mathscr{H}(\Lambda_2)$ and $\mathscr{H}(\Lambda_1)$, respectively, so that (c) results from Proposition 2.5.6.[23] Finally (b) results from (a) and (c).

7.2.11 PROPOSITION.

(a) *Let* ρ *be* \mathbf{Z}^{ν}-*invariant ; then the following limit exists*

$$s(\rho) = \lim_{n \to \infty} V(a)^{-1} S(\Lambda(a)) \tag{2.30}$$

and $0 \le s(\rho) \le \log(N + 1)$.

(b) *The functional* $s(\cdot)$ *is affine upper semicontinuous on* $E \cap \mathscr{L}^{\perp}_{\mathbf{Z}^{\nu}}$.

The proof of these results is essentially the same as for classical lattice systems (Proposition 7.2.3). We remark only that (2.14) is replaced by

$$- \operatorname{Tr}[\alpha \mu_{1\Lambda} \log \mu_{1\Lambda} + (1 - \alpha)\mu_{2\Lambda} \log \mu_{2\Lambda}]$$
$$\le - \operatorname{Tr}[\alpha \mu_{1\Lambda} + (1 - \alpha)\mu_{2\Lambda}] \log[\alpha \mu_{1\Lambda} + (1 - \alpha)\mu_{2\Lambda}]$$
$$\le - \operatorname{Tr}[\alpha \mu_{1\Lambda} \log(\alpha \mu_{1\Lambda}) + (1 - \alpha)\mu_{2\Lambda} \log((1 - \alpha)\mu_{2\Lambda})]$$
$$\le - \operatorname{Tr}[\alpha \mu_{1\Lambda} \log \mu_{1\Lambda} + (1 - \alpha)\mu_{2\Lambda} \log \mu_{2\Lambda}] + \log 2 \tag{2.31}$$

where we have in the first step used Proposition 2.5.7 and in the second step Proposition 2.5.8.

[22] Remember that $N + 1$ is the dimension of the Hilbert space \mathscr{H}_x at every $x \in \mathbf{Z}^{\nu}$.
[23] It is not known at present if *strong* subadditivity holds.

7.2.12 INTEGRAL REPRESENTATION OF THE MEAN ENTROPY. We mention without proof that, in each of the cases considered above, the mean entropy of an invariant state ρ may be written as the average of the entropies of the pure phases which constitute ρ. More precisely, if μ_ρ is the measure introduced in Theorem 6.4.1, then

$$s(\rho) = \int d\mu_\rho(\sigma)s(\sigma) = \mu_\rho(s) \tag{2.32}$$

7.2.13 QUANTUM CONTINUOUS SYSTEMS. If (μ_Λ) is a family of density distributions describing an \mathbf{R}^ν-invariant state of a boson or fermion system, we may define

$$S(\Lambda) = -\operatorname{Tr}_{\mathscr{H}_\varepsilon(\Lambda)} \mu_\Lambda \log \mu_\Lambda \tag{2.33}$$

Then $S(\Lambda) \geq 0$, and the subadditivity property (2.29) holds.[24] However no satisfactory proof has yet been obtained that $S(\Lambda)/V(\Lambda)$ tends to a limit when $\Lambda \to \infty$.

7.3 THERMODYNAMIC LIMIT OF STATES

We have studied in Chapter 4 the thermodynamic limit of the equilibrium state for systems at low activity. In this section we drop the condition of low activity; we are then still able to prove the existence of the thermodynamic limit, but only for "almost all" interactions in a sense which will be made precise later.

We shall restrict ourselves to classical and quantum lattice systems.[25] We let the Banach space \mathscr{B} be defined as in Section 2.2 or 2.4 and \mathscr{B}_0 be the subspace of finite range interactions. It will be convenient to work, in the rest of this chapter, with an arbitrary Banach space \mathscr{B}_1 such that $\mathscr{B}_0 \subset \mathscr{B}_1 \subset \mathscr{B}$, which has a norm larger than that of \mathscr{B} and in which \mathscr{B}_0 is dense. In particular, we may take $\mathscr{B}_1 = \mathscr{B}$.

7.3.1 CLASSICAL LATTICE SYSTEMS. We consider again lattice gases. The correlation function $\rho_\Lambda^\Phi(\cdot)$ corresponding to the interaction $\Phi \in \mathscr{B}_1$ and the finite set $\Lambda \subset \mathbf{Z}^\nu$ is given for $X \subset \Lambda$ by

$$\rho_\Lambda^\Phi(X) = Z(\Phi)^{-1} \sum_{Y \subset \Lambda \setminus X} \exp[-U_\Phi(X \cup Y)] \tag{3.1}$$

where

$$Z_\Lambda(\Phi) = \sum_{X \subset \Lambda} \exp[-U_\Phi(X)] \tag{3.2}$$

[24] This follows again from Proposition 2.5.6.

[25] An extension of the results of Sections 7.3–7.5 to classical continuous systems with hard cores has been given by Gallavotti and Miracle [3].

Let also $\rho_\Lambda{}^\Phi(X) = 0$ if $X \not\subset \Lambda$. By averaging over translations we get

$$\bar{\rho}_\Lambda{}^\Phi(X) = N(\Lambda)^{-1} \sum_{x \in \mathbf{Z}^\nu} \rho_\Lambda{}^\Phi(x + X) \tag{3.3}$$

For any $\Psi \in \mathscr{B}_1$ we have thus

$$\sum_{X \ni 0} \bar{\rho}_\Lambda{}^\Phi(X) \frac{\Psi(X)}{N(X)} = N(\Lambda)^{-1} \sum_{x \in \mathbf{Z}^\nu} \sum_{X \ni x} \rho_\Lambda{}^\Phi(X) \frac{\Psi(X)}{N(X)}$$

$$= N(\Lambda)^{-1} \sum_X \rho_\Lambda{}^\Phi(X) \Psi(X)$$

$$= Z_\Lambda^{-1} N(\Lambda)^{-1} \sum_X \sum_{Y: X \subset Y \subset \Lambda} \exp[-U_\Phi(Y)] \cdot \Psi(X)$$

$$= Z_\Lambda^{-1} N(\Lambda)^{-1} \sum_{Y \subset \Lambda} \exp[-U_\Phi(Y)] \sum_{X \subset Y} \Psi(X)$$

$$= Z_\Lambda^{-1} N(\Lambda)^{-1} \sum_{Y \subset \Lambda} \exp[-U_\Phi(Y)] U_\Psi(Y) \tag{3.4}$$

This formula asserts that $\bar{\rho}_\Lambda$ is up to the factor $-N(X)$ the functional derivative of $P_\Lambda(\Phi)$ with respect to Φ, where

$$P_\Lambda(\Phi) = N(\Lambda)^{-1} \log Z_\Lambda(\Phi) \tag{3.5}$$

It is therefore natural to try to obtain the infinite volume correlation function as functional derivative of the infinite volume thermodynamic function $P(\cdot)$.[26]

Let D be the set of all $\Phi \in \mathscr{B}_1$ such that the graph[27] of $P(\cdot)$ has a unique tangent plane at the point $(\Phi, P(\Phi))$. If $\Phi \in D$ there exists a unique element α^Φ in the dual $\mathscr{B}_1{}^*$ of \mathscr{B}_1 such that

$$P(\Phi + \Psi) \geq P(\Phi) - \alpha^\Phi(\Psi) \tag{3.6}$$

for all $\Psi \in \mathscr{B}_1$.

7.3.2 THEOREM.[28] *Let $\Phi \in D$; then when $\Lambda \to \infty$ in the sense of Van Hove,*

$$\lim_{\Lambda \to \infty} \sum_{X \ni 0} \bar{\rho}_\Lambda{}^\Phi(X) \frac{\Psi(X)}{N(X)} = \alpha^\Phi(\Psi) \tag{3.7}$$

for all $\Psi \in \mathscr{B}_1$. In particular for every finite $X \subset \mathbf{Z}^\nu$ the following limit exists

$$\lim_{\Lambda \to \infty} \bar{\rho}_\Lambda{}^\Phi(X) = \bar{\rho}^\Phi(X) \tag{3.8}$$

defining the infinite volume correlation function $\bar{\rho}^\Phi$.

[26] This point of view has in particular been used by Fisher [2].
[27] The graph of $P(\cdot)$ is the set $\{(\Phi, P(\Psi)) \in \mathscr{B}_1 \times \mathbf{R} : \Phi \in \mathscr{B}_1\}$.
[28] See Gallavotti–Miracle [1].

For finite Λ, the graph of the convex function $P_\Lambda(\cdot)$ has a unique tangent plane at each point $(\Phi, P(\Phi))$:

$$P_\Lambda(\Phi + \Psi) \geq P_\Lambda(\Phi) - \alpha_\Lambda{}^\Phi(\Psi) \tag{3.9}$$

where, according to (3.4),

$$\alpha_\Lambda{}^\Phi(\Psi) = Z_\Lambda^{-1} N(\Lambda)^{-1} \sum_{X \subset \Lambda} \exp[-U_\Phi(X)] U_\Psi(X) = \sum_{X \ni 0} \bar{\rho}_\Lambda{}^\Phi(X) \frac{\Psi(X)}{N(X)} \tag{3.10}$$

Let $\Phi \in D$, $\Psi \in \mathscr{B}_1$ and the point $(\Phi + \Psi, \xi)$ be strictly above the graph of $P(\cdot)$ (see Fig. 18). For sufficiently large Λ, $(\Phi + \Psi, \xi)$ is also above the graph of $P_\Lambda(\cdot)$; hence, above the tangent plane to $P_\Lambda(\cdot)$ at $(\Phi, P_\Lambda(\Phi))$:

$$\xi \geq P_\Lambda(\Phi + \Psi) \geq P_\Lambda(\Phi) - \alpha_\Lambda{}^\Phi(\Psi) \tag{3.11}$$

This ensures that the tangent plane to $P_\Lambda(\cdot)$ tends to the tangent plane to $P(\cdot)$ because of the uniqueness of the latter:

$$\lim_{\Lambda \to \infty} \alpha_\Lambda{}^\Phi(\Psi) = \alpha^\Phi(\Psi) \tag{3.12}$$

for all $\Psi \in \mathscr{B}$. From (3.12) and (3.10) we obtain (3.7).

We have thus proved the existence of the thermodynamic limit for the correlation function when $\Phi \in D$. We want to show now that D is a "large" set, that "almost all" interactions belong to it. It can be proved that, with respect to many measures on \mathscr{B}_1, the convex continuous function $P(\cdot)$ is almost everywhere differentiable. For simplicity we indicate only the following results.

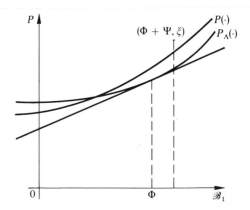

FIGURE 18 The graphs of $P(\cdot)$ and $P_\Lambda(\cdot)$.

7.3.3 PROPOSITION.

(a) *The set D contains a countable intersection of dense open subsets of \mathscr{B}_1 and is therefore dense in \mathscr{B}_1.*[29]

(b) *There exists a subset \tilde{D} of the Banach space $\tilde{\mathscr{B}}$ defined by (4.10) of Chapter 2 such that for $\tilde{\varphi} \in \tilde{D}$ and for (Lebesgue) almost every $(\beta, \mu) \in \mathbf{R}^2$, the point $\beta(-\mu, \tilde{\varphi})$ belongs to D.*

Let Σ be the boundary of a convex set with an interior point in a separable Banach space, and let T be the set of points of Σ at which there is a unique tangent plane, then T contains a countable intersection of dense open sets. This result is obtained in Dunford and Schwartz [1], theorem V.9.8, and proves (a). We omit the proof of (b).[30]

7.3.4 QUANTUM LATTICE SYSTEMS.

The extension of the results discussed above to quantum lattice systems is straightforward and left to the interested reader. We remark only that we start with the definition

$$\rho_\Lambda{}^\Phi(X) = Z(\Phi)^{-1} \operatorname{Tr}_{\mathscr{H}(\Lambda \setminus X)} \exp[-H_\Phi(\Lambda)] \tag{3.13}$$

and that we may write at the end $\bar{\rho}^\Phi(X) = \mu_X{}^\Phi$ where $(\mu_\Lambda{}^\Phi)$ is a system of density matrices satisfying the conditions (a), (b) of 7.1.3. Equation (3.7) is replaced by

$$\lim_{\Lambda \to \infty} \sum_{X \ni 0} N(X)^{-1} \operatorname{Tr}_{\mathscr{H}(X)} \bar{\rho}_\Lambda{}^\Phi(X) \Psi(X)$$

$$= \lim_{\Lambda \to \infty} Z_\Lambda^{-1} N(\Lambda)^{-1} \operatorname{Tr}_{\mathscr{H}(\Lambda)} [\exp(-H_\Phi(\Lambda)) H_\Psi(\Lambda)]$$

$$= \alpha^\Phi(\Psi) \tag{3.14}$$

7.4 A VARIATIONAL PRINCIPLE

According to a well-known principle of thermodynamics, the equilibrium state of a system at a given energy and density makes its entropy maximum. We are going to prove a variant of this principle, which is applicable to infinite systems.

Before formulating our result we have to define the mean energy per site associated with an interaction $\Phi \in \mathscr{B}_1$ and an invariant state ρ of the system. We introduce first $A_\Phi \in \mathfrak{A}$ by

$$A_\Phi(X) = \sum_{Y:0 \in Y \subset X} \frac{\Phi(Y)}{N(Y)} \tag{4.1}$$

[29] Baire's theorem: In a complete metric space, a countable intersection of dense open sets is again dense.

[30] See Gallavotti and Miracle [1].

for classical lattice gases and

$$A_\Phi = \sum_{Y \ni 0} \frac{\Phi(Y)}{N(Y)} \qquad (4.2)$$

for quantum lattice systems. It follows from the definitions that the mapping $\Phi \to A_\Phi$ is norm-decreasing from \mathscr{B}_1 to \mathfrak{A}. For a classical lattice gas we have, using (1.2), for finite Λ,

$$U_\Phi(\Lambda) = \sum_{X \subset \Lambda} \Phi(X) = \sum_{x \in \Lambda} \sum_{X : x \in X \subset \Lambda} \frac{\Phi(X)}{N(X)}$$

$$= \sum_{x \in \Lambda} \tau_x A_\Phi(\Lambda) = \sum_{x \in \mathbf{Z}^\nu} \tau_x A_\Phi(\Lambda) \qquad (4.3)$$

This shows that A_Φ has the meaning of the contribution of the origin to the energy of the system. A similar interpretation is possible for quantum systems. We may thus consider $\rho(A_\Phi)$ as the expectation value of the energy per site for the interaction Φ in the invariant state ρ.

7.4.1 THEOREM.[31] *Let* $\Phi \in \mathscr{B}_1$; *then*

$$P(\Phi) = \sup_{\rho \in E \cap \mathscr{L}^\perp} [s(\rho) - \rho(A_\Phi)] \qquad (4.4)$$

Since $|\rho(A_\Phi) - \rho(A_\Psi)| \leq \|\Phi - \Psi\|$, both sides of (4.4) depend continuously on Φ. Therefore it is sufficient to prove the theorem for $\Phi \in \mathscr{B}_0$ (i.e., Φ of finite range). In that case there exists a finite set $\Delta \subset \mathbf{Z}^\nu$ such that $A_\Phi \in \mathfrak{A}_\Delta$. For definiteness we consider the case of a lattice gas.

First we show that for each $\rho \in E \cap \mathscr{L}^\perp$ we have

$$P(\Phi) \geq s(\rho) - \rho(A_\Phi) \qquad (4.5)$$

Using the system of density distributions $(\mu_\Lambda(\cdot))$ associated with ρ we have, if $x + \Delta \subset \Lambda(a)$,

$$\rho(A_\Phi) = \sum_{X \subset \Lambda(a)} \mu_{\Lambda(a)}(X) \tau_x A_\Phi(X) \qquad (4.6)$$

From this and (4.3) we obtain

$$\rho(A_\Phi) = \lim_{a \to \infty} V(a)^{-1} \sum_{x \in \Lambda(a)} \sum_{X \subset \Lambda(a)} \mu_{\Lambda(a)}(X) \tau_x A_\Phi(X)$$

$$= \lim_{a \to \infty} V(a)^{-1} \sum_{X \subset \Lambda(a)} \mu_{\Lambda(a)} U_\Phi(X) \qquad (4.7)$$

[31] See Ruelle [9] for classical systems and Robinson [1],I for the quantum case. See also O. Lanford and D. W. Robinson, Statistical Mechanics of Quantum Spin System III, *Commun. Math. Phys.* **9**, 327–338 (1968).

On the other hand Proposition 7.2.3 gives

$$s(\rho) = \lim_{a \to \infty} V(a)^{-1} \left[- \sum_{X \subset \Lambda(a)} \mu_{\Lambda(a)}(X) \log \mu_{\Lambda(a)}(X) \right] \qquad (4.8)$$

From (4.7), (4.8), and the concavity of the logarithm we get

$$s(\rho) - \rho(A_\Phi) = \lim_{a \to \infty} V(a)^{-1} \left[- \sum_{X \subset \Lambda(a)} \mu_{\Lambda(a)}(X)(U_\Phi(X) + \log \mu_{\Lambda(a)}(X)) \right]$$

$$= \lim_{a \to \infty} V(a)^{-1} \sum_{X \subset \Lambda(a)} \mu_{\Lambda(a)}(X) \log \frac{\exp[-U_\Phi(X)]}{\mu_{\Lambda(a)}(X)}$$

$$\leq \lim_{a \to \infty} V(a)^{-1} \log \sum_{X \subset \Lambda(a)} \exp[-U(X)]$$

$$= P(\Phi) \qquad (4.9)$$

which proves (4.5).

We show now that, given $\varepsilon > 0$, we can find $\rho \in E \cap \mathscr{L}^1$ such that

$$P(\Phi) < s(\rho) - \rho(A_\Phi) + \varepsilon \qquad (4.10)$$

The state ρ is constructed such that it looks, as much as possible, like an equilibrium state for the interaction Φ. We choose first a such that

$$|P(\Phi) - V(a)^{-1} \log Z_{\Lambda(a)}(\Phi)| < \varepsilon/2 \qquad (4.11)$$

The translates Λ_n of $\Lambda(a)$ defined by (1.4) of Chapter 2 form a partition of \mathbf{Z}^ν. For every finite union Λ of the Λ_n:

$$\Lambda = \bigcup_{i=1}^{N} \Lambda_{n_i}$$

and $X \subset \Lambda$; we let

$$\tilde{\mu}_\Lambda(X) = \prod_{i=1}^{n} [Z_{\Lambda(a)}^{-1} \exp(-U_\Phi(X \cap \Lambda_{n_i}))] \qquad (4.12)$$

The density distributions $\tilde{\mu}_\Lambda(\cdot)$ define a state $\tilde{\rho}$ which reduces in each Λ_n to the finite system equilibrium state $\tilde{\mu}_{\Lambda_n}(\cdot)$. The state $\tilde{\rho}$ is periodic with periodicity cell $\Lambda(a)$; we obtain the invariant state ρ by averaging over translations:

$$\rho(A) = V(a)^{-1} \sum_{x \in \Lambda(a)} \tilde{\rho}(\tau_x A) \qquad (4.13)$$

It is easy to verify that

$$s(\rho) = \lim_{\Lambda \to \infty} N(\Lambda)^{-1} \left[- \sum_{X \subset \Lambda} \mu_\Lambda(X) \log \mu_\Lambda(X) \right]$$

$$= \lim_{\Lambda \to \infty} N(\Lambda)^{-1} \left[- \sum_{X \subset \Lambda} \tilde{\mu}_\Lambda(X) \log \tilde{\mu}_\Lambda(X) \right]$$

$$= V(a)^{-1}\left[-\sum_{X \subset \Lambda(a)} \tilde{\mu}_{\Lambda(a)}(X) \log \tilde{\mu}_{\Lambda(a)}(X)\right] \qquad (4.14)$$

Also, using (4.7), we see that if a is sufficiently large

$$\left|\rho(A_\Phi) - V(a)^{-1} \sum_{X \subset \Lambda_n} \tilde{\mu}_{\Lambda(a)}(X) U_\Phi(X)\right| < \varepsilon/2 \qquad (4.15)$$

From (4.14), (4.15), and (4.11) we obtain

$$s(\rho) - \rho(A_\Phi) + \varepsilon/2 > -V(a)^{-1} \sum_{X \subset \Lambda(a)} \tilde{\mu}_{\Lambda(a)}(X)[\log \tilde{\mu}_{\Lambda(a)}(X) + U_\Phi(X)]$$

$$= V(a)^{-1} \log Z_{\Lambda(a)}(\Phi) > P(\Phi) - \varepsilon/2$$

which proves (4.10) and therefore the theorem.

7.5 THE GIBBS PHASE RULE

For a one-component thermodynamic system, the *Gibbs phase rule* asserts that in the plane of two intensive thermodynamic parameters almost every point corresponds to a pure thermodynamic phase, that the points of coexistence of two phases form a set of dimension one, and the points of co-existence of three phases a set of dimension zero. For instance, using the variables p, T, we have the familiar phase diagram shown in Fig. 19.

It is conceivable that the Gibbs phase rule holds only for almost all interactions and breaks down for exceptional interactions.[32] Therefore we would hope to prove only a result of the following type: "For almost all interactions, chemical potentials, and temperatures, the equilibrium state is a pure phase." If μ and β are included in the interaction, the statement becomes: "For almost all interactions the equilibrium state is a pure phase."

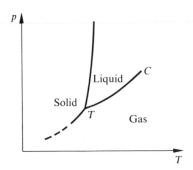

FIGURE 19 Phase diagram.

[32] Among the translation invariant interactions for continuous systems, those which are Euclidean invariant are probably exceptional in this sense since they presumably give rise to an unoriented crystal state in some region of the phase diagram. Such a state is not ergodic (see 6.5.2) and therefore does not count as a pure phase.

We now formulate and prove a weak version of the Gibbs phase rule.

7.5.1 THEOREM.[33] *Let $D \subset \mathscr{B}_1$ be the "large" set introduced in Section 7.3.*

(a) *If $\Phi \in D$, the function $\rho \to s(\rho) - \rho(A_\Phi)$ reaches its maximum $P(\Phi)$ at exactly one point $\rho^\Phi \in E \cap \mathscr{L}^\perp$.*

(b) *If $\Phi \in D$ and $\alpha^\Phi \in \mathscr{B}_1^*$ is defined by (3.6), then, for all $\Psi \in \mathscr{B}_1$,*

$$\rho^\Phi(A_\Psi) = \alpha^\Phi(\Psi) \tag{5.1}$$

so that ρ^Φ is the infinite volume equilibrium state corresponding to the interaction Φ.

(c) *If $\Phi \in D$, ρ^Φ is a \mathbf{Z}^ν-ergodic state and may thus be interpreted as pure thermodynamic phase.*

Let $\Phi \in \mathscr{B}_1$; then by Section 7.2 the functional $\rho \to s(\rho) - \rho(A_\Phi)$ is affine upper semicontinuous on $E \cap \mathscr{L}^\perp$. Therefore, if Δ_Φ is the set of points at which this functional reaches its maximum, Δ_Φ is a nonempty convex compact set. Let ρ^Φ be an extremal point[34] of Δ_Φ; then ρ^Φ is an extremal point of $E \cap \mathscr{L}^\perp$. (In the opposite case we could write $\rho^\Phi = \frac{1}{2}\rho_1 + \frac{1}{2}\rho_2$ with $\rho_1, \rho_2 \in E \cap \mathscr{L}^\perp$ and $\rho_2 \notin \Delta_\Phi$ in contradiction with the fact that an affine function reaches its maximum on Δ_Φ.)

If $\rho \in \Delta_\Phi$ we have by Theorem 7.4.1, for all $\Psi \in \mathscr{B}_1$,

$$P(\Phi + \Psi) \geq s(\rho) - \rho(A_{\Phi+\Psi}) = s(\rho) - \rho(A_\Phi) - \rho(A_\Psi)$$

$$= P(\Phi) - \rho(A_\Psi) \tag{5.2}$$

therefore $\Psi \to P(\Phi) - \rho(A_\Psi)$ is a tangent plane to the graph of $P(\cdot)$ at $(\Phi, P(\Phi))$. Notice that distinct elements of Δ_Φ yield distinct tangent planes. From (4.2) we find indeed in the quantum case

$$\rho(A_\Psi) = \sum_{X \ni 0} \frac{\rho(\Phi(X))}{N(X)} = \sum_{X \ni 0} N(X)^{-1} \operatorname{Tr}_{\mathscr{H}(X)} \rho_X \Phi(X)$$

so that (by \mathbf{Z}^ν-invariance) the functional $\Psi \to \rho(A_\Psi)$ determines the system (ρ_Λ). For a classical lattice gas, (4.1) and (1.8) give

$$\rho(A_\Psi) = \lim_{\Lambda \to \infty} \sum_{X \subset \Lambda} \mu_\Lambda(X) A_\Psi(X)$$

$$= \lim_{\Lambda \to \infty} \sum_{X \subset \Lambda} \mu_\Lambda(X) \sum_{Y: 0 \in Y \subset X} \frac{\Psi(Y)}{N(Y)}$$

$$= \lim_{\Lambda \to \infty} \sum_{Y: 0 \in Y \subset \Lambda} \frac{\Psi(Y)}{N(Y)} \sum_{X: Y \subset X \subset \Lambda} \mu_\Lambda(X) = \sum_{Y \ni 0} \frac{\Psi(Y)}{N(Y)} \rho(Y)$$

[33] See Ruelle [9].

[34] The fact that a nonempty convex compact set has an extremal point follows from the Krein–Milman theorem.

so that the functional $\Psi \to \rho(A_\Psi)$ determines the correlation functions and therefore [by (1.9)] the state ρ.

If $\Phi \in D$, the uniqueness of the tangent plane implies that Δ_Φ consists of one point ρ^Φ which is extremal in $E \cap \mathscr{L}^\perp$, proving (a) and (c). The proof of (b) results from comparison of (5.2) and (3.6); the interpretation of ρ^Φ as infinite volume equilibrium state is justified by Theorem 7.3.2 and its quantum counterpart.

7.6 TIME EVOLUTION OF QUANTUM LATTICE SYSTEMS AND THE KUBO–MARTIN–SCHWINGER BOUNDARY CONDITION

In this section we describe briefly some results on the time evolution of quantum lattice systems. These results remain essentially in the realm of equilibrium statistical mechanics, and the techniques used to prove them are close in spirit to those of the preceding sections.

A norm $\| \cdot \|_1$ on \mathscr{B}_0 is defined by

$$\|\Phi\|_1 = \sum_{X \ni 0} \|\Phi(X)\| \exp(N(X) - 1) \tag{6.1}$$

and the Banach space \mathscr{B}_1 is defined as the closure of \mathscr{B}_0 with respect to this norm.

The time evolution of the system occupying the finite region $\Lambda \subset \mathbf{Z}^\nu$, with respect to the interaction Φ, is defined by a one-parameter group of automorphisms of \mathfrak{A}_Λ which associates to the "observable" A at time 0 the following "observable" at time t:

$$\exp[itH_\Phi(\Lambda)]A \exp[-itH_\Phi(\Lambda)] = \sum_{n=0}^{\infty} i^n \frac{t^n}{n!} [H_\Phi(\Lambda), A]^{(n)} \tag{6.2}$$

where

$$[B, A]^0 = A \qquad [B, A]^{(n)} = [B, [B, A]^{(n-1)}] \tag{6.3}$$

7.6.1 LEMMA. *If $\Phi \in \mathscr{B}_1$ and $A \in \mathfrak{A}_{\Lambda_1}$, $\Lambda_1 \subset \Lambda$, then*

$$\|[H_\Phi(\Lambda), A]^{(n)}\| \le \|A\| \exp(N(\Lambda_1)) \cdot n!(2\|\Phi\|_1)^n \tag{6.4}$$

In the formula

$$[H_\Phi(\Lambda), A]^{(n)} = \sum_{X_1 \subset \Lambda} \cdots \sum_{X_n \subset \Lambda} [\Phi(X_n), [\cdots [\Phi(X_1), A]\cdots]] \tag{6.5}$$

we write

$$\sum_{X_i} = \sum_{k_i=0}^{\infty} \sum_{X_i : N(X_i) = k_i + 1}$$

and we remark that we may restrict the summations by the conditions $X_i \cap S_i \neq \emptyset$ where $S_1 = \Lambda_1$ and $S_{i+1} = X_i \cup S_i$ for $i = 1, \ldots, n - 1$. Notice that S_i contains at most $N(\Lambda_1) + k_1 + \cdots + k_{i-1}$ elements; therefore,

$$\| [H_\Phi(\Lambda), A]^{(n)} \|$$

$$\leq 2^n \|A\| \sum_{k_1, \ldots, k_n} \prod_{i=1}^{n} \left[(N(\Lambda_1) + k_1 + \cdots + k_{i-1}) \sum_{X_i \ni 0 : N(X_i) = k_i + 1} \|\Phi(X_i)\| \right]$$

(6.6)

We now use the fact that

$$\frac{1}{n!} \prod_{i=1}^{n} (N(\Lambda_1) + k_1 + \cdots + k_{i-1})$$

$$\leq \frac{1}{n!} (N(\Lambda_1) + k_1 + \cdots + k_n)^n \leq \exp(N(\Lambda_1) + k_1 + \cdots + k_n)$$

$$= \exp(N(\Lambda_1)) \cdot \prod_{i=1}^{n} \exp k_i$$

Inserting this into (6.6) yields (6.4).

7.6.2 THEOREM.[35] *If* $\Phi \in \mathcal{B}_1$, \mathfrak{A} *has a strongly continuous*[36] *one-parameter group of automorphisms* $\bar{\tau} : \mathbf{R} \to \mathrm{aut} \, \mathfrak{A}$ *defined as follows. If* $A \in \mathfrak{A}_{\Lambda_1}$ *for some finite* Λ_1 *and* Λ *tends to infinity in the sense that it eventually contains every finite subset of* \mathbf{Z}^ν, *then*

$$\lim_{\Lambda \to \infty} \exp[it H_\Phi(\Lambda)] A \exp[-it H_\Phi(\Lambda)] = \bar{\tau}_t A$$

(6.7)

The proof results from the following points

(a) We have term by term

$$\lim_{\Lambda \to \infty} \sum_{n=0}^{\infty} i^n \frac{t^n}{n!} [H_\Phi(\Lambda), A]^{(n)}$$

$$= \sum_{n=0}^{\infty} i^n \frac{t^n}{n!} \sum_{X_1 \subset \mathbf{Z}^\nu} \cdots \sum_{X_n \subset \mathbf{Z}^\nu} [\Phi(X_n), [\cdots [\Phi(X_1), A] \cdots]]$$

(6.8)

and if $|t| < (2\|\Phi\|_1)^{-1}$, (6.4) and (6.2) show that we may write

$$\lim_{\Lambda \to \infty} \exp[it H_\Phi(\Lambda)] A \exp[-it H_\Phi(\Lambda)] = \bar{\tau}_t A$$

(6.9)

defining $\bar{\tau}_t A$.

(b) $\bar{\tau}_t$ extends by continuity to an automorphism of \mathfrak{A}.

[35] Adapted from Robinson [1],II.
[36] This means that for all $A \in \mathfrak{A}$, $\lim_{t \to 0} \| A - \bar{\tau}_t A \| = 0$.

(c) For $t_1, t_2, t_1 + t_2$ in the interval $(-(2\|\Phi\|_1)^{-1}, +(2\|\Phi\|_1)^{-1})$ we have

$$\bar{\tau}_{t_1+t_2}A = \bar{\tau}_{t_1}\bar{\tau}_{t_2}A \qquad (6.10)$$

We may define $\bar{\tau}_t A$ for all real t so that (6.10) holds whenever $t_1, t_2 \in \mathbf{R}$. Then (6.9) holds for all $t \in \mathbf{R}$.

(d) In view of (6.8) and (6.4), $\bar{\tau}_t A$ is given for $A \in \mathfrak{A}_{\Lambda_1}$ and small t by a convergent power series. Therefore

$$\lim_{t\to 0} \|A - \bar{\tau}_t A\| = 0 \qquad (6.11)$$

and this remains true for all $A \in \mathfrak{A}$, concluding the proof of the theorem.

7.6.3 THEOREM.[37] *Let* $\Phi \in D \subset \mathscr{B}_1$ *and* ρ^Φ *be as in Theorem 7.5.1; then we have*

(a) *Invariance*: $\rho^\Phi(\bar{\tau}_t A) = \rho^\Phi(A)$ *for all* $A \in \mathfrak{A}$, $t \in \mathbf{R}$.

(b) *Kubo–Martin–Schwinger (KMS) boundary condition*: *If* $A, B \in \mathfrak{A}$ *there exists a bounded continuous function* F *on* $\{z : 0 \le \operatorname{Im} z \le 1\}$, *analytic for* $0 < \operatorname{Im} z < 1$ *and such that for all real* t

$$\rho^\Phi(B \cdot \bar{\tau}_t A) = F(t) \qquad \rho^\Phi(\bar{\tau}_t A \cdot B) = F(t + i) \qquad (6.12)$$

It is sufficient to prove invariance for $A \in \mathfrak{A}_{\Lambda_1}$, Λ_1 finite. We have by 7.3.4

$$\rho^\Phi(A) = \lim_{\Lambda\to\infty} \bar{\rho}_\Lambda{}^\Phi(A)$$

$$= \lim_{\Lambda\to\infty} N(\Lambda)^{-1} Z_\Lambda(\Phi)^{-1} \sum_{x:x+\Lambda_1\subset\Lambda} \operatorname{Tr}_{\mathscr{H}(\Lambda)}[\tau_x A \exp(-H_\Phi(\Lambda))] \qquad (6.13)$$

when $\Lambda \to \infty$ in the sense of Van Hove. It is useful to notice at this point that (6.13) remains true if Λ_1 is replaced by any nonempty finite subset of \mathbf{Z}^ν and A by any element of \mathfrak{A}, provided $\tau_x A$ is suitably approximated in the right-hand side by an element of $\mathscr{H}(\Lambda)$.

In view of Theorem 7.6.2, given $t \in \mathbf{R}$ and $\varepsilon > 0$, we can find a finite $\Lambda_0 \subset \mathbf{Z}^\nu$ such that

$$\Lambda \supset \Lambda_0 \quad \Rightarrow \quad \|e^{itH(\Lambda)}A\,e^{-itH(\Lambda)} - \bar{\tau}_t A\| < \varepsilon$$

and therefore also

$$\Lambda \supset x + \Lambda_0 \quad \Rightarrow \quad \|e^{itH(\Lambda)}\tau_x A\,e^{-itH(\Lambda)} - \tau_x\bar{\tau}_t A\| < \varepsilon \qquad (6.14)$$

where we have used $\tau_x\bar{\tau}_t = \bar{\tau}_t\tau_x$. If we replace in (6.13) Λ_1 by Λ_0, A by $\bar{\tau}_t A$, and make use of (6.14), we find

$$\left| \lim_{\Lambda\to\infty} N(\Lambda)^{-1} Z_\Lambda^{-1} \sum_{x:x+\Lambda_0\subset\Lambda} \operatorname{Tr}_{\mathscr{H}(\Lambda)}[e^{itH(\Lambda)}\tau_x A\,e^{-itH(\Lambda)}\,e^{-H(\Lambda)}] \right.$$

$$\left. - \rho^\Phi(\bar{\tau}_t A) \right| < \varepsilon$$

[37] See Robinson [1],II. An interesting study of the KMS boundary condition has been made for continuous quantum systems by Haag, Hugenholtz, and Winnink [1].

hence

$$\rho^{\Phi}(\bar{\tau}_t A) = \lim_{\Lambda \to \infty} N(\Lambda)^{-1} Z_\Lambda^{-1} \sum_{x:x+\Lambda_0 \subset \Lambda} \mathrm{Tr}_{\mathscr{H}(\Lambda)}[\tau_x A \exp(-H(\Lambda))]$$

$$= \rho^{\Phi}(A)$$

proving (a).

To prove the KMS boundary condition we consider the entire functions F_Λ defined by

$$F_\Lambda(z) = N(\Lambda)^{-1} Z_\Lambda^{-1} \sum_{x:x+\Lambda_1 \subset \Lambda} \mathrm{Tr}_{\mathscr{H}(\Lambda)}[\tau_x B\, e^{izH(\Lambda)} \tau_x A\, e^{-izH(\Lambda)}\, e^{-H(\Lambda)}]$$

where $A \in \mathfrak{A}_{\Lambda_1}$, $B \in \mathfrak{A}$. Using (6.14) as above we obtain for all real t

$$\lim_{\Lambda \to \infty} F_\Lambda(t) = \rho^{\Phi}(B \cdot \bar{\tau}_t A) \qquad \lim_{\Lambda \to \infty} F_\Lambda(t+i) = \rho^{\Phi}(\bar{\tau}_t A \cdot B)$$

The functions F_Λ and their derivatives F'_Λ are clearly individually bounded on the strip $\{z: 0 \le \mathrm{Im}\, z \le 1\}$. Furthermore, for $\mathrm{Im}\, z = 0$ and $\mathrm{Im}\, z = 1$ we have bounds which are uniform in Λ:

$$|F_\Lambda| \le \|A\| \cdot \|B\| \qquad |F'_\Lambda| \le \|[A, H(\Lambda)]\| \cdot \|B\| \tag{6.15}$$

and

$$\|[A, H(\Lambda)]\| \le 2\|A\| \sum_{X:X\cap\Lambda_1 \ne \phi} \|\Phi(X)\| \le 2N(\Lambda_1)\|A\| \cdot \|\Phi\| \tag{6.16}$$

Therefore the F_Λ, F'_Λ are bounded on the strip $\{z: 0 \le \mathrm{Im}\, z \le 1\}$ uniformly in Λ and we can choose a sequence $\Lambda \to \infty$ such that F_Λ converges uniformly on the compact subsets of the strip. The limit F of the F_Λ then satisfies the KMS boundary condition.

We still have to get rid of the condition $A \in \mathfrak{A}_{\Lambda_1}$: If a sequence A_n in $\bigcup_\Lambda \mathfrak{A}_\Lambda$ tends in norm to any $A \in \mathfrak{A}$, then the functions $t \to \rho^{\Phi}(B \cdot \bar{\tau}_t A_n)$, $\rho^{\Phi}(\bar{\tau}_t A_n \cdot B)$ tend uniformly to their limits and therefore F does also (the functions F being bounded, the maximum principle holds[38]).

We indicate now some consequences of the KMS boundary condition.

7.6.4 PROPOSITION.[39] Let $\rho = \rho^{\Phi}$. The center of $[\pi_\rho(\mathfrak{A})]''$ is contained in the commutant of $\bar{U}_\rho(\mathbf{R})$:

$$[\pi_\rho(\mathfrak{A})]' \cap [\pi_\rho(\mathfrak{A})]'' \subset [\bar{U}_\rho(\mathbf{R})]' \tag{6.17}$$

Let C be in the center of $[\pi_\rho(\mathfrak{A})]''$, and choose $B_n \in \mathfrak{A}$ such that $\pi_\rho(B_n)$ tends strongly to C and $\|B_n\| \le \|C\|$; this is possible by the simplicity of \mathfrak{A}

[38] If a function is holomorphic and bounded on a strip, then the supremum of its modulus on the strip is equal to the supremum of its modulus on the boundary.
[39] Due to Hugenholtz and Winnink (private communication). We denote by $\bar{U}_\rho(t)$ the unitary operator corresponding to the automorphism $\bar{\tau}_t$.

and Kaplansky's density theorem A.4.6. Let F_n be the function analytic in the strip $\{z:0 < \operatorname{Im} z < 1\}$ which has the boundary values $\rho(B_n \cdot \bar{\tau}_t A)$ for $z = t$ and $\rho(\bar{\tau}_t A \cdot B_n)$ for $z = t + i$. The derivatives of the functions F_n are uniformly bounded in the strip $\{z:0 < \operatorname{Im} z < 1\}$ as follows from (6.15) and (6.16). Therefore, $F_n \to \tilde{F}$ where \tilde{F} is analytic in the strip $\{z:0 < \operatorname{Im} z < 1\}$ and has boundary values $(\Omega_\rho, C\pi_\rho(\bar{\tau}_t A)\Omega_\rho)$ for $z = t$ and $(\Omega_\rho, \pi_\rho(\bar{\tau}_t A)C\Omega_\rho) = (\Omega_\rho, C\pi_\rho(\bar{\tau}_t A)\Omega_\rho)$ for $z = t + i$. This implies that \tilde{F} may be continued to a periodic analytic function [such that $\tilde{F}(z) = \tilde{F}(z + i)$ in the whole complex plane] and since \tilde{F} is bounded, it is a constant K:

$$(\Omega_\rho, \pi_\rho(\bar{\tau}_t A)C\Omega_\rho) = K$$

Taking $A = A_1 \cdot A_2$, we get

$$(\Omega_\rho, \pi_\rho(A_1)\bar{U}_\rho(t)^{-1}C\bar{U}_\rho(t)\pi_\rho(A_2)\Omega_\rho) = K$$

hence $\bar{U}_\rho(t)^{-1}C\bar{U}_\rho(t)$ is independent of t, proving the proposition.

7.6.5 COROLLARY. *If ρ^Φ is ergodic for time translations, then ρ^Φ is a factor state* (see Exercise 6.E).

This follows from 6.3.3,(b) and 7.6.3.

Finally, we indicate the following result which relates $[\pi_\rho(\mathfrak{A})]'$ and $[\pi_\rho(\mathfrak{A})]''$.

7.6.6 THEOREM.[40] *Let $\rho = \rho^\Phi$. There is an antiunitary operator J in \mathfrak{H}_ρ such that $J^2 = 1$ and*

$$J\Omega_\rho = \Omega_\rho \tag{6.18}$$

$$J\bar{U}_\rho(t)J = \bar{U}_\rho(t) \tag{6.19}$$

$$J[\pi_\rho(\mathfrak{A})]''J = [\pi_\rho(\mathfrak{A})]' \tag{6.20}$$

The proof uses the theory of quasi-Hilbert algebras.[41]

For a discussion of the type of $[\pi_\rho(\mathfrak{A})]''$ in the case that it is a factor, see Hugenholtz [1], Størmer [2].

BIBLIOGRAPHICAL NOTE

The results described in this chapter are largely due to physicists with a background of "axiomatic" relativistic quantum field theory. This imparts a somewhat special flavor to the subject.

[40] See Haag, Hugenholtz, and Winnink [1].
[41] See Dixmier [1],I,§5.

The study of the entropy in Section 7.2 is due to Robinson and Ruelle [2] for classical systems, Lanford and Robinson [1] for quantum systems. For the most part the results of Section 7.3 are due to Gallavotti and Miracle [1], those of Sections 7.4 and 7.5 to Ruelle [9], those of Section 7.6 to Robinson [1] and Haag, Hugenholtz, and Winnink [1].

EXERCISES

7.A. In the situation of 7.1.1 (classical lattice gas), a state $\rho \in \mathscr{E}(E \cap \mathscr{L}^-)$ is a \mathbf{Z}^ν-ergodic measure on K. Use the pointwise ergodic theorem[42] to show that ρ is carried by the configurations $X \in K$ such that for all $A \in \mathfrak{A}$

$$\lim_{a \to \infty} V(a)^{-1} \sum_{x \in \Lambda(a)} A(x + X) = \rho(A)$$

7.B. In the situation of 7.1.1, show that $\mathscr{E}(E \cap \mathscr{L}^\perp)$ is dense in $E \cap \mathscr{L}^\perp$ for the weak topology. This is an example of a simplex where the extremal points are dense.

[Given $\hat{\rho} \in E \cap \mathscr{L}^\perp$, construct ρ by analogy with (4.12), (4.13), replacing (4.12) by

$$\tilde{\mu}_\Lambda(X) = \prod_{i=1}^n \hat{\mu}_{\Lambda_{n_i}}(X \cap \Lambda_{n_i})$$

Check that ρ is ergodic and tends to $\hat{\rho}$ when $a \to \infty$.]

7.C. We have indicated in 7.1.2 that the elements σ of $\mathscr{E}(E) \cap \mathscr{F}$ may be identified with infinite configurations X_σ (defined as countable subsets of \mathbf{R}^ν where the points X_σ are attributed a finite multiplicity and $X_\sigma \cap \Lambda$ is finite for every bounded Λ):

$$\sigma(\varphi(Sf_1, \ldots, Sf_q)) = \varphi(Sf_1(X_\sigma), \ldots, Sf_q(X_\sigma))$$

Show that the weak topology on E induces on $\mathscr{E}(E) \cap \mathscr{F}$ a topology with the subbasis[43] $(\mathcal{O}^n_{\Lambda K})$ for the open sets; here K runs over the compact subsets of \mathbf{R}^ν, Λ over the open sets such that $\Lambda \subset K$, n over the integers greater than or equal to zero and $\mathcal{O}^n_{\Lambda K}$ is the set of configurations X_σ with n particles in Λ and no particle in $K \setminus \Lambda$.

7.D.[44] Using the notation and results of Theorem 7.6.3, show that $\rho^\Phi(A^*A) = 0$ implies $A = 0$. [Show that $\rho^\Phi(A^*A) = 0 \Rightarrow \rho^\Phi(A^*\bar{\tau}_t A) = 0 \Rightarrow \rho^\Phi(AA^*) = 0$ and use the fact that \mathfrak{A} is simple.]

[42] See Dunford and Schwartz [1], Theorem VIII.6.9.

[43] $(\mathcal{O}^n_{\Lambda K})$ is a subbasis if every open set in $\mathscr{E}(E) \cap \mathscr{F}$ is a union of finite intersections of the sets $\mathcal{O}^n_{\Lambda K}$. Lanford [1] has shown that $\mathscr{E}(E) \cap \mathscr{F}$ is a separable complete metric space for the topology considered.

[44] See Robinson [1],II.

APPENDIX

SOME MATHEMATICAL TOOLS

A.1 SOME TERMINOLOGY

A.1.1 NOTATION. For real $a \leq b$, (a, b) is an open interval, $[a, b]$ a closed interval, and $(a, b] = \{x : a < x \leq b\}$.

We denote by \mathbf{Z} the ring of integers, by \mathbf{R}, \mathbf{C} the real and complex fields; \mathbf{Z}^v (resp. \mathbf{R}^v, \mathbf{C}^v) are the spaces of v-tuples of integers (resp. real, complex numbers). The origin of \mathbf{Z}^v (resp. \mathbf{R}^v, \mathbf{C}^v) is $0 = (0, \ldots, 0)$.

If A and B are subsets of E, $A \setminus B$ is the complement in A of $A \cap B$.

A.1.2 EQUICONTINUITY. A family (f_α) of (real or complex) functions on a topological space E is said to be *equicontinuous* if, for all $x \in E$ and $\varepsilon > 0$, there is a neighborhood \mathcal{N}_x of x (independent of α) such that

$$y \in \mathcal{N}_x \quad \Rightarrow \quad |f_\alpha(y) - f_\alpha(x)| < \varepsilon \tag{1.1}$$

If the family (f_α) is equicontinuous and if the functions f_α converge on a dense subset of E, then they converge everywhere on E.

A.1.3 UPPER SEMICONTINUITY. A function f with values in $\mathbf{R} \cup \{-\infty\}$ on the topological space E is *upper semicontinuous* (u.s.c) if for every $x \in E$ and $a > f(x)$ there is a neighborhood \mathcal{N}_x of x such that

$$y \in \mathcal{N}_x \quad \Rightarrow \quad f(y) < a \tag{1.2}$$

It is equivalent to require that, for every real a, $\{x \in E : f(x) < a\}$ is open or $\{x \in E : f(x) \geq a\}$ closed.

The lower bound of a family of real continuous functions on E is u.s.c. If E is compact, and f is u.s.c., there exists $x \in E$ such that

$$f(x) = \sup_{y \in E} f(y) \tag{1.3}$$

Thus, on a compact set, a u.s.c. function reaches its supremum.

198

A.1.4 CONVEXITY. Let E be a real vector space. A set $C \subset E$ is *convex* if

$$x_1, x_2 \in C \quad \text{and} \quad \alpha \in [0, 1] \Rightarrow \alpha x_1 + (1 - \alpha)x_2 \in C \tag{1.4}$$

The convex hull of a set $S \subset E$ is the smallest convex set containing S, i.e., the set of all finite linear combinations $\Sigma_i \lambda_i x_i$ with $x_i \in S$, $\lambda_i \geq 0$, $\Sigma_i \lambda_i = 1$.
 A real function f defined on the convex set C is *convex* if

$$x_1, x_2 \in C \quad \text{and} \quad \alpha \in [0, 1] \Rightarrow f(\alpha x_1 + (1 - \alpha)x_2) \leq \alpha f(x_1) + (1 - \alpha)f(x_2)$$

$$\tag{1.5}$$

We say that f is *concave* if $-f$ is convex; we say that f is *affine* if f is convex and concave. If C is a convex open subset of \mathbf{R}^n, then every convex function on C is continuous.

A.2 GENERALITIES ON B*–ALGEBRAS

The elements of the theory of B*-algebras are reviewed in this and the next section. Corresponding to the fact that proofs are not given, the order of exposition is somewhat unconventional. For a more detailed introduction to B*-algebras, see the first two sections of the book of Dixmier [2] (referred to as [D] in what follows).

A.2.1 INVOLUTION OF AN ALGEBRA OVER **C**. Let \mathfrak{A} be an algebra over the field **C** of complex numbers. A mapping $A \rightarrow A^*$ of \mathfrak{A} into itself will be called an involution if it is

(a) involutory, i.e., $(A^*)^* = A$,
(b) conjugate linear, i.e., $(A_1 + \lambda A_2)^* = A_1^* + \lambda^* A_2^*$,
(c) an antiautomorphism, i.e., $(A_1 A_2)^* = A_2^* A_1^*$.

A.2.2 B*-ALGEBRAS. An algebra \mathfrak{A} over **C** with an involution $A \rightarrow A^*$ and a norm $A \rightarrow \|A\|$ will be called a B*-algebra if the following conditions are satisfied

(a) $\|A^*\| \quad = \|A\|$
(b) $\|A_1 A_2\| \leq \|A_1\| \cdot \|A_2\|$
(c) \mathfrak{A} is complete (with respect to the norm).
(d) $\|A^* A\| = \|A\|^2$

A.2.3 REMARK. Condition (c) above expresses that, with respect to its linear structure and norm, \mathfrak{A} is a Banach space.

A.2.4 ALGEBRAS OF FUNCTIONS. Let L be a locally compact topological space and $\mathscr{C}_0(L)$ the space of complex continuous functions vanishing at

infinity on L. If $f \in \mathscr{C}_0(L)$, let f^* be its complex conjugate and

$$\|f\| = \sup_{x \in L} |f(x)| \tag{2.1}$$

With respect to the involution $f \to f^*$ and norm $f \to \|f\|$, $\mathscr{C}_0(L)$ is an Abelian B*-algebra. Conversely (see A.3.9) every Abelian B*-algebra is isomorphic to some $\mathscr{C}_0(L)$.

A.2.5 C*-ALGEBRAS. Let \mathfrak{H} be a complex Hilbert space and $\mathfrak{B}(\mathfrak{H})$ the algebra (over \mathbf{C}) of the bounded operators on \mathfrak{H}. If $A \in \mathfrak{B}(\mathfrak{H})$, let A^* be its adjoint and

$$\|A\| = \sup_{\|\Phi\| \leq 1} \|A\Phi\| \tag{2.2}$$

its norm. A subalgebra \mathfrak{A} of $\mathfrak{B}(\mathfrak{H})$ which is self-adjoint (i.e., $A \in \mathfrak{A} \Rightarrow A^* \in \mathfrak{A}$) and norm-closed (i.e., complete with respect to the norm) is called a C*-algebra. With respect to the involution $A \to A^*$ and norm $A \to \|A\|$, every C*-algebra is a B*-algebra. Conversely (see A.3.6), every B*-algebra is isomorphic to a C*-algebra of operators on some Hilbert space.

A.2.6 EXISTENCE OF AN IDENTITY. A B*-algebra \mathfrak{A} may or may not have an identity. For instance the algebra $\mathscr{C}_0(L)$ (see A.2.4) has an identity if and only if L is compact. If \mathfrak{A} has an identity 1, then $\|1\| = 1$. Let \mathfrak{A} have an identity; then an element U of \mathfrak{A} is said to be unitary if $UU^* = U^*U = 1$; the unitaries generate \mathfrak{A} (in fact every element of \mathfrak{A} is a linear combination of unitaries).

A.2.7 ADJUNCTION OF AN IDENTITY. From any algebra \mathfrak{A} over \mathbf{C} we construct as follows an algebra $\tilde{\mathfrak{A}}$ with an identity. An element of $\tilde{\mathfrak{A}}$ is a pair (λ, A) with $\lambda \in \mathbf{C}$, $A \in \mathfrak{A}$, and we write

$$(\lambda_1, A_1) + (\lambda_2, A_2) = (\lambda_1 + \lambda_2, A_1 + A_2) \tag{2.3}$$

$$\lambda(\lambda_1, A_1) = (\lambda\lambda_1, \lambda A_1) \tag{2.4}$$

$$(\lambda_1, A_1) \cdot (\lambda_2, A_2) = (\lambda_1\lambda_2, \lambda_1 A_2 + \lambda_2 A_1 + A_1 A_2) \tag{2.5}$$

If \mathfrak{A} has an involution $A \to A^*$ then $\tilde{\mathfrak{A}}$ has an involution

$$(\lambda, A) \to (\lambda, A)^* = (\lambda^*, A^*) \tag{2.6}$$

If \mathfrak{A} is a B*-algebra then ([D],1.3.8) there exists a norm which makes $\tilde{\mathfrak{A}}$ a B*-algebra, and this norm is determined by (2.10).

There is an isomorphism $A \to (0, A)$ of \mathfrak{A} onto a two-sided ideal of co-dimension one of $\tilde{\mathfrak{A}}$; this isomorphism preserves the involution if \mathfrak{A} has an involution and the norm if \mathfrak{A} is a B*-algebra.

A.2.8 SPECTRUM OF AN ELEMENT. Let \mathfrak{A} be an algebra over \mathbf{C} with identity 1 and $A \in \mathfrak{A}$; then

$$\mathrm{Sp}_{\mathfrak{A}}A = \{\lambda \in \mathbf{C} : A - \lambda 1 \quad \text{has no inverse in } \mathfrak{A}\} \tag{2.7}$$

is the spectrum of A. If \mathfrak{A} is an algebra over \mathbf{C} and $A \in \mathfrak{A}$ we write

$$\mathrm{Sp}'_{\mathfrak{A}}A = \mathrm{Sp}_{\tilde{\mathfrak{A}}}(0, A) \tag{2.8}$$

Let \mathfrak{A} be a B*-algebra and define the spectral norm $\|A\|_{\mathrm{Sp}} = \sup|z|$ for $z \in \mathrm{Sp}'\, A$. If A is self-adjoint ($A = A^*$) then ([D],1.3.7)

$$\|A\| = \|A\|_{\mathrm{Sp}} \tag{2.9}$$

For all $A \in \mathfrak{A}$, A^*A is self-adjoint and, by A.2.2(d),

$$\|A\|^2 = \|A^*A\| = \|A^*A\|_{\mathrm{Sp}} \tag{2.10}$$

A.2.9 MORPHISMS. A sub-B*-algebra of a B*-algebra \mathfrak{A} is by definition a norm-closed self-adjoint subalgebra of \mathfrak{A}.

Given two B*-algebras \mathfrak{A}_1, \mathfrak{A}_2, a morphism (or *-homomorphism) $\pi : \mathfrak{A}_1 \to \mathfrak{A}_2$ is by definition a homomorphism of algebras over \mathbf{C} such that $\pi(A^*) = (\pi(A))^*$. It can be proved ([D],1.3.7) that a morphism reduces norms ($\|\pi(A)\| \leq \|A\|$) and is therefore continuous. The kernel of a morphism is thus a closed two-sided ideal.

It can be proved ([D],1.8.1) that a morphism of B*-algebras is injective (i.e., one-to-one into) if and only if it is isometric (i.e., norm-preserving). In particular, an automorphism of \mathfrak{A} is an isometry of \mathfrak{A} onto itself.

If \mathfrak{A} is a B*-algebra and \mathfrak{I} a closed two-sided ideal, then \mathfrak{I} is self-adjoint, $\mathfrak{A}/\mathfrak{I}$ with the quotient norm is a B*-algebra, and the quotient mapping $\mathfrak{A} \to \mathfrak{A}/\mathfrak{I}$ is a morphism ([D],1.8.2). In particular the image $\pi(\mathfrak{A}_1)$ of a morphism $\pi : \mathfrak{A}_1 \to \mathfrak{A}_2$ is closed and therefore a B*-algebra.

A.2.10 REMARK. The B*-algebras, with the morphisms introduced in A.2.9 form a category.[1] It would appear from A.2.2 that the definition of B*-algebras introduces an essentially nonalgebraic notion (the norm). It follows however from (2.10) that the norm of a B*-algebra is determined by its structure of algebra over \mathbf{C} with an involution. Notice that the morphisms are defined without reference to norms.

A.3 STATES ON B*–ALGEBRAS

A.3.1 POSITIVE ELEMENTS ([D],1.6). An element A of a B*-algebra is said to be positive ($A \geq 0$) if one of the following equivalent conditions is satisfied:

[1] See, e.g., MacLane [1].

(a) $A = A^*$ and $\mathrm{Sp}'\, A \geq 0$,
(b) $A = B^*B$ for some $B \in \mathfrak{A}$.

An order relation is thus defined on \mathfrak{A} (with $A \leq B \Leftrightarrow B - A \geq 0$) and the set $P = \{A \in \mathfrak{A} : A \geq 0\}$ is a closed convex cone in \mathfrak{A}.

A.3.2 EXISTENCE OF AN APPROXIMATE IDENTITY ([D],1.7). Every B*-algebra \mathfrak{A} has an increasing approximate identity (U_α) in the following sense.

(a) The set $\{\alpha\}$ of indices is ordered and such that, given α_1, α_2, there exists $\alpha \geq \alpha_1, \alpha_2$.

(b) $\qquad\qquad\qquad 0 \leq U_\alpha$ and $\|U_\alpha\| \leq 1$ \qquad for all $\quad \alpha$

(c) $\qquad\qquad\qquad\qquad \alpha_1 \leq \alpha_2 \Rightarrow U_{\alpha_1} \leq U_{\alpha_2}$

(d) $\quad \lim_\alpha \|U_\alpha A - A\| = 0 \quad$ and $\quad \lim_\alpha \|A U_\alpha - A\| = 0 \qquad$ for all $\quad A \in \mathfrak{A}$

If \mathfrak{A} is separable (i.e., contains a countable dense set), we may choose the integers greater than or equal to zero as set of indices.

A.3.3 STATES. A continuous linear functional ρ on \mathfrak{A} is called a state if it is positive ($A \geq 0 \Rightarrow \rho(A) \geq 0$) and has norm 1. If \mathfrak{A} has an identity then $\rho(1) = 1$. Conversely, let ρ be a positive linear form on a B*-algebra \mathfrak{A} with identity; if $\rho(1) = 1$, then ρ is a state ([D],2.1). If (U_α) is an approximate identity of \mathfrak{A} and ρ is a state, then $\lim \rho(U_\alpha) = 1$[[D],2.1.5,(v)].

If h is a self-adjoint continuous linear form on \mathfrak{A}, there exist (uniquely defined) continuous positive linear forms f, f' on \mathfrak{A} such that $h = f - f'$, $\|h\| = \|f\| + \|f'\|$ ([D],12.3.4).

The set E of states on a B*-algebra \mathfrak{A} is a convex subset of the unit ball of the dual \mathfrak{A}'. If \mathfrak{A} has an identity, E is weakly compact because E is the intersection of the unit ball (weakly compact by the Alaoglu–Bourbaki theorem[2]) and the plane $\{f \in \mathfrak{A}' : f(1) = 1\}$.

A state of \mathfrak{A} is said to be a pure state if it is an extremal point of E, i.e., if it cannot be written in the form $\rho = \alpha\rho_1 + (1 - \alpha)\rho_2$ with $\rho_1, \rho_2 \in E$, $\rho_1 \neq \rho_2, 0 < \alpha < 1$. We shall denote by $\mathscr{E}(E)$ the set of pure states (extremal points of E). If \mathfrak{A} has an identity the Krein–Milman theorem[3] applied to the convex (weakly) compact set E shows that the finite convex linear combinations of pure states are (weakly) dense in E.

If \mathfrak{A}_1 is a sub-B*-algebra of \mathfrak{A}, then any state ρ_1 on \mathfrak{A}_1 extends to a state ρ on \mathfrak{A}; if ρ_1 is pure, ρ may be chosen pure. If $\mathfrak{A}_1 = \mathfrak{I}$ is a closed two-sided ideal, the extension is unique ([D],2.10.1–2 and 2.11.7). In particular a state on a B*-algebra has a unique extension to a state on the algebra \mathfrak{A} obtained by adjunction of an identity.

[2,3] For the Alaoglu–Bourbaki and Krein–Milman theorems see Köthe [1], pp. 250, 335.

A.3.4 REPRESENTATIONS. A representation of a B*-algebra \mathfrak{A} is a pair (\mathfrak{H}, π) where \mathfrak{H} is a complex Hilbert space and $\pi: \mathfrak{A} \to \mathfrak{B}(\mathfrak{H})$ is a morphism. The representation is said to be essential (or nondegenerate) if $\Phi \in \mathfrak{H}$ and $\pi(A)\Phi = 0$ for all $A \in \mathfrak{A}$ imply $\Phi = 0$.

The representation is said to be irreducible if one of the following equivalent conditions is satisfied ([D],2.8.4).

(a) *Topological irreducibility*: If $C \in \mathfrak{B}(\mathfrak{H})$ and $[C, \pi(A)] = 0$ for all $A \in \mathfrak{A}$ then C is a multiple of the identity operator on \mathfrak{H}.

(b) *Algebraic irreducibility*: The only subspaces of \mathfrak{H} stable under $\pi(\mathfrak{A})$ are $\{0\}$ and \mathfrak{H}.

A cyclic representation of \mathfrak{A} is a triple $(\mathfrak{H}, \pi, \Omega)$ where \mathfrak{H}, π form a representation and $\Omega \in \mathfrak{H}$ is such that $\|\Omega\| = 1$ and $\pi(\mathfrak{A})\Omega$ is dense in \mathfrak{H}.

A.3.5 THE GEL'FAND–SEGAL CONSTRUCTION ([D],2.4). If $(\mathfrak{H}, \pi, \Omega)$ is a cyclic representation of the B*-algebra \mathfrak{A}, $A \to \rho(A) = (\Omega, \pi(A)\Omega)$ defines a state ρ on \mathfrak{A}. Conversely if ρ is a state on \mathfrak{A}, there exists a cyclic representation $(\mathfrak{H}_\rho, \pi_\rho, \Omega_\rho)$ such that

$$\rho(\cdot) = (\Omega_\rho, \pi_\rho(\cdot)\Omega_\rho) \tag{3.1}$$

We sketch now the construction of this cyclic representation (Gel'fand–Segal construction).

We assume that \mathfrak{A} has an identity (this can be achieved by adjunction of 1). Define

$$\mathcal{N} = \{A \in \mathfrak{A}: \rho(A^*A) = 0\}$$

then \mathcal{N} is a left ideal of \mathfrak{A}. With respect to the scalar product defined by ρ, \mathfrak{A}/\mathcal{N} is a Hausdorff pre-Hilbert space; let \mathfrak{H}_ρ be the completed Hilbert space. If $A, A_1 \in \mathfrak{A}$, the mapping $A_1 \to AA_1$ defines a mapping $\mathfrak{A}/\mathcal{N} \to \mathfrak{A}/\mathcal{N}$ (because \mathcal{N} is a left ideal) and it can be shown that it has norm less than or equal to $\|A\|$; it extends thus by continuity to a unique mapping $\pi_\rho(A): \mathfrak{H}_\rho \to \mathfrak{H}_\rho$. We check readily that π_ρ is a morphism $\mathfrak{A} \to \mathfrak{B}(\mathfrak{H})$. Let Ω_ρ be the class of 1 in \mathfrak{A}/\mathcal{N}; then (3.1) holds.

The Gel'fand–Segal construction is unique in the sense that any cyclic representation $(\mathfrak{H}, \pi, \Omega)$ such that $(\Omega, \pi(\cdot)\Omega) = \rho(\cdot)$ is unitarily equivalent to $(\mathfrak{H}_\rho, \pi_\rho, \Omega_\rho)$, i.e., there is an isometry V of \mathfrak{H}_ρ onto \mathfrak{H} such that $\pi = V\pi_\rho V^{-1}, \Omega = V\Omega_\rho$.

If ρ is a state on \mathfrak{A} and f a positive linear form on \mathfrak{A} such that $f \leq \rho$ then ([D],2.5.1) there exists a unique self-adjoint operator T on \mathfrak{H}_ρ, commuting with $\pi_\rho(\mathfrak{A})$ and such that

$$0 \leq T \leq 1 \qquad f(\cdot) = (T\Omega_\rho, \pi_\rho(\cdot)T\Omega_\rho) \tag{3.2}$$

A.3.6 Universal Representation and Enveloping von Neumann Algebra. Let \mathfrak{A} be a B*-algebra, E the set of states on \mathfrak{A}, (\mathfrak{H}, π) the representation of \mathfrak{A} such that $\mathfrak{H} = \bigoplus_{\rho \in E} \mathfrak{H}_\rho$, $\pi = \bigoplus_{\rho \in E} \pi_\rho$; then (\mathfrak{H}, π) is called the universal representation of \mathfrak{A}. The morphism $\pi \colon \mathfrak{A} \to \mathfrak{B}(\mathfrak{H})$ is an isomorphism of \mathfrak{A} onto a C*-algebra $\pi(\mathfrak{A}) \subset \mathfrak{B}(\mathfrak{H})$ ([D], 2.6.1). This shows in particular that every B*-algebra is isomorphic to a C*-algebra.

Let \mathfrak{B} be the weak closure of $\pi(\mathfrak{A})$; \mathfrak{B} is called the enveloping von Neumann algebra of \mathfrak{A}. If $f \in \mathfrak{A}'$ (f is a continuous linear functional on \mathfrak{A}), then there is a unique weakly continuous linear functional \tilde{f} on \mathfrak{B} such that $\tilde{f}(\pi(\cdot)) = f(\cdot)$ on \mathfrak{A}. Every element C of \mathfrak{B} determines in this way a continuous linear functional $f \to \tilde{f}(C)$ on \mathfrak{A}', and it can be shown that this identifies \mathfrak{B} and the bidual \mathfrak{A}'' of \mathfrak{A} as Banach spaces ([D], 12.1.3).

A.3.7 Irreducible Representations. Let ρ be a state on \mathfrak{A}, $(\mathfrak{H}_\rho, \pi_\rho, \Omega_\rho)$ the corresponding cyclic representation. The following conditions are equivalent ([D]. 2.5.4 and 2.9.5).

(a) $\rho \in \mathscr{E}(E)$: ρ is pure.
(b) The representation $(\mathfrak{H}_\rho, \pi_\rho)$ is irreducible.
(c) The left ideal $\mathscr{N} = \{A \in \mathfrak{A} : \rho(A^*A) = 0\}$ is maximal regular (regular means that there exists $U \in \mathfrak{A}$ such that for all $A \in \mathfrak{A}$, $UA - A \in \mathscr{N}$; if \mathfrak{A} has an identity every left ideal is regular).

If they are satisfied, $\mathfrak{H}_\rho = \mathfrak{A}/\mathscr{N}$ with the quotient norm ([D], 2.12.18).

A.3.8 Kadison's Transitivity Theorem ([D], 2.8). Let (\mathfrak{H}, π) be an irreducible representation of the B*-algebra \mathfrak{A}. If $\Phi_1, \ldots, \Phi_n, \Psi_1, \ldots, \Psi_n \in \mathfrak{H}$ and $B \in \mathfrak{B}(\mathfrak{H})$ is such that

$$B\Phi_i = \Psi_i \qquad i = 1, \ldots, n \qquad (3.3)$$

then there exists $A \in \mathfrak{A}$ such that

$$\pi(A)\Phi_i = \Psi_i \qquad i = 1, \ldots, n \qquad (3.4)$$

If B is self-adjoint, then A may be chosen self-adjoint; if \mathfrak{A} has an identity and B is unitary, then A may be chosen unitary (i.e., $A^*A = AA^* = 1$). In particular, if $\Omega \in \mathfrak{H}$, $\|\Omega\| = 1$, then $(\mathfrak{H}, \pi, \Omega)$ is a cyclic representation.

Let ρ be a pure state on \mathfrak{A} and $(\mathfrak{H}_\rho, \pi_\rho, \Omega_\rho)$ the corresponding cyclic representation. If $A \in \mathfrak{A}$ and $\rho(A) = 0$ there exist A_1, A_2 such that $\pi_\rho(A_1)\Omega = 0$, $\pi_\rho(A_2)^*\Omega = 0$, and $A = A_1 + A_2$ ([D], 2.9.1).

A.3.9 Commutative B*-Algebras and the Gel'fand Isomorphism.[4] If \mathfrak{A} is an Abelian B*-algebra the following sets are in natural one-to-one correspondence:

[4] See Loomis [1].

(a) $\mathscr{E}(E)$: the pure states;

(b) the homomorphisms of \mathfrak{A} onto \mathbf{C};

(c) the maximal regular ideals of \mathfrak{A} (the maximal ideals if \mathfrak{A} has an identity).

The set $\mathscr{E}(E)$, with the weak topology, is called the spectrum of \mathfrak{A}; it is a locally compact space. For $A \in \mathfrak{A}$, let \hat{A} be the function on $\mathscr{E}(E)$ defined by $\hat{A}(\rho) = \rho(A)$; then the mapping $A \to \hat{A}$ is an isomorphism (Gel'fand isomorphism) of the B*-algebra \mathfrak{A} onto the B*-algebra of continuous complex functions tending to zero at infinity on $\mathscr{E}(E)$.

The spectrum $\mathscr{E}(E)$ is compact if and only if \mathfrak{A} has an identity. The states on \mathfrak{A} are put in one-to-one correspondence with the probability measures on $\mathscr{E}(E)$ by the Gel'fand isomorphism. In particular, if $\rho \in E$, let μ be the corresponding measure on $\mathscr{E}(E)$, and $(\mathfrak{H}_\rho, \pi_\rho, \Omega_\rho)$ the Gel'fand–Segal cyclic representation; we have the identifications $\mathfrak{H}_\rho = L^2(\mu), \pi_\rho \colon A \to \hat{A} \in \mathfrak{B}(L^2(\mu))$, $\Omega_\rho = 1 \in L^2(\mu)$.

Let \mathfrak{A}'_h be the space of self-adjoint continuous linear forms on a B*-algebra \mathfrak{A}; then \mathfrak{A}'_h is a lattice for its natural order if and only if \mathfrak{A} is Abelian ([D], 2.12.17).

A.4 VON NEUMANN ALGEBRAS

The theory of von Neumann algebras can be found in the book of J. Dixmier [1]. A convenient summary of the results is given as Appendix A of [D].

A.4.1 COMMUTANT. Let M be a subset of $\mathfrak{B}(\mathfrak{H})$; then the commutant M' of M is defined by

$$M' = \{A \in \mathfrak{B}(\mathfrak{H}) : B \in M \Rightarrow [A, B] = 0\} \qquad (4.1)$$

If M' consists of the multiples of $\mathbf{1}$, we say that M is irreducible. The commutant $M'' = (M')'$ of M' is called the bicommutant of M.

A.4.2 VON NEUMANN ALGEBRAS. Let \mathfrak{H} be a complex Hilbert space, a self-adjoint subalgebra \mathscr{B} of $\mathfrak{B}(\mathfrak{H})$ is called von Neumann algebra if it satisfies one of the following equivalent conditions.

(a) \mathscr{B} contains the identity operator $\mathbf{1}$ on \mathfrak{H} and is weakly closed in $\mathfrak{B}(\mathfrak{H})$.

(b) \mathscr{B} contains $\mathbf{1}$ and is strongly closed in $\mathfrak{B}(\mathfrak{H})$.

(c) \mathscr{B} is equal to its bicommutant: $\mathscr{B} = \mathscr{B}''$.

In particular, a von Neumann algebra is a C*-algebra with identity.

If M is a self-adjoint subset of $\mathfrak{B}(\mathfrak{H})$, then M' is a von Neumann algebra and M'' is the smallest von Neumann algebra containing M (i.e., the von Neumann algebra generated by M).

A.4.3 RESTRICTION TO A SUBSPACE OF \mathfrak{H}. If a self-adjoint subalgebra \mathscr{B} of $\mathfrak{B}(\mathfrak{H})$ is weakly (= strongly) closed, then \mathscr{B} contains a largest projection P; $PB = B$ for all $B \in \mathscr{B}$; and the restriction of \mathscr{B} to the range of P is a von Neumann algebra.

Let $\mathscr{B} \subset \mathfrak{B}(\mathfrak{H})$ be a von Neumann algebra. If $P' \in \mathscr{B}'$ is a projection, the restriction of \mathscr{B} to the range of P' is a von Neumann algebra $\mathscr{B}_{P'}$. If $P \in \mathscr{B}$ is a projection, the restriction of $P\mathscr{B}P$ to the range of P is a von Neumann algebra \mathscr{B}_P and $(\mathscr{B}_P)' = (\mathscr{B}')_P$.

A.4.4 CYCLIC AND SEPARATING VECTORS. Let M be a self-adjoint subset of $\mathfrak{B}(\mathfrak{H})$ and $\Phi \in \mathfrak{H}$. We say that Φ is cyclic for M if $M\Phi$ is dense in \mathfrak{H}; that Φ is separating for M if $B \in M$ and $B\Phi = 0 \Rightarrow B = 0$. If Φ is cyclic for M, then Φ is separating for M'. If Φ is separating for M', then Φ is cyclic for M''.

A.4.5 FACTORS AND ABELIAN VON NEUMANN ALGEBRAS. A von Neumann algebra \mathscr{B} is said to be a factor if $\mathscr{B} \cap \mathscr{B}'$ consists of the multiples of the identity operator. A von Neumann algebra \mathscr{B} is Abelian if and only if $\mathscr{B} \subset \mathscr{B}'$. If $\mathscr{B} = \mathscr{B}'$, then \mathscr{B} is said to be maximal Abelian. If an Abelian von Neumann algebra has a cyclic vector, then it is maximal Abelian.

A.4.6 KAPLANSKY'S DENSITY THEOREM. Let \mathscr{B}_1 and \mathscr{B}_2 be self-adjoint subalgebras of $\mathfrak{B}(\mathfrak{H})$ and \mathscr{B}_{1h} (resp. \mathscr{B}_{2h}) the subspace of self-adjoint elements in \mathscr{B}_1 (resp. \mathscr{B}_2). If $\mathscr{B}_1 \subset \mathscr{B}_2$ and \mathscr{B}_1 is dense in \mathscr{B}_2 for the strong topology, then the unit ball of \mathscr{B}_1 (resp. \mathscr{B}_{1h}) is strongly dense in the unit ball of \mathscr{B}_2 (resp. \mathscr{B}_{2h}).

A.5 INTEGRAL REPRESENTATIONS ON CONVEX COMPACT SETS

Proofs of the results in this section will be found in Bourbaki [1] and a paper by Choquet and Meyer [1] referred to in what follows as [C.–M.].[5]

A.5.1 RESULTANT OF A MEASURE. Let F be a locally convex topological vector space and K a convex, compact subset of F. The dual \mathscr{M} of $\mathscr{C}(K)$ consists of the measures on K; we denote by \mathscr{M}_+ the convex cone of positive measures and by \mathscr{M}_1 the set of positive measures of norm 1 on K [\mathscr{M}_1 is the set of probability measures on K and may be identified with the set of states on $\mathscr{C}(K)$]. If $\rho \in K$, we denote by $\delta_\rho \in \mathscr{M}_1$ the unit mass (= Dirac measure) at ρ [the δ_ρ are the pure states on $\mathscr{C}(K)$].

If $\mu \in \mathscr{M}_1$ there exists $\rho \in K$ such that, for all $f \in F'$,

$$f(\rho) = \int f(\sigma)\, d\mu(\sigma) \tag{5.1}$$

[5] See also Phelps [1].

ρ is called the resultant of μ.[6] If $\mu \in \mathcal{M}_1$ has resultant ρ then μ can be approximated weakly by measures $\mu' \in \mathcal{M}_1$ with resultant ρ and finite support (i.e., $\mu' = \Sigma_{i=1}^n \lambda_i \delta_{\rho_i}$, $\lambda_i \geq 0$, $\Sigma_{i=1}^n \lambda_i = 1$, $\rho_i \in K$, $\Sigma_{i=1}^n \lambda_i \rho_i = \rho$).[7] If $\mu \in \mathcal{M}_1$ has resultant ρ and the function f is affine upper-semicontinuous on K, then $\mu(f) = f(\rho)$ ([C.–M.], Lemme 10).

A.5.2 MAXIMAL MEASURES. With the same notation as above, let $S \subset \mathscr{C}(K)$ be the convex cone of convex continuous functions on K. An order \prec on \mathcal{M}_+ is introduced (Bishop and de Leeuw) by

$$\mu_1 \prec \mu_2 \quad \Leftrightarrow \quad \text{for all } f \in S : \mu_1(f) \leq \mu_2(f)$$

If $\mu_1 \prec \mu_2$ and f is an affine continuous function, then $\mu_1(f) = \mu_2(f)$; in particular, $\|\mu_1\| = \|\mu_2\|$ and (if $\mu_1 \in \mathcal{M}_1$) μ_1, μ_2 have the same resultant. If $\mu \in \mathcal{M}_1$ and $\rho \in K$, the conditions "μ has resultant ρ" and "$\mu \succ \delta_\rho$" are equivalent.

We shall say that $\mu \in \mathcal{M}_+$ is maximal if it is maximal for the order \prec. If $\mu \in \mathcal{M}_+$, there exists a maximal measure $\succ \mu$ ([C.–M.], Théorème 3). In particular, if $\rho \in K$, there exists a maximal measure with resultant ρ.

A.5.3 UNIQUENESS PROBLEM. We use the same notation as above and we assume for convenience that K is a basis of a convex cone C with apex 0. (K is the intersection of C with a closed hyperplane $H \not\ni 0$ of F which meets all generating lines of C. This situation may always be arranged, replacing F by $\mathbf{R} \times F$ and imbedding K in $\mathbf{R} \times F$ as $\{1\} \times K$). The cone C defines an order in F; if C is a lattice for this order, we say that K is a *simplex* (Choquet). This definition does not depend upon the choice of C.

The following conditions are equivalent ([C.–M.], Théorème 11).

(a) K is a simplex.

(b) If $\rho \in K$ there exists a unique maximal measure $\mu_\rho \succ \delta_\rho$ (i.e., every $\rho \in K$ is the resultant of a unique maximal measure).

If K is a simplex the mapping $\rho \to \mu_\rho$ is affine (this is implicit in the proof of Théorème 11 of [C.–M.]).

A.5.4 MAXIMAL MEASURES AND EXTREMAL POINTS. Let $\mathscr{E}(K)$ be the set of extremal points on K; the Krein–Milman theorem says that K is the closure of the convex hull of $\mathscr{E}(K)$. If $\mu \in \mathcal{M}_+$ is carried[8] by $\mathscr{E}(K)$, then μ is maximal ([C.–M.], Proposition 15). Conversely, if K is metrizable and $\mu \in \mathcal{M}_+$ is maximal, then μ is carried by $\mathscr{E}(K)$ ([C.–M.], Lemme 13).

[6] See Bourbaki [1], p. 216, Corollaire.

[7] See Bourbaki [1], p. 217, Proposition 3.

[8] A measure μ is carried by a set S if S is μ-measurable and $|\mu|(S') = 0$ where S' is the complement of S.

Therefore if K is metrizable and $\mu \in \mathcal{M}_+$,

$$\mu \text{ maximal} \Leftrightarrow \mu \text{ carried by } \mathcal{E}(K)$$

In particular every $\rho \in K$ is the resultant of a measure μ_ρ carried by $\mathcal{E}(K)$ and, if K is a simplex, $\rho \to \mu_\rho$ is one-to-one from K onto the probability measures on K carried by $\mathcal{E}(K)$. In this case we may say that every $\rho \in K$ has a unique integral representation on $\mathcal{E}(K)$ by a measure μ_ρ such that $f(\rho) = \mu_\rho(f)$ for any affine continuous function f on K.

A.5.5 EXAMPLE: STATES ON B*-ALGEBRAS. Let \mathfrak{A} be a B*-algebra with identity; then E is convex and weakly compact. According to A.3.9 and A.5.3, every state ρ on \mathfrak{A} will be the resultant of a unique maximal measure μ_ρ if and only if \mathfrak{A} is Abelian. In fact, if \mathfrak{A} is Abelian and $\rho \in E$, μ_ρ is concentrated on $\mathcal{E}(E)$, whether or not E is metrizable (μ_ρ is just the measure on the spectrum of \mathfrak{A} associated to ρ by the Gel'fand isomorphism).

A.6 GROUPS WITH AN INVARIANT MEAN[9]

Let G be a locally compact group and $\mathcal{C}_B(G)$ be the Abelian B*-algebra of bounded continuous complex functions on G. If $h \in G$, we define the right translate f_h and the left translate ${}_h f$ of $f \in \mathcal{C}_B(G)$ by

$$f_h(g) = f(gh) \qquad {}_h f(g) = f(h^{-1}g)$$

A state \mathfrak{M} on $\mathcal{C}_B(G)$ is called a right-invariant (resp. a left-invariant) mean if, for all $h \in G$,

$$\mathfrak{M}(f_h) = \mathfrak{M}(f) \qquad [\text{resp. } \mathfrak{M}({}_h f) = \mathfrak{M}(f)]$$

If there exists a right-invariant mean on G, there exists also a left-invariant mean and a two-sided invariant mean; we say that G is a *group with an invariant mean*.

G has an invariant mean if it is Abelian, or compact, or admits a composition series consisting of such groups (example: the Euclidean group). On the other hand, a connected semisimple Lie group has an invariant mean only if it is compact.

A.6.1 M-NETS. We say that a family (χ_α) of functions on a locally compact group G is an *M-net* if the following conditions are satisfied.

(a) The set $\{\alpha\}$ of indices is ordered and such that, given α_1, α_2, there exists $\alpha \geq \alpha_1, \alpha_2$.

[9] For a detailed study of these groups (also called *amenable groups*), see Greenleaf [1].

(b) $0 \leq \chi_\alpha$ and $\int dg\chi_\alpha(g) = 1$ for all α (here dg is a right Haar measure).

(c) $$\lim_\alpha \int dg|\chi_\alpha(gh) - \chi_\alpha(g)| = 0 \qquad \text{for all} \quad h \in G$$

A necessary and sufficient condition for the existence of an M-net on G is that G has an invariant mean. If G has a countable basis, we may choose the integers greater than or equal to zero as set of indices.

REFERENCES

Araki, H. and Woods, E. J.
 [1] Representations of the Canonical Commutation Relations Describing a Nonrelativistic Infinite Free Bose Gas, *J. Math. Phys.* **4**, 637–662 (1963).

Araki, H. and Wyss, W.
 [1] Representations of Canonical Anticommutation Relations, *Helv. Phys. Acta.* **37**, 136–159 (1964).

Arnold, V. I. et Avez, A.
 Problèmes ergodiques de la Mécanique classique (Gauthier-Villars, Paris, 1967).

Bendat, J. and Sherman, S.
 [1] Monotone and Convex Operator Functions, *Trans. Am. Math. Soc.* **79**, 58–71 (1955).

Berezin, F. A., Dobrushin, R. L., Minlos, R. A., Povzner, A. Ya., and Sinai, Ya. G.
 [1] Some Facts about General Properties of Thermodynamical Potentials and Phase Transitions of the First Kind by Low Temperatures (Unpublished Report, 1966).

Borchers, H. J. and Zimmermann, W.
 [1] On the Self-Adjointness of Field Operators, *Nuovo Cim.* **31**, 1047–1059 (1964).

Bourbaki, N.
 [1] *Eléments de Mathématique. Intégration* (Hermann, Paris, 1965), 2nd ed., Chapitres 1, 2, 3, et 4.

Choquet, G. et Meyer, P.-A.
 [1] Existence et unicité des représentations intégrales dans les convexes compacts quelconques, *Ann. Inst. Fourier.* **13**, 139–154 (1963).

Dell'Antonio, G.-F., Doplicher, S., and Ruelle, D.
 [1] A Theorem on Canonical Commutation and Anticommutation Relations, *Commun. Math. Phys.* **2**, 223–230 (1966).

Dixmier, J.
[1] *Les algèbres d'opérateurs dans l'espace hilbertien (Algèbres de von Neumann)* (Gauthier-Villars, Paris, 1957).
[2] *Les C*-algèbres et leurs représentations* (Gauthier-Villars, Paris, 1964).

Dobrushin, R. L.
[1] Investigation of the Conditions of the Asymptotic Existence of the Configuration Integral of the Gibbs Distribution, *Teorija Verojatn. i ee Prim.* **9**, 626–643 (1964).
[2] The Existence of a Phase Transition in the Two- and Three-Dimensional Ising Models, *Teorija Verojatn. i ee Prim.* **10**, 209–230 (1965).

Doplicher, S., Kadison, R. V., Kastler, D., and Robinson, D. W.
[1] Asymptotically Abelian Systems, *Commun. Math. Phys.* **6**, 101–120 (1967).

Doplicher, S., Kastler, D., and Robinson, D. W.
[1] Covariance Algebras in Field Theory and Statistical Mechanics, *Commun. Math. Phys.* **3**, 1–28 (1966).

Dunford, N. and Schwartz, J. T.
[1] *Linear Operators, I, General Theory* (Interscience, New York, 1958).

Dyson, F. J.
[1] Stability of Matter (Lectures at Brandeis Summer Institute, 1966).
[2] Ground-State Energy of a Finite System of Charged Particles, *J. Math. Phys.* **8**, 1538–1545 (1967).

Dyson, F. J. and Lenard, A.
[1] Stability of Matter, I, *J. Math. Phys.* **8**, 423–434 (1967); Stability of Matter, II, *J. Math. Phys.* **9**, 698–711 (1968).

Fisher, M. E.
[1] The Free Energy of a Macroscopic System, *Arch. Rat. Mech. Anal.* **17**, 377–410 (1964).
[2] Correlation Functions and the Coexistence of Phases, *J. Math. Phys.* **6**, 1643–1653 (1965).
[3] The Theory of Condensation and the Critical Point, *Physics* **3**, 255–283 (1967).
[4] Critical Temperatures of Anisotropic Ising Lattices II General Upper Bounds (To Appear).

Fisher, M. E. and Ruelle, D.
[1] The Stability of Many-Particle Systems, *J. Math. Phys.* **7**, 260–270 (1966).

Gallavotti, G. and Miracle-Sole, S.
[1] Statistical Mechanics of Lattice Systems, *Commun. Math. Phys.* **5**, 317–323 (1967).

[2] Correlation Functions of a Lattice System, *Commun. Math. Phys.* **7**, 274–288 (1968).

[3] A Variational Principle for the Equilibrium of Hard Sphere Systems, *Ann. Inst. Henri Poincaré* **8**, 287–299 (1968).

Gallavotti, G., Miracle-Sole, S., and Robinson, D. W.

[1] Analyticity Properties of a Lattice Gas, *Phys. Letters* **25** A, 493–494 (1967).

Gantmacher, F. R.

[1] *The Theory of Matrices*, 2 Vol (Chelsea, New York, 1964).

Gårding, L. and Wightman, A. S.

[1] Representations of the Anticommutation Relations, *Proc. Nat. Acad. Sci.* **40**, 617–621 (1954).

[2] Representations of the Commutation Relations, *Proc. Nat. Acad. Sci.* **40**, 622–626 (1954).

Ginibre, J.

[1] Reduced Density Matrices of Quantum Gases I. Limit of Infinite Volume, *J. Math. Phys.* **6**, 238–251 (1965). II. Cluster Property, *J. Math. Phys.* **6**, 252–262 (1965). III. Hard-Core Potentials, *J. Math. Phys.* **6**, 1432–1446 (1965).

[2] Rigorous Lower Bound on the Compressibility of a Classical System, *Phys. Letters* **24** A, 223–224 (1967).

[3] On the Asymptotic Exactness of the Bogoliubov Approximation for Many Boson Systems, *Commun. Math. Phys.* **8**, 26–51 (1968).

Ginibre, J., Grossmann, A., and Ruelle, D.

[1] Condensation of Lattice Gases, *Commun. Math. Phys.* **3**, 187–193 (1966).

Greenleaf, F. P.

[1] Invariant Means on Topological Groups and Their Applications (Dept. of Math., UC Berkeley, 1966).

Griffiths, R. B.

[1] A Proof that the Free Energy of a Spin System is Extensive, *J. Math. Phys.* **5**, 1215–1222 (1964).

[2] Peierls Proof of Spontaneous Magnetization in a Two-Dimensional Ising Ferromagnet, *Phys. Rev.* **136** A, 437–439 (1964).

[3] Microcanonical Ensemble in Quantum Statistical Mechanics, *J. Math. Phys.* **6**, 1447–1461 (1965).

[4] Correlations in Ising Ferromagnets, I. *J. Math. Phys.* **8**, 478–483 (1967). II. External Magnetic Fields, *J. Math. Phys.* **8**, 484–489 (1967). III. A Mean-Field Bound for Binary Correlations, *Commun. Math. Phys.* **6**, 121–127 (1967).

Groeneveld, J.

[1] Two Theorems on Classical Many-Particle Systems, *Phys. Letters* **3** 50–51 (1962).

[2] Estimation Methods for Mayer's Graphical Expansions (Thesis, Amsterdam, 1967).

Haag, R., Hugenholtz, N. M., and Winnink, M.
[1] On the Equilibrium States in Quantum Statistical Mechanics, *Commun. Math. Phys.* **5**, 215–236 (1967).

Hill, T. L.
[1] *Statistical Mechanics* (McGraw-Hill, New York, 1956).

Hugenholtz, N. M.
[1] On the Factor Type of Equilibrium States in Quantum Statistical Mechanics, *Commun. Math. Phys.* **6**, 189–193 (1967).

Hurst, C. A. and Green, H. S.
[1] New Solution of the Ising Problem for a Rectangular Lattice, *J. Chem. Phys.* **33**, 1059–1062 (1960).

Jacobs, K.
[1] *Lecture Notes on Ergodic Theory* (Aarhus Universitet, Aarhus, 1963).

Jentzsch, R.
[1] Über Integralgleichungen mit positivem Kern, *Crelles J.* **141**, 235–244 (1912).

Kastler, D. and Robinson, D. W.
[1] Invariant States in Statistical Mechanics, *Commun. Math. Phys.* **3**, 151–180 (1966).

Kato, T.
[1] Fundamental properties of Hamiltonian operators of Schrödinger type, *Trans. Am. Math. Soc.* **70**, 195–211 (1951).
[2] *Perturbation Theory for Linear Operators* (Springer, Berlin, 1966).

Kelly, D. G. and Sherman, S.
[1] General Griffiths Inequalities on Correlations in Ising Ferromagnets, *J. Math. Phys.* **9**, 466–484 (1968).

Khinchin, A. I.
[1] *Mathematical Foundations of Information Theory* (Dover, New York, 1957).

Köthe, G.
[1] *Topologische Lineare Räume I* (Springer, Berlin, 1960).

Lanford, O.
[1] The Classical Mechanics of One-Dimensional Systems of Infinitely Many Particles, I. An Existence Theorem, *Commun. Math. Phys.* **9**, 176–191 (1968). II. Kinetic Theory, *Commun. Math. Phys.* (To Appear).

Lanford, O. and Robinson, D. W.
[1] Mean Entropy of States in Quantum Statistical Mechanics, *J. Math. Phys.* **9**, 1120–1125 (1968).

Lanford, O. and Ruelle, D.
[1] Integral Representations of Invariant States on B* Algebras, *J. Math. Phys.* **8**, 1460–1463 (1967).

Lebowitz, J. L. and Penrose, O.
[1] Convergence of Virial Expansions, *J. Math. Phys.* **5**, 841–847 (1964).
[2] Analytic and Clustering Properties of Thermodynamic Functions and Distribution Functions for Classical Lattice and Continuum Systems (To Appear).

Lee, T. D. and Yang, C. N.
[1] Statistical Theory of Equations of State and Phase Transitions, II, Lattice Gas and Ising Model, *Phys. Rev.* **87**, 410–419 (1952).

Lieb, E. H.
[1] A New Method in the Theory of Imperfect Gases and Liquids, *J. Math. Phys.* **4**, 671–678 (1963).

Loomis, L. H.
[1] *An Introduction to Abstract Harmonic Analysis* (D. Van Nostrand, Princeton, 1953).

Maclane, S.
[1] Categorical Algebra, *Bull. Am. Math. Soc.* **71**, 40–106 (1965).

Mayer, J. E.
[1] *J. Chem. Phys.* **15**, 187 (1947).

Mermin, N. D.
[1] Absence of Ordering in Certain Classical Systems, *J. Math. Phys.* **8**, 1061–1064 (1967).

Mermin, N. D. and Wagner, H.
[1] Absence of Ferromagnetism or Antiferromagnetism in One- or Two-Dimensional Isotropic Heisenberg Models, *Phys. Rev. Letters* **17**, 1133–1136 (1966).

Minlos, R. A. and Sinai, Ya. G.
[1] Phenomenon of "Phase Separation" at low temperatures in some models of lattice gases I, *Mat. Sbornik* **73**, 375–448 (1967).

Nelson, E.
[1] Analytic Vectors, *Ann. of Math.* **70**, 572–615 (1959).
[2] Feynman Integrals and the Schrödinger Equation, *J. Math. Phys.* **5**, 332–343 (1964).

Nettleton, R. E. and Green, M. S.
[1] Expression in Terms of Molecular Distribution Functions for the Entropy Density in an Infinite System, *J. Chem. Phys.* **29**, 1365–1370 (1958).

Onsager, L.
[1] Electrostatic Interaction of Molecules, *J. Phys. Chem.* **43**, 189–196 (1939).
[2] Crystal Statistics, I, A Two-Dimensional Model with an Order-Disorder Transition, *Phys. Rev.* **65**, 117–149 (1944).

Peierls, R.
[1] On Ising's Model of Ferromagnetism, *Proc. Camb. Phil. Soc.* **32**, 477–481 (1936).

Penrose, O.
[1] Convergence of Fugacity Expansions for Fluids and Lattice Gases, *J. Math. Phys.* **4**, 1312–1320 (1963).
[2] The Remainder in Mayer's Fugacity Series, *J. Math. Phys.* **4**, 1488–1494 (1963).
Phelps, R. R.
[1] *Lectures on Choquet's Theorem* (D. Van Nostrand, Princeton, 1966).
Polya, G. and Szegö, G.
[1] *Aufgaben und Lehrsätze aus der Analysis* (Springer, Berlin, 1964), 2 Vol. 3rd ed.
Powers, R.
[1] Representations of the Canonical Anticommutation Relations (Thesis, Princeton, 1967).
Riesz, F. et Sz–Nagy, B.
[1] *Leçons d'Analyse Fonctionelle* (Académie des Sciences de Hongrie, 1955), 3rd ed.
Robinson, D. W.
[1] Statistical Mechanics of Quantum Spin Systems I. *Commun. Math. Phys.* **6**, 151–160 (1967). II. *Commun. Math. Phys.* **7**, 337–348 (1968).
Robinson, D. W. and Ruelle, D.
[1] Extremal Invariant States, *Ann. Inst. Henri Poincaré* **6**, 299–310 (1967).
[2] Mean Entropy of States in Classical Statistical Mechanics, *Commun. Math. Phys.* **5**, 288–300 (1967).
Ruelle, D.
[1] Classical Statistical Mechanics of a System of Particles, *Helv. Phys. Acta* **36**, 183–197 (1963).
[2] Statistical Mechanics of Quantum Systems of Particles, *Helv. Phys. Acta* **36**, 789–799 (1963).
[3] Correlation Functions of Classical Gases, *Ann. Phys.* **25**, 109–120 (1963).
[4] Cluster Property of the Correlation Functions of Classical Gases, *Rev. Mod. Phys.* **36**, 580–584 (1964).
[5] Rigorous Results in Statistical Mechanics in *Lectures in Theoretical Physics* (University of Colorado Press, Boulder, 1964), Vol. VI.
[6] Correlation Functionals, *J. Math. Phys.* **6**, 201–220 (1965).
[7] States of Physical Systems, *Commun. Math. Phys.* **3**, 133–150 (1966).
[8] States of Classical Statistical Mechanics, *J. Math. Phys.* **8**, 1657–1668 (1967).
[9] A Variational Formulation of Equilibrium Statistical Mechanics and the Gibbs Phase Rule, *Commun. Math. Phys.* **5**, 324–329 (1967).
[10] Statistical Mechanics of a One-Dimensional Lattice Gas, *Commun. Math. Phys.* **9**, 267–278 (1968).
Schultz, T. D., Mattis, D. C., and Lieb, E. H.
[1] Two-Dimensional Ising Model as a Soluble Problem of Many Fermions, *Rev. Mod. Phys.* **36**, 856–871 (1964).

Segal, I. E.

[1] A Class of Operator Algebras which Are Determined by Groups, *Duke Math. J.* **18**, 221–265 (1951).

[2] *Mathematical Problems of Relativistic Physics* (Am. Math. Soc., Providence, 1963).

Shohat, J. A. and Tamarkin, J. D.

[1] *The Problem of Moments* (Am. Math. Soc., Providence, 1963).

Sinai, Ya. G.

[1] Dynamical Systems with Countably-Multiple Lebesgue Spectrum
 I. *Izv. Akad. Nauk SSSR Ser. Mat.* **25**, 899–924 (1961).
 Amer. Math. Soc. Transl. (2) **39**, 83–110 (1964).
 II. *Izv. Akad. Nauk SSSR Ser. Mat.* **30**, 15–68 (1966).

Størmer, E.

[1] Large Groups of Automorphisms of C*-Algebras, *Commun. Math. Phys.* **5**, 1–22 (1967).

[2] Types of von Neumann Algebras Associated with Extremal Invariant States, *Commun. Math. Phys.* **6**, 194–204 (1967).

[3] Symmetric States of Infinite Tensor Products of C*-Algebras (To Appear).

Uhlenbeck, G. E. and Ford, G. W.

[1] *Lectures in Statistical Mechanics* (Amer. Math. Soc., Providence, 1963).

Van Hove, L.

[1] Quelques propriétés générales de l'intégrale de configuration d'un système de particules avec interaction, *Physica* **15**, 951–961 (1949).

[2] Sur l'intégrale de configuration pour les systèmes de particules à une dimension, *Physica* **16**, 137–143 (1950).

Weng, C.-Y., Griffiths, R. B., and Fisher, M. E.

[1] Critical Temperatures of Anisotropic Ising Lattices, I. Lower bounds (To Appear).

Wightman, A. S.

[1] Quantum Field Theory in Terms of Vacuum Expectation Values, *Phys. Rev.* **101**, 860–866 (1956).

Wils, W.

[1] De Vrije Energie in de Thermodynamische Limiet van de Statistische Mechanica (Unpublished).

Yang, C. N.

[1] The Spontaneous Magnetization of a Two-Dimensional Ising Model, *Phys. Rev.* **85**, 808–816 (1952).

Yang, C. N. and Lee, T. D.

[1] Statistical Theory of Equations of State and Phase Transitions, I. Theory of Condensation, *Phys. Rev.* **87**, 404–409 (1952).

INDEX